Lecture Notes in Physics

Volume 876

For further volumes:
www.springer.com/series/5304

The Lecture Notes in Physics

The series Lecture Notes in Physics (LNP), founded in 1969, reports new developments in physics research and teaching—quickly and informally, but with a high quality and the explicit aim to summarize and communicate current knowledge in an accessible way. Books published in this series are conceived as bridging material between advanced graduate textbooks and the forefront of research and to serve three purposes:

- to be a compact and modern up-to-date source of reference on a well-defined topic
- to serve as an accessible introduction to the field to postgraduate students and nonspecialist researchers from related areas
- to be a source of advanced teaching material for specialized seminars, courses and schools

Both monographs and multi-author volumes will be considered for publication. Edited volumes should, however, consist of a very limited number of contributions only. Proceedings will not be considered for LNP.

Volumes published in LNP are disseminated both in print and in electronic formats, the electronic archive being available at springerlink.com. The series content is indexed, abstracted and referenced by many abstracting and information services, bibliographic networks, subscription agencies, library networks, and consortia.

Proposals should be sent to a member of the Editorial Board, or directly to the managing editor at Springer:

Christian Caron
Springer Heidelberg
Physics Editorial Department I
Tiergartenstrasse 17
69121 Heidelberg/Germany
christian.caron@springer.com

Gustavo E. Romero · Gabriela S. Vila

Introduction to Black Hole Astrophysics

 Springer

Gustavo E. Romero
Instituto Argentino de Radioastronomía
Villa Elisa, Argentina

Gabriela S. Vila
Instituto Argentino de Radioastronomía
Villa Elisa, Argentina

ISSN 0075-8450
ISSN 1616-6361 (electronic)
Lecture Notes in Physics
ISBN 978-3-642-39595-6
ISBN 978-3-642-39596-3 (eBook)
DOI 10.1007/978-3-642-39596-3
Springer Heidelberg New York Dordrecht London

Library of Congress Control Number: 2013949608

Printed on acid-free paper

Springer is part of Springer Science+Business Media (www.springer.com)

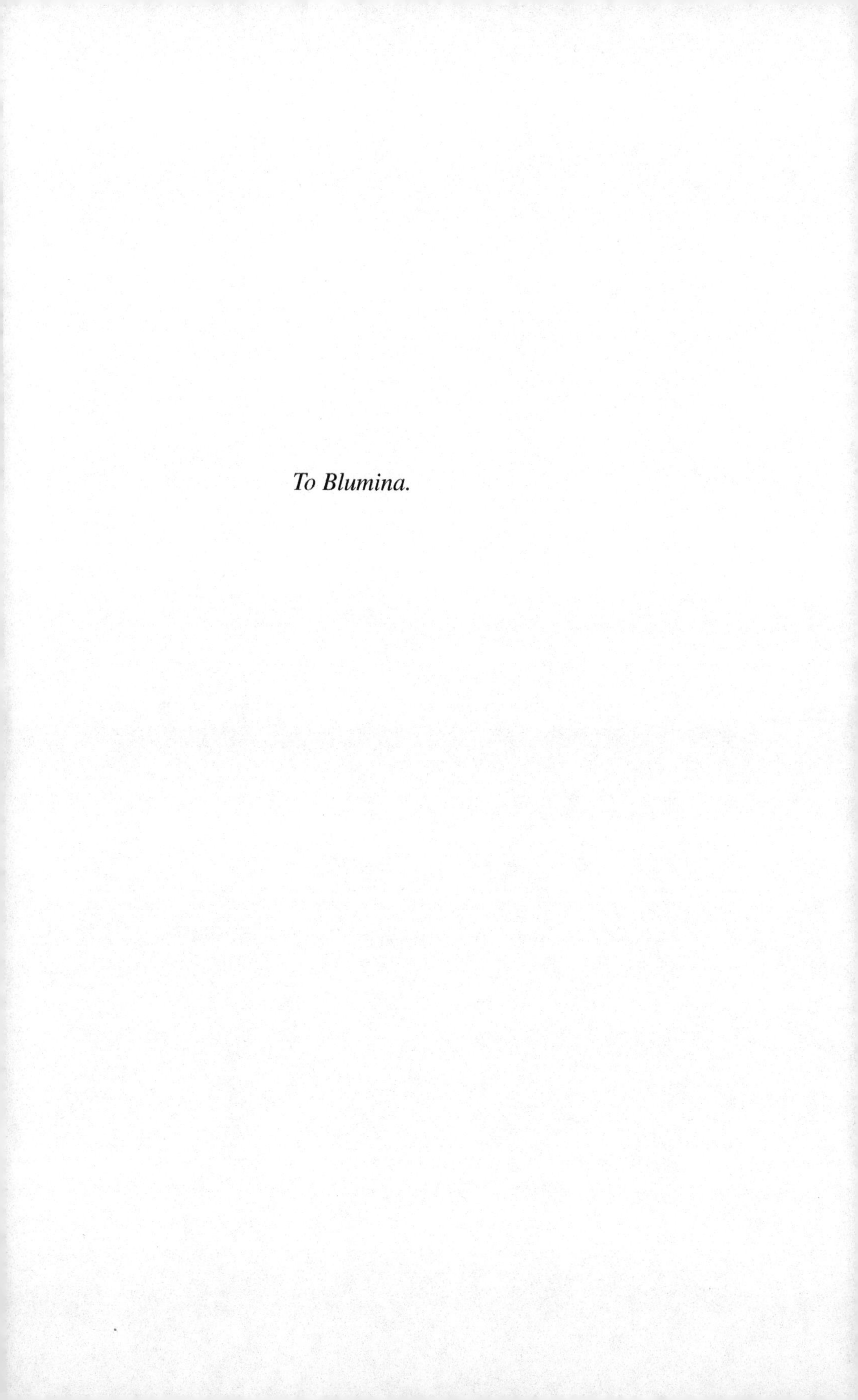

To Blumina.

Preface

This book is based on the lecture notes of the course on black hole astrophysics given at the University of La Plata, Argentina, by one of us (GER). The material aims at advanced undergraduate and graduate students with interest in astrophysics. The course takes one semester and is usually complemented by a course on high-energy astrophysics. The material included goes beyond what is found in classic textbooks like that by Shapiro and Teukolsky and is focused exclusively on black holes. We do not consider applications to other compact objects, such as neutron stars and white dwarfs. Instead, we provide more details on astrophysical manifestations of black holes. In particular, we include abundant material on jet physics and accounts of objects such as microquasars, active galactic nuclei, gamma-ray bursts, and ultraluminous X-ray sources. Other topics, normally not covered in introductory texts, like black holes in alternative theories of gravity, are discussed since we have found that they are highly stimulating for the students. Obviously, in a book of this kind, completeness is not possible, and some selection criterion must be applied to the material. Ours has been quite personal: we selected the topics on which GER has been working for around 20 years now, and we think that these topics form the core and starting point for basic research in this fascinating area of astrophysics.

In writing the book, we tried to avoid unnecessary technicalities, and to some degree the book is self-contained. Some previous knowledge of General Relativity would be desirable, but the reader will find the basic tools in Chap. 1. The appendices provide some additional mathematical details that will be useful to pursue the study and a guide to the bibliography on the subject.

La Plata   Gustavo E. Romero
May 2013   Gabriela S. Vila

Acknowledgements

I want to thank the many students that have taken my courses over the years. Their questions and interest have shaped the material of this book. I am also grateful to my former teacher assistants Drs. Sofia Cora, Mariana Orellana and Anabella T. Araudo. I have learned many things from discussions on the topics of this book with friends and colleagues. I want to mention here to Santiago Perez-Bergliaffa, Mario Novello, Felipe Tobar Falciano, Nelson Pinto Nieto, Josep Maria Paredes, Felix Mirabel, Felix Aharonian, Valenti Bosch-Ramon, Brian Punsly, Frank Rieger, Mario Bunge, Jorge Combi, Paula Benaglia, Sergio Cellone, Dmitri Khangulyan, Leonardo Pellizza, Ernesto Eiroa, Atsuo Okazaki, Stan Owocki, Wolfgang Kundt, Orlando Peres, Jun-Hui Fan, K.S. Cheng, Tiberio Harko, Marc Ribó, Pol Bordas, Diego Torres, Luis Anchordoqui, Margarita Safonova, and Daniela Pérez. I am also very grateful to some of my present and former PhD students for questions and discussions, especially to Marina Kaufman Bernadó, Mariana Orellana, Ernesto Eiroa, Ileana Andruchow, Matías Reynoso, Anabella T. Araudo, Gabriela S. Vila, Daniela Pérez, Florencia Vieyro, and María Victoria del Valle.

Thanks also to my family, for patience and encouragement.

Gustavo E. Romero

My first and deepest acknowledgment goes to my co-author, Gustavo E. Romero. Over the last seven years he has thought me most of what I know about many fascinating areas of astrophysics. He has also been a permanent source of encouragement and a good friend.

I thank as well Valenti Bosch-Ramon, Matías Reynoso, Leonardo Pellizza, Florencia Vieyro, María Victoria del Valle, and Daniela Pérez for many fruitful discussions and collaborations. I am also very grateful to Drs. Felix Aharonian and Josep Maria Paredes for their support.

Finally, I want to thank Nicolás Casco for his endless patience and unconditional support, and also for having helped me with some technical aspects of the writing of this book.

Gabriela S. Vila

We both are grateful to Ramon Khanna for encouraging us to write this book, and for his guidance during each stage of its preparation.

We are pleased to thank those colleagues that have kindly granted us permission to illustrate the book with figures taken from their works: Felix Aharonian, Maxim Barkov, Andrei Beloborodov, John Blondin, Alfio Bonanno, Pol Bordas, Avery Broderick, Heino Falcke, Stefan Gillessen, Alexander Heger, Pablo Jimeno Romero, Yoshi Kato, Shinji Koide, Serguei Komissarov, Nicholas McConnell, Yosuke Mizuno, Ramesh Narayan, Brian Punsly, Mark Reynolds, Ken Rice, and Alexander Tchekhovskoy. We also acknowledge the kind collaboration of the following publishers and editorials: American Astronomical Society, American Institute of Physics, American Physical Society, EDP Sciences, Elsevier, and Oxford University Press.

This work was supported, in part, by a research grant from CONICET (PIP 2010-0078) and ANPCyT (PICT 2007-00848 / PICT 2012-00878), as well as by Spanish Ministerio de Ciencia e Innovación (MICINN) under grant AYA2010-21782-C03-01.

Contents

Acronyms

ADAF	Advection-Dominated Accretion Flow
ADIOS	Advection-Dominated Inflow Outflow Solution
AGN	Active Galactic Nucleus (Nuclei)
ANEC	Average Null Energy Condition
AWEC	Average Weak Energy Condition
BATSE	Burst and Transient Source Experiment
BH	Black Hole
BLRG	Broad Line Radio Galaxy
BLR	Broad Line Region
CDAF	Convection-Dominated Accretion Flow
CDI	Current-Driven Instability
CMB	Cosmic Microwave Background
CTA	Cherenkov Telescope Array
CTC	Closed Time-like Curve
DECIGO	DECI-Hertz Interferometer Gravitational wave Observatory
EGRET	Energetic Gamma-Ray Experiment Telescope
FLRW	Friedmann-Lemaitre-Robertson-Walker (metric)
FR	Fanaroff-Riley (radio galaxy)
FSRQ	Flat Spectrum Radio Quasar
GRB	Gamma-Ray Burst
HMXRB	High-Mass X-Ray Binary
HS	High-Soft (X-ray state)
IC	Inverse Compton (scattering)
IGM	Inter Galactic Medium
IMBH	Intermediate-Mass Black Hole
ISM	Inter Stellar Medium
KHI	Kelvin-Helmholtz Instability
LAT	Large Area Telescope
LED	Large Extra Dimension(s)
LH	Low-Hard (X-ray state)
LIGO	Laser Interferometer Gravitational-wave Observatory

LISA	Laser Interferometer Space Antenna
LMXRB	Low-Mass X-Ray Binary
MAGIC	Major Atmospheric Gamma-ray Imaging Cherenkov Telescopes
MDAF	Magnetically-Dominated Accretion Flow
MHD	Magneto-Hydro-Dynamics
MQ	MicroQuasar
NGO	New Gravitational wave Observatory
NLRG	Narrow Line Radio Galaxy
NLR	Narrow Line Region
NS	Neutron Star
PSC	Principle of Self-Consistency
PTA	Pulsar Timing Array
QPO	Quasi-Periodic Oscillation
QSO	Quasi-Stellar Object
QSRS	Quasi-Stellar Radio Source
RIAF	Radiatively Inefficient Accretion Flow
SLS	Stationary Limit Surface
SPH	Smoothed Particle Hydrodynamics
ULX	Ultra Luminous X-ray (source)
VH	Very High (X-ray state)
VLBI	Very Large Baseline Interferometry
XRB	X-Ray Binary

Chapter 1
Space-Time and Gravitation

1.1 Space-Time

Strictly speaking, black holes do not exist. Moreover, holes, of any kind, do not exist. You can talk about holes of course. For instance you can say: "there is a hole in the wall". You can give many details of the hole: it is big, it is round-shaped, light comes in through it. Even, perhaps, the hole could be such that you can go through it to the outside. But we are sure that you do not think that there is a thing made out of nothingness in the wall. Certainly not. To talk about the hole is an indirect way of talking about the wall. What really exists is the wall. The wall is made out of bricks, atoms, protons and leptons, whatever. To say that there is a hole in the wall is just to say that the wall has certain topology, a topology such that not every closed curve on the surface of the wall can be continuously contracted to a single point. The hole is not a thing. The hole is a property of the wall.

Let us come back to black holes. What are we talking about when we talk about black holes? *Space-time*. What is space-time?

Space-time is the ontological sum of all events of all things.

A thing is an individual endowed with physical properties. An event is a change in the properties of a thing. An ontological sum is an aggregation of things or physical properties, i.e. a physical entity or an emergent property. An ontological sum should not be confused with a set, which is a mathematical construct and has only mathematical (i.e. fictional) properties.

Everything that has happened, everything that happens, everything that will happen, is just an element, a "point", of space-time. Space-time is not a thing, it is just the relational property of all things.[1]

As seems to happen with every physical property, we can represent space-time with some mathematical structure, in order to describe it. We shall adopt the following mathematical structure for space-time:

---

[1] For more details on this view see Perez-Bergliaffa et al. (1998) and Romero (2013).

G.E. Romero, G.S. Vila, *Introduction to Black Hole Astrophysics*,
Lecture Notes in Physics 876, DOI 10.1007/978-3-642-39596-3_1,
© Springer-Verlag Berlin Heidelberg 2014

Space-time can be represented by a C^∞-differentiable, 4-dimensional, real manifold.

A real 4-D manifold[2] is a set that can be covered completely by subsets whose elements are in a one-to-one correspondence with subsets of $\mathfrak{R}^4$, the 4-dimensional space of real numbers. Each element of the manifold represents an event. We adopt 4 dimensions because it seems enough to give 4 real numbers to localize an event. For instance, a lightning has beaten the top of the building at 25 m above the sea level, located in the 38th Av., between streets 20 and 21, La Plata city, at 4:35 am, local time, March 22nd, 2012 (this is Romero's home at the time of writing). We see now why we choose a manifold to represent space-time: we can always provide a set of 4 real numbers for every event, and this can be done independently of the intrinsic geometry of the manifold. If there is more than a single characterization of an event, we can always find a transformation law between the different coordinate systems of each characterization. This is a basic property of manifolds.

Now, if we want to calculate distances between two events, we need more structure on our manifold: we need a geometric structure. We can get this introducing a metric tensor that allows to calculate distances. For instance, consider an Euclidean metric tensor $\delta_{\mu\nu}$ (indices run from 0 to 3):

$$\delta_{\mu\nu} = \begin{pmatrix} 1 & 0 & 0 & 0 \\ 0 & 1 & 0 & 0 \\ 0 & 0 & 1 & 0 \\ 0 & 0 & 0 & 1 \end{pmatrix}. \tag{1.1}$$

Then, adopting the Einstein convention of sum, we have that the distance ds between two arbitrarily close events is:

$$ds^2 = \delta_{\mu\nu}dx^\mu dx^\nu = \left(dx^0\right)^2 + \left(dx^1\right)^2 + \left(dx^3\right)^2 + \left(dx^3\right)^2. \tag{1.2}$$

Restricted to 3 coordinates, this is the way distances have been calculated since Pythagoras. The world, however, seems to be a little more complicated. After the introduction of the Special Theory of Relativity by Einstein (1905), the German mathematician Hermann Minkowski introduced the following pseudo-Euclidean metric which is consistent with Einstein's theory (Minkowski 1907, 1909):

$$ds^2 = \eta_{\mu\nu}dx^\mu dx^\nu = \left(dx^0\right)^2 - \left(dx^1\right)^2 - \left(dx^3\right)^2 - \left(dx^3\right)^2. \tag{1.3}$$

The Minkowski metric tensor $\eta_{\mu\nu}$ has rank 2 and trace -2. We call the coordinates with the same sign *spatial coordinates* (adopting the convention $x^1 = x$, $x^2 = y$, and $x^3 = z$) and the coordinate $x^0 = ct$ is called *temporal coordinate*. The constant c is introduced to make the units uniform.

There is an important fact about Eq. (1.3): contrary to what was thought by Kant and others, it is not a necessary statement. Things might have been different. We can

---

[2]See Appendix A.

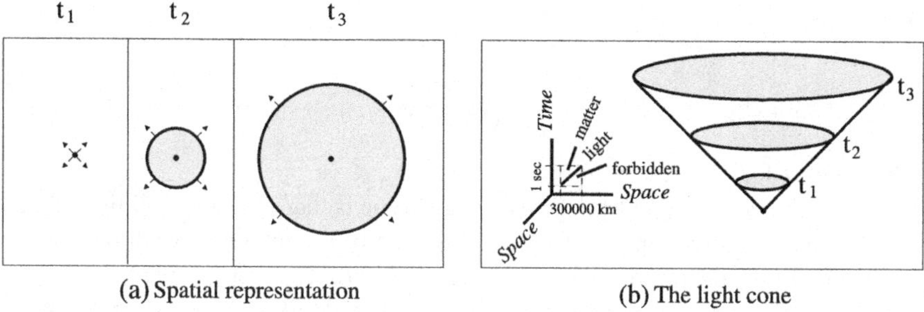

Fig. 1.1 Light cone. From J-P. Luminet (1998)

easily imagine possible worlds with other metrics. This means that the metric tensor has empirical information about the real universe.

Once we have introduced a metric tensor we can separate space-time at each point in three regions according to $ds^2 < 0$ (space-like region), $ds^2 = 0$ (light-like or null region), and $ds^2 > 0$ (time-like region). Particles that go through the origin can only reach time-like regions. The null surface $ds^2 = 0$ can be inhabited only by particles moving at the speed of light, like photons. Points in the space-like region cannot be reached by material objects from the origin of the *light cone* that can be formed at any space-time point. A light cone is shown in Fig. 1.1.

The introduction of the metric allows to define the future and the past of a given event. Once this is done, all events can be classified by the relation "earlier than" or "later than". The selection of a "present" event - or "now" - is entirely conventional. To be present is not an intrinsic property of any event. Rather, it is a secondary, relational property that requires interaction with a conscious being. The extinction of the dinosaurs will always be earlier than the beginning of World War II. But the latter was present only to some human beings at some physical state. The present is a property like a scent or a color. It emerges from the interaction of self-conscious individuals with changing things and has not existence independently of them (for more about this, see Grünbaum 1973 Chap. X, and Romero 2012).

Let us consider the unitary vector $T^\nu = (1, 0, 0, 0)$. A vector x^μ points to the future if $\eta_{\mu\nu}x^\mu T^\nu > 0$. In a similar way, the vector points towards the past if $\eta_{\mu\nu}x^\mu T^\nu < 0$.

We define the *proper time* τ of a physical system as the time of a co-moving system, i.e. $dx = dy = dz = 0$, and hence:

$$d\tau^2 = \frac{1}{c^2}ds^2. \tag{1.4}$$

Since the interval is an invariant (i.e. it has the same value in all coordinate systems), it is easy to show that

$$d\tau = \frac{dt}{\gamma}, \tag{1.5}$$

where

$$\gamma = \frac{1}{\sqrt{1 - (\frac{v}{c})^2}} \tag{1.6}$$

is the Lorentz factor of the system.

A basic characteristic of Minskowski space-time is that it is "flat": all light cones point in the same direction, i.e. the local direction of the future does not depend on the coefficients of the metric since these are constant. More general space-times are possible. If we want to describe gravity in the framework of space-time, we have to introduce a pseudo-Riemannian space-time, whose metric can be flexible, i.e. a function of the material properties (mass-energy and momentum) of the physical systems that produce the events of space-time.

1.2 Tetrads: Orthogonal Unit Vector Fields

Let us consider a scalar product

$$\mathbf{v} \bullet \mathbf{w} = \left(v^{\mu}\hat{e}_{\mu}\right) \bullet \left(w^{\nu}\hat{e}_{\nu}\right) = (\hat{e}_{\mu} \bullet \hat{e}_{\nu})v^{\mu}w^{\nu} = g_{\mu\nu}v^{\mu}w^{\nu},$$

where

$$\hat{e}_{\mu} = \lim_{\delta x^{\mu} \to 0} \frac{\delta \mathbf{s}}{\delta x^{\mu}},$$

and we have defined

$$\hat{e}_{\mu}(x) \bullet \hat{e}_{\nu}(x) = g_{\mu\nu}(x).$$

Similarly,

$$\hat{e}^{\mu}(x) \bullet \hat{e}^{\nu}(x) = g^{\mu\nu}(x).$$

We call $\hat{e}_{\mu}$ a *coordinate basis vector* or a *tetrad*. Here $\delta \mathbf{s}$ is an infinitesimal displacement vector between a point P on the manifold (see Fig. 1.2) and a nearby point Q whose coordinate separation is δx^{μ} along the x^{μ} coordinate curve. $\hat{e}_{\mu}$ is the tangent vector to the x^{μ} curve at P. We can write

$$d\mathbf{s} = \widehat{e}_{\mu}dx^{\mu}$$

and then

$$ds^2 = d\mathbf{s} \bullet d\mathbf{s} = \left(dx^{\mu}\hat{e}_{\mu}\right) \bullet \left(dx^{\nu}\hat{e}_{\nu}\right) = (\hat{e}_{\mu} \bullet \hat{e}_{\nu})dx^{\mu}dx^{\nu} = g_{\mu\nu}dx^{\mu}dx^{\nu}.$$

At any given point P the manifold is flat, so

$$g_{\mu\nu}(P) = \eta_{\mu\nu}.$$

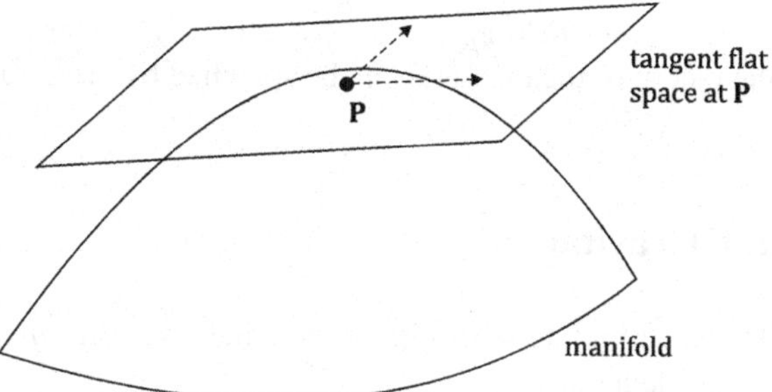

Fig. 1.2 Tangent flat space at a point P of a curved manifold

A manifold with such a property is called *pseudo-Riemannian*. If $g_{\mu\nu}(P) = \delta_{\mu\nu}$ the manifold is called strictly *Riemannian*.

The basis is called *orthonormal* when $\hat{e}^{\mu} \bullet \hat{e}_{\nu} = \eta_{\nu}^{\mu}$ at any given point P. Notice that since the tetrads are 4-dimensional we can write

$$e_{\mu a}(x)e_{\nu}^{a}(x) = g_{\mu\nu}(x),$$

and

$$e_{\mu a}(P)e_{\nu}^{a}(P) = \eta_{\mu\nu}.$$

The tetrads can vary along a given world-line, but always satisfying

$$e_{\mu a}(\tau)e_{\nu}^{a}(\tau) = \eta_{\mu\nu}.$$

We can also express the scalar product $\mathbf{v} \bullet \mathbf{w}$ in the following ways:

$$\mathbf{v} \bullet \mathbf{w} = \left(v_{\mu}\hat{e}^{\mu}\right) \bullet \left(w_{\nu}\hat{e}^{\nu}\right) = \left(\hat{e}^{\mu} \bullet \hat{e}^{\nu}\right)v_{\mu}w_{\nu} = g^{\mu\nu}v_{\mu}w_{\nu},$$

$$\mathbf{v} \bullet \mathbf{w} = \left(v^{\mu}\hat{e}_{\mu}\right) \bullet \left(w_{\nu}\hat{e}^{\nu}\right) = \left(\hat{e}_{\mu} \bullet \hat{e}^{\nu}\right)v^{\mu}w_{\nu} = v^{\mu}w_{\nu}\delta_{\mu}^{\nu} = v^{\mu}w_{\mu}$$

and

$$\mathbf{v} \bullet \mathbf{w} = \left(v_{\mu}\hat{e}^{\mu}\right) \bullet \left(w^{\nu}\hat{e}_{\nu}\right) = \left(\hat{e}^{\mu} \bullet \hat{e}_{\nu}\right)v_{\mu}w^{\nu} = \delta_{\nu}^{\mu}v_{\mu}w^{\nu} = v_{\mu}w^{\mu}.$$

By comparing these expressions for the scalar product of two vectors we see that

$$g_{\mu\nu}w^{\nu} = w_{\mu},$$

so the quantities $g_{\mu\nu}$ can be used to lower and raise indices. Similarly,

$$g^{\mu\nu}w_{\nu} = w^{\mu}.$$

We also have that

$$g_{\mu\nu}w^{\nu}g^{\mu\nu}w_{\nu} = g_{\mu\nu}g^{\mu\nu}w^{\nu}w_{\nu} = w_{\mu}w^{\mu}.$$

From here it follows

$$g^{\mu\nu}g_{\mu\sigma} = \delta_{\sigma}^{\nu}.$$

The tensor field $g_{\mu\nu}(x)$ is called the *metric tensor* of the manifold. Alternative, the metric of the manifold can be specified by the tetrads $e_\mu^a(x)$.

1.3 Gravitation

The key to relate space-time to gravitation is the *equivalence principle* introduced by Einstein (1907):

> At every space-time point in an arbitrary gravitational field it is possible to choose a locally inertial coordinate system such that, within a sufficiently small region of the point in question, the laws of nature take the same form as in unaccelerated Cartesian coordinate systems in absence of gravitation (formulation by Weinberg 1972).

This is equivalent to state that at every point P of the manifold that represents space-time there is a flat tangent surface. Einstein called the idea that gravitation vanishes in free-falling systems "the happiest thought of my life" (Pais 1982).

In order to introduce gravitation in a general space-time we define a metric tensor $g_{\mu\nu}$, such that its components can be related to those of a locally Minkowski space-time defined by $ds^2 = \eta_{\alpha\beta}d\xi^\alpha d\xi^\beta$ through a general transformation:

$$ds^2 = \eta_{\alpha\beta}\frac{\partial\xi^\alpha}{\partial x^\mu}\frac{\partial\xi^\beta}{\partial x^\nu}dx^\mu dx^\nu = g_{\mu\nu}dx^\mu dx^\nu. \tag{1.7}$$

In the absence of gravity we can always find a global coordinate system (ξ^α) for which the metric takes the form given by Eq. (1.3) everywhere. With gravity, on the contrary, such a coordinate system can represent space-time only in an infinitesimal neighborhood of a given point. This situation is represented in Fig. 1.2, where the tangent flat space to a point P of the manifold is shown. The curvature of space-time means that it is not possible to find coordinates in which $g_{\mu\nu} = \eta_{\mu\nu}$ at *all points on the manifold*. However, it is always possible to represent the event (point) P in a system such that $g_{\mu\nu}(P) = \eta_{\mu\nu}$ and $(\partial g_{\mu\nu}/\partial x^\sigma)_P = 0$.

To find the equation of motion of a free particle (i.e. only subject to gravity) in a general space-time of metric $g_{\mu\nu}$ let us consider a freely falling coordinate system ξ^α. In such a system

$$\frac{d^2\xi^\alpha}{ds^2} = 0, \tag{1.8}$$

where $ds^2 = (cd\tau)^2 = \eta_{\alpha\beta}d\xi^\alpha d\xi^\beta$.

Let us consider now any other coordinate system x^μ. Then,

$$0 = \frac{d}{ds}\left(\frac{\partial\xi^\alpha}{\partial x^\mu}\frac{dx^\mu}{ds}\right), \tag{1.9}$$

$$0 = \frac{\partial\xi^\alpha}{\partial x^\mu}\frac{d^2 x^\mu}{ds^2} + \frac{\partial^2\xi^\alpha}{\partial x^\mu\partial x^\nu}\frac{dx^\mu}{ds}\frac{dx^\nu}{ds}. \tag{1.10}$$

Multiplying both sides by $\partial x^\lambda / \partial \xi^\alpha$ and using

$$\frac{\partial \xi^\alpha}{\partial x^\mu} \frac{\partial x^\lambda}{\partial \xi^\alpha} = \delta^\lambda_\mu, \tag{1.11}$$

we get

$$\frac{d^2 x^\lambda}{ds^2} + \Gamma^\lambda_{\mu\nu} \frac{dx^\mu}{ds} \frac{dx^\nu}{ds} = 0, \tag{1.12}$$

where $\Gamma^\lambda_{\mu\nu}$ is the *affine connection* of the manifold:

$$\Gamma^\lambda_{\mu\nu} \equiv \frac{\partial x^\lambda}{\partial \xi^\alpha} \frac{\partial^2 \xi^\alpha}{\partial x^\mu \partial x^\nu}. \tag{1.13}$$

The affine connection can be expressed in terms of derivatives of the metric tensor (see, e.g., Weinberg 1972):

$$\Gamma^\lambda_{\mu\nu} = \frac{1}{2} g^{\lambda\alpha} (\partial_\mu g_{\nu\alpha} + \partial_\nu g_{\mu\alpha} - \partial_\alpha g_{\mu\nu}). \tag{1.14}$$

Here we have used that $g^{\mu\alpha} g_{\alpha\nu} = \delta^\mu_\nu$, and the notation $\partial_\nu f = \partial f / \partial x^\nu$. Notice that under a coordinate transformation from $x^\mu \to x'^\mu$ the affine connection is not transformed as a tensor, despite the metric $g_{\mu\nu}$ is a tensor of second rank.

The coefficients $\Gamma^\lambda_{\mu\nu}$ are said to define a *connection* on the manifold. What is connected are the tangent spaces at different points of the manifold, in such a way that it is possible to connect a vector in the tangent space at point P with the vector parallel to it at another point Q. There is some degree of freedom in the specification of the affine connection, so we demand symmetry in the last two indices:

$$\Gamma^\lambda_{\mu\nu} = \Gamma^\lambda_{\nu\mu}$$

or

$$\Gamma^\lambda_{[\mu\nu]} = \Gamma^\lambda_{\mu\nu} - \Gamma^\lambda_{\nu\mu} = 0.$$

In general space-times this requirement is not necessary, and a tensor can be introduced such that

$$T^\lambda_{\mu\nu} = \Gamma^\lambda_{[\mu\nu]}. \tag{1.15}$$

This tensor represents the *torsion* of space-time. In General Relativity space-time is always considered as torsionless, but in the so-called *teleparallel* theory of gravity (e.g. Einstein 1928; Arcos and Pereira 2004) torsion represents the gravitational field instead of curvature, which is nil.

In a pseudo-Riemannian space-time the usual partial derivative is not a meaningful quantity since we can give it different values through different choices of coordinates. This can be seen in the way the derivative transforms under a coordinate change. Defining $f_{,v} = \partial_v f$, we have

$$A'^{\mu}_{,v} = \frac{\partial}{\partial x'^v}\left(\frac{\partial x'^\mu}{\partial x^\mu} A^\mu\right) = \frac{\partial x'^\mu}{\partial x^\mu}\frac{\partial x^v}{\partial x'^v} A^\mu_{,v} + \frac{\partial^2 x'^\mu}{\partial x^\mu \partial x^v}\frac{\partial x^v}{\partial x'^v} A^\mu. \tag{1.16}$$

We can define a covariant differentiation $A_{\mu;v}$ through the condition of parallel transport:

$$A_{\mu;v} = \frac{\partial A_\mu}{\partial x^v} - \Gamma^\lambda_{\mu v} A_\lambda. \tag{1.17}$$

A useful, alternative notation, is

$$\nabla_v A_\mu = \frac{\partial A_\mu}{\partial x^v} - \Gamma^\lambda_{\mu v} A_\lambda. \tag{1.18}$$

A covariant derivative of a vector field is a rank 2 tensor of type $(1, 1)$. The covariant divergence of a vector field yields a scalar field:

$$\nabla_\mu A^\mu = \partial_\mu A^\mu(x) - \Gamma^\mu_{\alpha\mu} A^\alpha(x) = \phi(x). \tag{1.19}$$

A tangent vector satisfies $V^v V_{v;\mu} = 0$. If there is a vector ζ^μ pointing in the direction of a symmetry of space-time, then it can be shown that (e.g. Weinberg 1972)

$$\zeta_{\mu;v} + \zeta_{v;\mu} = 0, \tag{1.20}$$

or

$$\nabla_v \zeta_\mu + \nabla_\mu \zeta_v = 0. \tag{1.21}$$

This equation is called Killing's equation. A vector field ζ^μ satisfying such a relation is called a Killing field.

If there is a curve γ on the manifold, such that its tangent vector is $u^\alpha = dx^\alpha/d\lambda$ and a vector field A^α is defined in a neighborhood of γ, we can introduce a derivative of A^α along γ as

$$\ell_u A^\alpha = A^\alpha_{,\beta} u^\beta - u^\alpha_{,\beta} A^\beta = A^\alpha_{;\beta} u^\beta - u^\alpha_{;\beta} A^\beta. \tag{1.22}$$

This derivative is a tensor, and it is usually called *Lie derivative*. It can be defined for tensors of any type. A Killing vector field is such that

$$\ell_\zeta g_{\mu v} = 0. \tag{1.23}$$

From Eq. (1.12) we can recover the classical Newtonian equations if

$$\Gamma^0_{i,j} = 0, \qquad \Gamma^i_{0,j} = 0, \qquad \Gamma^i_{0,0} = \frac{\partial \Phi}{\partial x^i},$$

where $i,\ j = 1, 2, 3$ and Φ is the Newtonian gravitational potential. Then

$$x^0 = ct = c\tau,$$

$$\frac{d^2 x^i}{d\tau^2} = -\frac{\partial \Phi}{\partial x^i}.$$

We see, hence, that the metric represents the gravitational potential and the affine connection the gravitational field.

The presence of gravity is indicated by the curvature of space-time. The Riemann tensor, or curvature tensor, provides a measure of this curvature:

$$R^\sigma_{\mu\nu\lambda} = \Gamma^\sigma_{\mu\lambda,\nu} - \Gamma^\sigma_{\mu\nu,\lambda} + \Gamma^\sigma_{\alpha\nu}\Gamma^\alpha_{\mu\lambda} - \Gamma^\sigma_{\alpha\lambda}\Gamma^\alpha_{\mu\nu}. \tag{1.24}$$

The form of the Riemann tensor for an affine-connected manifold can be obtained through a coordinate transformation $x^\mu \to \bar{x}^\mu$ that makes the affine connection vanish everywhere, i.e.

$$\bar{\Gamma}^\sigma_{\mu\nu}(\bar{x}) = 0, \quad \forall \bar{x},\ \sigma,\ \mu,\ \nu. \tag{1.25}$$

The coordinate system $\bar{x}^\mu$ exists if

$$\Gamma^\sigma_{\mu\lambda,\nu} - \Gamma^\sigma_{\mu\nu,\lambda} + \Gamma^\sigma_{\alpha\nu}\Gamma^\alpha_{\mu\lambda} - \Gamma^\sigma_{\alpha\lambda}\Gamma^\alpha_{\mu\nu} = 0 \tag{1.26}$$

for the affine connection $\Gamma^\sigma_{\mu\nu}(x)$. The left hand side of Eq. (1.26) is the Riemann tensor $R^\sigma_{\mu\nu\lambda}$. In such a case the metric is flat, since its derivatives are zero. If $R^\sigma_{\mu\nu\lambda} > 0$ the metric has a positive curvature.

The Ricci tensor is defined as

$$R_{\mu\nu} = g^{\lambda\sigma} R_{\lambda\mu\sigma\nu} = R^\sigma_{\mu\sigma\nu}. \tag{1.27}$$

Finally, the Ricci scalar is

$$R = g^{\mu\nu} R_{\mu\nu}. \tag{1.28}$$

1.4 Field Equations

The key issue to determine the geometric structure of space-time, and hence to specify the effects of gravity, is to find the law that fixes the metric once the source of the gravitational field is given. The source of the gravitational field is the energy-momentum tensor $T_{\mu\nu}$ that represents the physical properties of material things. This was Einstein's fundamental intuition: the curvature of space-time at any event is related to the energy-momentum content at that event. For example, for the simple case of a perfect fluid the energy-momentum tensor takes the form

$$T_{\mu\nu} = (\rho + P)u_\mu u_\nu - P g_{\mu\nu}, \tag{1.29}$$

Fig. 1.3 Albert Einstein during a lecture in Vienna in 1921, by the Austrian photographer Ferdinand Schmutzer. From Wikimedia Commons, http://commons.wikimedia.org

where ρ is the mass-energy density, P is the pressure, and $u^{\mu} = dx^{\mu}/ds$ is the 4-velocity.

The field equations were found by Einstein (1915) and Hilbert (1915) on November 1915.[3] We can write Einstein's physical intuition in the following form:

$$K_{\mu\nu} = \kappa T_{\mu\nu}, \tag{1.30}$$

where $K_{\mu\nu}$ is a rank-2 tensor related to the curvature of space-time and κ is a constant. Since the curvature is expressed by $R_{\mu\nu\sigma\rho}$, $K_{\mu\nu}$ must be constructed from this tensor and the metric tensor $g_{\mu\nu}$. The tensor $K_{\mu\nu}$ has to satisfy the following properties:

(i) the Newtonian limit $\nabla^2 \Phi = 4\pi G\rho$ suggests that it should contain terms no higher than linear in the second order derivatives of the metric tensor;
(ii) since $T_{\mu\nu}$ is symmetric then $K_{\mu\nu}$ must be symmetric as well.

---

[3]Recent scholarship has arrived to the conclusion that Einstein (Fig. 1.3) was the first to find the equations and that Hilbert incorporated the final form of the equations in the proof reading process, after Einstein's communication (Corry et al. 1997).

Since $R_{\mu\nu\sigma\rho}$ is already linear in the second order derivatives of the metric, the most general form of $K_{\mu\nu}$ is

$$K_{\mu\nu} = aR_{\mu\nu} + bRg_{\mu\nu} + \lambda g_{\mu\nu}, \tag{1.31}$$

where a, b, and λ are constants.

If *every* term in $K_{\mu\nu}$ must be linear in the second order derivatives of $g_{\mu\nu}$, then $\lambda = 0$. Hence

$$K_{\mu\nu} = aR_{\mu\nu} + bRg_{\mu\nu}. \tag{1.32}$$

The conservation of energy-momentum requires that $T^{\mu\nu}_{\ \ ;\mu} = 0$. So,

$$\left(aR^{\mu\nu} + bRg^{\mu\nu}\right)_{;\mu} = 0. \tag{1.33}$$

Also, it happens that (Bianchi identities)

$$\left(R^{\mu\nu} - \frac{1}{2}Rg^{\mu\nu}\right)_{;\mu} = 0. \tag{1.34}$$

From here we get $b = -a/2$ and $a = 1$. We can then re-write the field equations as

$$\left(R_{\mu\nu} - \frac{1}{2}Rg_{\mu\nu}\right) = \kappa T_{\mu\nu}. \tag{1.35}$$

In order to fix κ, we must compare the weak-field limit of these equations with the Poisson's equations of Newtonian gravity. This requires that $\kappa = -8\pi G/c^4$.

The Einstein field equations can then be written in the simple form

$$R_{\mu\nu} - \frac{1}{2}g_{\mu\nu}R = -\frac{8\pi G}{c^4}T_{\mu\nu}. \tag{1.36}$$

This is a set of ten non-linear partial differential equations for the metric coefficients. In Newtonian gravity, otherwise, there is only one gravitational field equation. General Relativity involves numerous non-linear differential equations. In this fact lies its complexity, and its richness.

The conservation of mass-energy and momentum can be derived from the field equations:

$$T^{\mu\nu}_{\ \ ;\nu} = 0 \quad \text{or} \quad \nabla_\nu T^{\mu\nu} = 0. \tag{1.37}$$

Contrary to classical electrodynamics, here the field equations entail the energy-momentum conservation and the equations of motion for free particles (i.e. for particles moving in the gravitational field, treated here as a background pseudo-Riemannian space-time).

Let us consider, for example, a distribution of dust (i.e. a pressureless perfect fluid) for which the energy-momentum tensor is

$$T^{\mu\nu} = \rho u^\mu u^\nu, \tag{1.38}$$

with u^μ the 4-velocity. Then,

$$T^{\mu\nu}_{\ ;\mu} = \left(\rho\, u^\mu u^\nu\right)_{;\mu} = \left(\rho\, u^\mu\right)_{;\mu} u^\nu + \rho\, u^\mu u^\nu_{\ ;\mu} = 0. \tag{1.39}$$

Contracting with u_ν,

$$c^2\left(\rho\, u^\mu\right)_{;\mu} + \left(\rho\, u^\mu\right) u_\nu u^\nu_{\ ;\mu} = 0, \tag{1.40}$$

where we used $u^\nu u_\nu = c^2$. Since the second term on the left is zero, we have

$$\left(\rho u^\mu\right)_{;\mu} = 0. \tag{1.41}$$

Replacing in Eq. (1.39) we obtain

$$u^\mu u^\nu_{\ ;\nu} = 0, \tag{1.42}$$

which is the equation of motion for the dust distribution in the gravitational field.

Einstein's equations (1.36) can be cast in the form

$$R^\mu_\nu - \frac{1}{2}\delta^\mu_\nu R = -\frac{8\pi G}{c^4}\, T^\mu_\nu. \tag{1.43}$$

Contracting by setting $\mu = \nu$ we get

$$R = -\frac{16\pi G}{c^4}\, T, \tag{1.44}$$

where $T = T^\mu_\mu$. Replacing the curvature scalar in Eqs. (1.36) we obtain the alternative form

$$R_{\mu\nu} = -\frac{8\pi G}{c^4}\left(T_{\mu\nu} - \frac{1}{2}T g_{\mu\nu}\right). \tag{1.45}$$

In a region of empty space, $T_{\mu\nu} = 0$ and then

$$R_{\mu\nu} = 0, \tag{1.46}$$

i.e. the Ricci tensor vanishes. The curvature tensor, which has 20 independent components, does not necessarily vanish. This means that a gravitational field can exist in empty space only if the dimensionality of space-time is 4 or higher. For space-times with lower dimensionality, the curvature tensor vanishes if $T_{\mu\nu} = 0$. The components of the curvature tensor that are not zero in empty space are contained in the Weyl tensor (see Sect. 1.10 below for a definition of the Weyl tensor). Hence, the Weyl tensor describes the curvature of empty space. Absence of curvature (flatness) demands that both the Ricci and Weyl tensors should be zero.

1.5 The Cosmological Constant

The set of Einstein's equations is not unique: we can add any constant multiple of $g_{\mu\nu}$ to the left member of (1.36) and still obtain a *consistent* set of equations. It is usual to denote this multiple by Λ, so the field equations can also be written as

$$R_{\mu\nu} - \frac{1}{2}g_{\mu\nu}R + \Lambda g_{\mu\nu} = -\frac{8\pi G}{c^4}\,T_{\mu\nu}. \qquad (1.47)$$

The constant Λ is a new universal constant called, because of historical reasons, the *cosmological constant*. If we consider some kind of "substance" with equation of state given by $P = -\rho c^2$, then its energy-momentum tensor would be

$$T_{\mu\nu} = -P\,g_{\mu\nu} = \rho\,c^2 g_{\mu\nu}. \qquad (1.48)$$

Notice that the energy-momentum tensor of this substance depends only on the space-time metric $g_{\mu\nu}$, so it describes a property of the "vacuum" itself. We can call ρ the energy density of the vacuum field. Then, we rewrite Eq. (1.47) as

$$R_{\mu\nu} - \frac{1}{2}g_{\mu\nu}R = -\frac{8\pi G}{c^4}\left(T_{\mu\nu} + T_{\mu\nu}^{\mathrm{vac}}\right), \qquad (1.49)$$

in such a way that

$$\rho_{\mathrm{vac}}c^2 = \frac{\Lambda c^4}{8\pi G}. \qquad (1.50)$$

There is evidence (e.g. Reiss et al. 1998; Perlmutter et al. 1999) that the energy density of the vacuum is different from zero. This means that Λ is small, but not zero.[4] The negative pressure seems to be driving a "cosmic acceleration".

There is a simpler interpretation of the repulsive force that produces the accelerate expansion: there is not a dark field. The only field is gravity, represented by $g_{\mu\nu}$. What is different is the law of gravitation: instead of being given by Eqs. (1.36), it is expressed by Eqs. (1.47); gravity can be repulsive under some circumstances.

Despite the complexity of Einstein's field equations a large number of exact solutions have been found. They are usually obtained imposing symmetries on the space-time in such a way that the metric coefficients can be calculated. The first and most general solution to Eqs. (1.36) was obtained by Karl Schwarzschild in 1916, short before he died in the Eastern Front of the Great War. This solution, as we shall see, describes a non-rotating black hole of mass M.

---

[4]The current value is around 10^{-29} g cm^{-3}.

1.6 Relativistic Action

Let us consider a mechanical system whose configuration can be uniquely defined by generalized coordinates q^a, $a = 1, 2, \ldots, n$. The *action* of such a system is

$$S = \int_{t_1}^{t_2} L\left(q^a, \dot{q}^a, t\right) dt, \tag{1.51}$$

where t is the time. The *Lagrangian* L is defined in terms of the kinetic energy T of the system and the potential energy U:

$$L = T - U = \frac{1}{2} m\, g_{ab}\, \dot{q}^a\, \dot{q}^b - U, \tag{1.52}$$

where g_{ab} is the metric of the configuration space, $ds^2 = g_{ab}\, dq^a\, dq^b$. Hamilton's principle states that for arbitrary variations such as

$$q^a(t) \to q'^a(t) = q^a(t) + \delta q^a(t), \tag{1.53}$$

the variation of the action δS vanishes. Assuming that $\delta q^a(t) = 0$ at the endpoints t_1 and t_2 of the trajectory, it can be shown that the Lagrangian must satisfy the Euler-Lagrange equations:

$$\frac{\partial L}{\partial q^a} - \frac{d}{dt}\left(\frac{\partial L}{\partial \dot{q}^a}\right) = 0, \quad a = 1, 2, \ldots, n. \tag{1.54}$$

These are the equations of motion of the system.

In the case of the action of a set of fields[5] defined on some general four dimensional space-time manifold, we can introduce a *Lagrangian density* $\mathcal{L}$ of the fields and their derivatives:

$$S = \int_{\mathcal{R}} \mathcal{L}\left(\Phi^a,\, \partial_\mu \Phi^a,\, \partial_\mu \partial_\nu \Phi^a, \ldots\right) d^4 x, \tag{1.55}$$

where $d^4 x = dx^0 dx^1 dx^2 dx^3$, Φ^a is a field on the manifold, and $\mathcal{R}$ is a region of the manifold. The action should be a scalar, then we should use the element of volume in a coordinate system x^μ written in the invariant form $\sqrt{-g}\, d^4 x$, where $g = \| g^{\mu\nu} \|$ is the determinant of the metric tensor in that coordinate system. The corresponding action is

$$S = \int_{\mathcal{R}} L \sqrt{-g}\, d^4 x, \tag{1.56}$$

where the *Lagrangian field L* is related to the Lagrangian density by

$$\mathcal{L} = L \sqrt{-g}. \tag{1.57}$$

---

[5]A field is a physical system with infinite degrees of freedom.

The field equations for Φ^a can be derived demanding that the action (1.55) is invariant under small variations in the fields:

$$\Phi^a(x) \to \Phi'^a(x) = \Phi^a + \delta\Phi^a(x). \tag{1.58}$$

No coordinate has been changed here, just the form of the fields in a fixed coordinate system. Assuming for simplicity that derivatives of order higher than first can be neglected we have:

$$\partial_\mu \Phi^a \to \partial_\mu \Phi'^a = \partial_\mu \Phi^a + \partial_\mu(\delta\Phi^a). \tag{1.59}$$

Using these variations we obtain the variation of the action $S \to S + \delta S$, where

$$\delta S = \int_{\mathcal{R}} \delta\mathcal{L} \, d^4x = \int_{\mathcal{R}} \left[\frac{\partial\mathcal{L}}{\partial\Phi^a} \delta\Phi^a + \frac{\partial\mathcal{L}}{\partial(\partial_\mu\Phi^a)} \delta(\partial_\mu\Phi^a) \right] d^4x. \tag{1.60}$$

After some math (see, e.g., Hobson et al. 2007), we get:

$$\frac{\delta\mathcal{L}}{\delta\Phi^a} = \frac{\partial\mathcal{L}}{\partial\Phi^a} - \partial_\mu \left[\frac{\partial\mathcal{L}}{\partial(\partial_\mu\Phi^a)} \right] = 0. \tag{1.61}$$

These are the Euler-Lagrange equations for the local field theory defined by the action (1.55).

If the field theory is General Relativity, we need to define a Lagrangian density which is a scalar under general coordinate transformations and which depends on the components of the metric tensor $g_{\mu\nu}$, which represents the dynamical potential of the gravitational field. The simplest scalar that can be constructed from the metric and its derivatives is the Ricci scalar R. The simplest possible action is the so-called *Einstein-Hilbert* action:

$$S_{\mathrm{EH}} = \int_{\mathcal{R}} R \sqrt{-g} \, d^4x. \tag{1.62}$$

The Lagrangian density is $\mathcal{L} = R\sqrt{-g}$. Introducing a variation in the metric

$$g_{\mu\nu} \to g_{\mu\nu} + \delta g_{\mu\nu}, \tag{1.63}$$

we can arrive, after significant algebra, to

$$\delta S_{\mathrm{EH}} = \int_{\mathcal{R}} \left(R_{\mu\nu} - \frac{1}{2} g_{\mu\nu} R \right) \delta g^{\mu\nu} \sqrt{-g} \, d^4x. \tag{1.64}$$

By demanding that $\delta S_{\mathrm{EH}} = 0$ and considering that $\delta g_{\mu\nu}$ is arbitrary, we get

$$G_{\mu\nu} \equiv R_{\mu\nu} - \frac{1}{2} g_{\mu\nu} R = 0. \tag{1.65}$$

These are the Einstein's field equations in vacuum. The tensor $G_{\mu\nu}$ is called the *Einstein tensor*. This variational approach was used by Hilbert in November 1915 to derive Einstein's equations from simplicity and symmetry arguments.

If there are non-gravitational fields present the action will have and additional component:

$$S = \frac{1}{2\kappa} S_{\text{EH}} + S_{\text{M}} = \int_{\mathcal{R}} \left(\frac{1}{2\kappa} \mathcal{L}_{\text{EH}} + \mathcal{L}_{\text{M}} \right) d^4 x, \qquad (1.66)$$

where S_{M} is the non-gravitational action and $\kappa = -8\pi G/c^4$. If we vary the action with respect to the inverse of the metric tensor we get:

$$\frac{1}{2\kappa} \frac{\delta \mathcal{L}_{\text{EH}}}{\delta g^{\mu\nu}} + \frac{\delta \mathcal{L}_{\text{M}}}{\delta g^{\mu\nu}} = 0. \qquad (1.67)$$

Since $\delta S_{\text{EH}} = 0$,

$$\frac{\delta \mathcal{L}_{\text{EH}}}{\delta g^{\mu\nu}} = \sqrt{-g}\, G_{\mu\nu}. \qquad (1.68)$$

Then, if we identify the energy-momentum tensor of the non-gravitational fields in the following way

$$T_{\mu\nu} = \frac{2}{\sqrt{-g}} \frac{\delta \mathcal{L}_{\text{M}}}{\delta g^{\mu\nu}}, \qquad (1.69)$$

we obtain the full Einstein's equations,

$$G_{\mu\nu} = -\frac{8\pi G}{c^4} T_{\mu\nu}.$$

1.7 Math Note: Invariant Volume Element

Let us calculate the N-dimensional invariant volume element $d^N V$ in an N-dimensional pseudo-Riemannian manifold. In an orthogonal coordinate system this volume element is:

$$d^N V = \sqrt{|g_{11}\, g_{22} \cdots g_{NN}|}\, dx^1 dx^2 \cdots dx^N.$$

In such a system the determinant of the metric tensor is

$$\|g_{ab}\| = g_{11}\, g_{22} \cdots g_{NN},$$

i.e. the product of the diagonal elements.

Using the notation adopted above for the determinant we can write:

$$d^N V = \sqrt{|g|}\, dx^1 dx^2 \cdots dx^N.$$

It is not difficult to show that this result remains valid in an arbitrary coordinate system (see Hobson et al. 2007).

1.8 The Cauchy Problem

The Cauchy problem concerns the solution of a partial differential equation that satisfies certain side conditions which are given on a hypersurface in the domain of the functions. It is an extension of the initial value problem. In the case of the Einstein field equations, the hypersurface is given by the condition $x^0/c = t$. If it were possible to obtain from the field equations an expression for $\partial^2 g_{\mu\nu}/\partial(x^0)^2$ everywhere at t, then it would be possible to compute $g_{\mu\nu}$ and $\partial g_{\mu\nu}/\partial x^0$ at a time $t + \delta t$, and repeating the process the metric could be calculated for all x^μ. This is the problem of finding the causal development of a physical system from initial data.

Let us prescribe initial data $g_{\mu\nu}$ and $g_{\mu\nu,0}$ on S defined by $x^0/c = t$. The dynamical equations are the six equations defined by

$$G^{i,j} = -\frac{8\pi G}{c^4} T^{ij}. \tag{1.70}$$

When these equations are solved for the 10 second derivatives $\partial^2 g_{\mu\nu}/\partial(x^0)^2$, there appears a fourfold ambiguity, i.e. four derivatives are left indeterminate. In order to completely fix the metric it is necessary to impose four additional conditions. These conditions are usually imposed upon the affine connection:

$$\Gamma^\mu \equiv g^{\alpha\beta}\Gamma^\mu_{\alpha\beta} = 0. \tag{1.71}$$

The condition $\Gamma^\mu = 0$ implies $\Box^2 x^\mu = 0$, so the coordinates are known as harmonic. With such conditions it can be shown the existence, uniqueness and stability of the solutions. But the result is in no way general and this is an active field of research. The fall of predictability posits a serious problem for the space-time interior of black holes and for multiply connected space-times, as we shall see.

1.9 The Energy-Momentum of Gravitation

Taking the covariant derivative to both sides of Einstein's equations and using Bianchi identities we get

$$\left(R^{\mu\nu} - \frac{1}{2} g^{\mu\nu} R \right)_{;\mu} = 0, \tag{1.72}$$

and then $T^{\mu\nu}_{;\mu} = 0$. This means the conservation of energy and momentum of matter and non-gravitational fields, but it is not strictly speaking a full conservation law, since the energy-momentum of the gravitational field is not included. Because of the Equivalence Principle, it is always possible to choose a coordinate system where the gravitational field locally vanishes. Hence, its local energy is zero. Energy is the more general property of things: the potential to change. This property, however, cannot be associated with a pure gravitational field at any point according to General Relativity. Therefore, it is not possible to associate a tensor with the energy-momentum of the gravitational field. Nonetheless, extended regions with gravitational field have energy-momentum since it is impossible to make the field null in all points of the region just through a coordinate change. We can then define a quasi-tensor for the energy-momentum of gravity. Quasi-tensors are objects that under *global* linear transformations behave like tensors.

We can define a quasi-tensor of energy-momentum such that

$$\Theta^{\mu\nu}{}_{,\nu} = 0. \tag{1.73}$$

In the absence of gravitational fields it satisfies $\Theta^{\mu\nu} = T^{\mu\nu}$. Hence, we can write:

$$\Theta^{\mu\nu} = \sqrt{-g}\left(T^{\mu\nu} + t^{\mu\nu}\right) = \Lambda^{\mu\nu\alpha}{}_{,\alpha}. \tag{1.74}$$

An essential property of $t^{\mu\nu}$ is that it is not a tensor, since in the *superpotential* on the right side appears the normal derivative, not the covariant one. Since $t^{\mu\nu}$ can be interpreted as the contribution of gravitation to the quasi-tensor $\Theta^{\mu\nu}$, we can expect that it should be expressed in geometric terms only, i.e. as a function of the affine connection and the metric. Landau and Lifshitz (1962) found an expression for $t^{\mu\nu}$ that contains only first derivatives and is symmetric:

$$
\begin{aligned}
t^{\mu\nu} = \frac{c^4}{16\pi G}\Big[&\left(2\Gamma^{\sigma}_{\rho\eta}\,\Gamma^{\gamma}_{\sigma\gamma} - \Gamma^{\sigma}_{\rho\gamma}\,\Gamma^{\gamma}_{\eta\sigma} - \Gamma^{\sigma}_{\rho\sigma}\,\Gamma^{\gamma}_{\eta\gamma}\right)\left(g^{\mu\rho}g^{\nu\eta} - g^{\mu\nu}g^{\rho\eta}\right) \\
&+ g^{\mu\rho}g^{\eta\sigma}\left(\Gamma^{\nu}_{\rho\gamma}\,\Gamma^{\gamma}_{\eta\sigma} + \Gamma^{\nu}_{\eta\sigma}\,\Gamma^{\gamma}_{\rho\gamma} + \Gamma^{\nu}_{\sigma\gamma}\,\Gamma^{\gamma}_{\rho\eta} + \Gamma^{\nu}_{\rho\eta}\,\Gamma^{\gamma}_{\sigma\gamma}\right) \\
&+ g^{\nu\rho}g^{\eta\sigma}\left(\Gamma^{\mu}_{\rho\gamma}\,\Gamma^{\gamma}_{\eta\sigma} + \Gamma^{\mu}_{\eta\sigma}\,\Gamma^{\gamma}_{\rho\gamma} + \Gamma^{\mu}_{\sigma\gamma}\,\Gamma^{\gamma}_{\rho\eta} + \Gamma^{\mu}_{\rho\eta}\,\Gamma^{\gamma}_{\sigma\gamma}\right) \\
&+ g^{\rho\eta}g^{\sigma\gamma}\left(\Gamma^{\mu}_{\rho\sigma}\,\Gamma^{\nu}_{\eta\gamma} - \Gamma^{\mu}_{\rho\eta}\,\Gamma^{\nu}_{\sigma\gamma}\right)\Big].
\end{aligned} \tag{1.75}
$$

It is possible to find in a curved space-time a coordinate system such that locally $t^{\mu\nu} = 0$. Similarly, an election of curvilinear coordinates in a flat space-time can yield non-vanishing values for the components of $t^{\mu\nu}$. We infer from this that the energy of the gravitational field is a global property, not a local one. There is energy in a region where there is a gravitational field, but in General Relativity it makes no sense to talk about the energy of a given point of the field. For other proposals of $t^{\mu\nu}$ see Maggiore (2008).

1.10 Weyl Tensor and the Entropy of Gravitation

The Weyl curvature tensor is the traceless component of the curvature (Riemann) tensor. In other words, it is a tensor that has the same symmetries as the Riemann tensor with the extra condition that metric contraction yields zero.

In 3 dimensions the Weyl curvature tensor vanishes identically. In dimensions ≥ 4 the Weyl curvature is generally nonzero. If the Weyl tensor vanishes, then there exists a coordinate system in which the metric tensor is proportional to a constant tensor.

The Weyl tensor can be obtained from the full curvature tensor by subtracting out various traces. This is most easily done by writing the Riemann tensor as a $(0, 4)$-valent tensor (by contracting with the metric). The Riemann tensor has 20 independent components, 10 of which are given by the Ricci tensor and the remaining 10 by the Weyl tensor.

The Weyl tensor is given in components by

$$C_{abcd} = R_{abcd} + \frac{2}{n-2}(g_{a[c}R_{d]b} - g_{b[c}R_{d]a}) + \frac{2}{(n-1)(n-2)} R\, g_{a[c}g_{d]b}, \quad (1.76)$$

where R_{abcd} is the Riemann tensor, R_{ab} is the Ricci tensor, R is the Ricci scalar and "[ ]" refers to the antisymmetric part. In 4 dimensions the Weyl tensor is

$$C_{abcd} = R_{abcd} + \frac{1}{2}(g_{ac}R_{db} - g_{bc}R_{da} - g_{ad}R_{cb} + g_{bd}R_{ca})$$

$$+ \frac{1}{6}(g_{ac}g_{db} - g_{ad}g_{cb})R. \quad (1.77)$$

In addition to the symmetries of the Riemann tensor, the Weyl tensor satisfies

$$C^{a}_{bad} \equiv 0. \quad (1.78)$$

Two metrics that are *conformally related* to each other, i.e.

$$\overline{g}_{ab} = \Omega^2 g_{ab}, \quad (1.79)$$

where $\Omega(x)$ is a non-zero differentiable function, have the same Weyl tensor:

$$\overline{C}^{a}_{bcd} = C^{a}_{bcd}. \quad (1.80)$$

The absence of structure in space-time (i.e. spatial isotropy and hence no gravitational principal null-directions) corresponds to the absence of Weyl conformal curvature $(C^2 = C^{abcd}C_{abcd} = 0)$. When clumping takes place, the structure is characterized by a non-zero Weyl curvature. In the interior of a black hole, as we shall see, the Weyl curvature is large and goes to infinity at the singularity. Actually, Weyl curvature goes faster to infinity than Riemann curvature (the former as r^{-3} and the latter as $r^{-3/2}$ for a Schwarzschild black hole). Since the initial conditions of the Universe seem highly uniform and the primordial state one of low-entropy, Penrose (1979) has proposed that the Weyl tensor gives a measure of the gravitational entropy and that the Weyl curvature vanishes at any initial singularity (this would be valid for white holes if they were to exist). In this way, despite the fact that matter was in local equilibrium in the early Universe, the global state was of low entropy, since the gravitational field was highly uniform and dominated the overall entropy.

1.11 Gravitational Waves

Before the development of General Relativity, Heinrik Lorentz had speculated that "gravitation can be attributed to actions which do not propagate with a velocity larger than that of the light" (Lorentz 1900). The term *gravitational waves* appeared for the first time in 1905 when H. Poincaré discussed the extension of Lorentz invariance to gravitation (Poincaré 1905, see Pais 1982 for further details). The idea that a perturbation in the source of the gravitational field can result in a wave that would manifest as a moving disturbance in the metric field was developed by Einstein in 1916, shortly after the final formulation of the field equations (Einstein 1916). Then, in 1918, Einstein presented the quadrupole formula for the energy loss of a mechanical system (Einstein 1918).

Einstein's approach was based on the weak-field approximation of the metric field:

$$g_{\mu\nu} = \eta_{\mu\nu} + h_{\mu\nu}, \tag{1.81}$$

where $\eta_{\mu\nu}$ is the Minkowski flat metric and $|h_{\mu\nu}| \ll 1$ is a small perturbation to the background metric. Since $h_{\mu\nu}$ is small, all products that involve it and its derivatives can be neglected. And because the metric is almost flat all indices can be lowered or raised through $\eta_{\mu\nu}$ and $\eta^{\mu\nu}$ instead of $g_{\mu\nu}$ and $g^{\mu\nu}$. We can then write

$$g^{\mu\nu} = \eta^{\mu\nu} - h^{\mu\nu}. \tag{1.82}$$

With this, we can compute the affine connection:

$$\Gamma^{\mu}_{\nu\sigma} = \frac{1}{2}\eta^{\mu\beta}(h_{\sigma\beta,\nu} + h_{\nu\beta,\sigma} - h_{\nu\sigma,\beta}) = \frac{1}{2}\left(h^{\mu}_{\sigma,\nu} + h^{\mu}_{\nu,\sigma} - h_{\nu\sigma}{}^{,\mu}\right). \tag{1.83}$$

Here, $h_{\nu\sigma}{}^{,\mu} = \eta^{\mu\beta}h_{\nu\sigma,\beta}$. Introducing $h = h^{\mu}_{\mu} = \eta^{\mu\nu}h_{\mu\nu}$, we can write the Ricci tensor and the curvature scalar as:

$$R_{\mu\nu} = \Gamma^{\alpha}_{\mu\alpha,\nu} - \Gamma^{\alpha}_{\mu\nu,\alpha} = \frac{1}{2}\left(h_{,\mu\nu} - h^{\alpha}_{\nu,\mu\alpha} - h^{\alpha}_{\mu,\nu\alpha} + h_{\mu\nu,}{}^{\alpha}{}_{\alpha}\right), \tag{1.84}$$

and

$$R \equiv g^{\mu\nu}R_{\mu\nu} = \eta^{\mu\nu}R_{\mu\nu} = h^{\alpha}{}_{,\alpha} - h^{\alpha\beta}{}_{,\alpha\beta}. \tag{1.85}$$

Then, the field equations (1.36) can be cast in the following way:

$$\bar{h}_{\mu\nu,\alpha}{}^{\alpha} + \left(\eta_{\mu\nu}\bar{h}^{\alpha\beta}{}_{,\alpha\beta} - \bar{h}^{\alpha}_{\nu,\mu\alpha} - \bar{h}^{\alpha}_{\mu,\nu\alpha}\right) = 2\kappa T_{\mu\nu}, \tag{1.86}$$

where

$$\bar{h}_{\mu\nu} \equiv h_{\mu\nu} - \frac{1}{2}h\,\eta_{\mu\nu}. \tag{1.87}$$

We can make further simplifications through a *gauge transformation*. A gauge transformation is a small change of coordinates

$$x'^{\mu} \equiv x^{\mu} + \xi^{\mu}(x^{\alpha}), \tag{1.88}$$

where the ξ^α are of the same order of magnitude as the perturbations of the metric. The matrix $\Lambda_\nu^\mu \equiv \partial x'^\mu / \partial x^\nu$ is given by

$$\Lambda_\nu^\mu = \delta_\nu^\mu + \xi^\mu_{,\nu}. \tag{1.89}$$

Under gauge transformations

$$\overline{h}'^{\mu\nu} = \overline{h}^{\mu\nu} - \xi^{\mu,\nu} - \xi^{\nu,\mu} + \eta^{\mu\nu}\xi^\alpha_{,\alpha}. \tag{1.90}$$

The gauge transformation can be chosen in such a way that

$$\overline{h}^{\mu\alpha}_{,\alpha} = 0. \tag{1.91}$$

Then, the field equations simplify to

$$\overline{h}_{\mu\nu,\alpha}{}^\alpha = 2\kappa T_{\mu\nu}. \tag{1.92}$$

The imposition of this gauge condition is analogous to what is done in electromagnetism with the introduction of the Lorentz gauge condition $A^\mu_{,\mu} = 0$, where A^μ is the electromagnetic 4-potential. A gauge transformation $A_\mu \to A_\mu - \psi_{,\mu}$ preserves the Lorentz gauge condition if $\psi^\mu_{,\mu} = 0$. In the gravitational case, we have $\xi^{\mu\alpha}_{,\alpha} = 0$.

Introducing the d'Alembertian

$$\Box^2 = \eta^{\mu\nu}\partial_\mu\partial_\nu = \frac{1}{c^2}\frac{\partial^2}{\partial t^2} - \nabla^2, \tag{1.93}$$

if $\overline{h}^{\mu\nu}_{,\nu} = 0$ we get

$$\Box^2\overline{h}^{\mu\nu} = 2\kappa T^{\mu\nu}. \tag{1.94}$$

The gauge condition can be expressed as

$$\Box^2\xi^\mu = 0. \tag{1.95}$$

Recalling the definition of κ, we can write the wave equations of the gravitational field, insofar as the amplitudes are small, as:

$$\Box^2\overline{h}^{\mu\nu} = -\frac{16\pi G}{c^4}T^{\mu\nu}. \tag{1.96}$$

In the absence of matter and non-gravitational fields, these equations become

$$\Box^2\overline{h}^{\mu\nu} = 0. \tag{1.97}$$

The simplest solution to Eq. (1.97) is

$$\overline{h}^{\mu\nu} = \Re\left[A^{\mu\nu}\exp\left(ik_\alpha x^\alpha\right)\right], \tag{1.98}$$

where $A^{\mu\nu}$ is the amplitude matrix of a plane wave that propagates with direction $k^\mu = \eta^{\mu\alpha}k_\alpha$ and $\mathfrak{R}$ indicates that just the real part of the expression should be considered. The 4-vector k^μ is null and satisfies

$$A^{\mu\nu}k_\nu = 0. \tag{1.99}$$

Since $\overline{h}^{\mu\nu}$ is symmetric the amplitude matrix has ten independent components. Equation (1.99) can be used to reduce this number to six. The gauge condition allows a further reduction, so finally we have only two independent components. Einstein realized this in 1918. These two components characterize two different possible polarization states for the gravitational waves. In the so-called *traceless and transverse gauge*—TT—we can introduce two linear polarization matrices defined as:

$$e_1^{\mu\nu} = \begin{pmatrix} 0 & 0 & 0 & 0 \\ 0 & 1 & 0 & 0 \\ 0 & 0 & -1 & 0 \\ 0 & 0 & 0 & 0 \end{pmatrix}, \tag{1.100}$$

and

$$e_2^{\mu\nu} = \begin{pmatrix} 0 & 0 & 0 & 0 \\ 0 & 0 & 1 & 0 \\ 0 & 1 & 0 & 0 \\ 0 & 0 & 0 & 0 \end{pmatrix}, \tag{1.101}$$

in such a way that the general amplitude matrix is

$$A^{\mu\nu} = \alpha e_1^{\mu\nu} + \beta e_2^{\mu\nu}, \tag{1.102}$$

with α and β complex constants.

The general solution of Eq. (1.96) is:

$$\overline{h}^{\mu\nu}\left(x^0, \mathbf{x}\right) = \frac{\kappa}{2\pi} \int \frac{T^{\mu\nu}(x^0 - |\mathbf{x} - \mathbf{x}'|, \mathbf{x}')}{|\mathbf{x} - \mathbf{x}'|} \, dV'. \tag{1.103}$$

In this integral we have considered only the effects of sources in the *past* of the space-time point $(x^0, \mathbf{x})$. The integral extends over the space-time region formed by the intersection the past half of the null cone at the field point with the world tube of the source.

If the source is small compared to the wavelength of the gravitational radiation, we can approximate (1.103) by

$$\overline{h}^{\mu\nu}(ct, \mathbf{x}) = \frac{4G}{c^4 r} \int T^{\mu\nu}\left(ct - r, \mathbf{x}'\right) dV'. \tag{1.104}$$

This approximation is valid in the *far zone*, where $r > l$, with l the typical size of the source. In this region the gravitational wave looks like a plane wave, and a simple expression for $\overline{h}^{ij}$ can be obtained (see Foster and Nightingale 2006).

1.12 Alternative Theories of Gravitation

1.12.1 Scalar-Tensor Gravity

Perhaps the most important alternative theory of gravitation is the Brans-Dicke theory of scalar-tensor gravity (Brans and Dicke 1961). The original motivation for this theory was to implement the idea of Mach that the phenomenon of inertia was due to the acceleration of a given system with respect to the general mass distribution of the universe. The masses of the different fundamental particles would not be basic intrinsic properties but a relational property originated in the interaction with some cosmic field. We can express this in the form:

$$m_i\left(x^\mu\right) = \lambda_i \phi\left(x^\mu\right).$$

Since the masses of the different particles can be measured only through the gravitational acceleration Gm/r^2, the gravitational constant G should be related to the average value of some cosmic scalar field ϕ, which is coupled with the mass density of the universe.

The simplest general covariant equation for a scalar field produced by matter is

$$\Box^2\phi = 4\pi\lambda\left(T^{\mathrm{M}}\right)^\mu_\mu, \tag{1.105}$$

where $\Box^2\phi = \phi^{;\mu}_{\;\;;\mu}$ is, again, the invariant d'Alembertian, λ is a coupling constant, and $(T^{\mathrm{M}})^{\mu\nu}$ is the energy-momentum tensor of everything but gravitation. The matter and non-gravitational fields generate the cosmic scalar field ϕ. This field is normalized such that

$$\langle\phi\rangle = \frac{1}{G}. \tag{1.106}$$

The scalar field, as anything else, also generates gravitation, so Einstein's field equations are re-written as:

$$R_{\mu\nu} - \frac{1}{2}g_{\mu\nu}R = -\frac{8\pi}{c^4\phi}\left(T^{\mathrm{M}}_{\mu\nu} + T^{\phi}_{\mu\nu}\right). \tag{1.107}$$

Here, $T^{\phi}_{\mu\nu}$ is the energy momentum tensor of the scalar field ϕ. Its explicit form is rather complicated (see Weinberg 1972, p. 159). Because of historical reasons the parameter λ is written as:

$$\lambda = \frac{2}{3 + 2\omega}.$$

In the limit $\omega \to \infty$, $\lambda \to 0$ and $T^{\phi}_{\mu\nu}$ vanishes, and hence the Brans-Dicke theory reduces to Einstein's.

One of the most interesting features of Brans-Dicke theory is that G varies with time because it is determined by the scalar field ϕ. A variation of G would affect the orbits of planets, the stellar evolution, and many other astrophysical phenomena. Experiments can constrain ω to $\omega > 500$. Hence, Einstein theory seems to be correct, at least at low energies.

1.12.2 Gravity with Extra Dimensions

The so-called hierarchy problem is the difficulty to explain why the characteristic energy scale of gravity, the Planck energy $M_{\mathrm{Pc}}c^2 \sim 10^{19}$ GeV,[6] is 16 orders of magnitude larger than the electro-weak scale, $M_{\mathrm{ew}}c^2 \sim 1$ TeV. A possible solution was presented in 1998 by Arkani-Hamed et al. (1998), with the introduction of gravity with large extra dimensions (LEDs). The idea of extra dimensions was, however, no new in physics. It was originally introduced by Kaluza (1921) with the aim of unifying gravitation and electromagnetism. In a different context, Nordstrøm (1914) also discussed the possibility of a fifth dimension.

Kaluza's fundamental insight was to write the action as

$$S = \frac{1}{16\pi \hat{G}} \int_{\mathcal{R}} \hat{R} \sqrt{-\hat{g}} \, d^4x \, dy, \tag{1.108}$$

instead of in the form given by expression (1.66). In Kaluza's action y is the coordinate of an extra dimension and the hats denote 5-dimensional (5-D) quantities. The interval results:

$$ds^2 = \hat{g}_{\mu\nu} dx^\mu dx^\nu, \tag{1.109}$$

with μ, ν running from 0 to 4, being $x^4 = y$ the extra dimension. The extra dimension should have no effect over the gravitation, hence Kaluza imposed the condition

$$\frac{\partial \hat{g}_{\mu\nu}}{\partial y} = 0. \tag{1.110}$$

Since gravitation manifests through the derivatives of the metric, condition (1.110) implies that the extra dimension does not affect the predictions of General Relativity. If we write the metric as

$$\hat{g}_{\mu\nu} = \phi^{-1/3} \begin{pmatrix} g_{\mu\nu} + \phi A_\mu A_\nu & \phi A_\mu \\ \phi A_\nu & \phi \end{pmatrix}, \tag{1.111}$$

then, the action becomes

$$S = \frac{1}{16\pi G} \int_{\mathcal{R}} \left(R - \frac{1}{4}\phi F_{ab} F^{ab} - \frac{1}{6\phi^2 \partial_a \phi} \partial^a \phi \right) \sqrt{-\hat{g}} \, d^4x, \tag{1.112}$$

---

[6] The Planck mass is $M_{\mathrm{P}} = \sqrt{\hbar c / G} = 2.17644(11) \times 10^{-5}$ g. The Planck mass is the mass of the Planck particle, a hypothetical minuscule particle whose effective radius equals the Planck length $l_{\mathrm{P}} = \sqrt{\hbar G / c^3} = 1.616252(81) \times 10^{-33}$ cm.

Fig. 1.4 Compactified extra dimensions in Kaluza-Klein and ADD braneworld theories. Adapted from Whisker (2006)

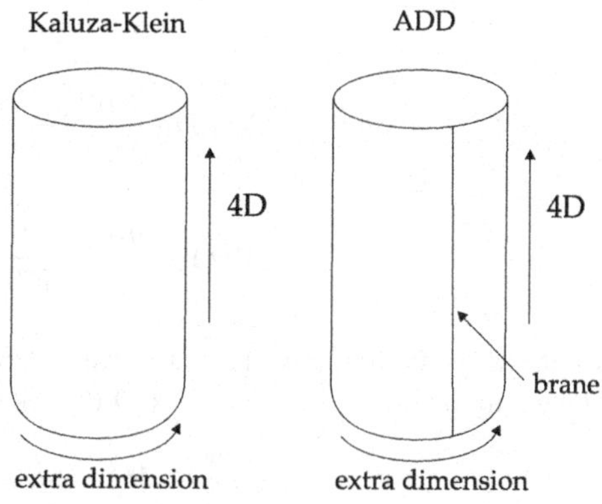

where $F_{ab} = \partial_a A_b - \partial_b A_a$ and

$$G = \frac{\hat{G}}{\int dy}.$$

The action (1.112) describes 4-D gravity along with electromagnetism. The price paid for this unification was the introduction of a scalar field ϕ called the *dilaton* (which was fixed by Kaluza to $\phi = 1$) and an extra fifth dimension which is not observed.

Klein (1926) suggested that the fifth dimension was not observable because it is *compactified* on a circle. This compactification can be achieved identifying y with $y + 2\pi R$. The quantity R is the size of the extra dimension. Such a size should be extremely small in order not to be detected in experiments. The only natural length of the theory is the Planck length: $R \approx l_P \sim 10^{-35}$ m.

A very interesting feature of the theory is that charge conservation can be interpreted as momentum conservation in the fifth dimension:

$$J^\mu = 2\alpha T^{\mu 5}, \tag{1.113}$$

where J^μ is the current density and α a constant. The variation of the action (1.112) yields both Einstein's and Maxwell's equations:

$$G_{\mu\nu} = \kappa T_{\mu\nu} \quad \text{and} \quad \partial_\mu F^{\mu\nu} = \frac{c^2\kappa}{2G} J^\nu.$$

Unfortunately, the Kaluza-Klein theory is not consistent with other observed features of particle physics as described by the Standard Model. This shortcoming is removed in the mentioned LED model by Arkani-Hamed et al. (1998), called ADD braneworld model. The model postulates n flat, compact extra dimensions of size R, but the Standard Model fields are *confined* to a 4-D *brane*, with only gravity propagating in the *bulk* (see Fig. 1.4). The effective potential for gravity behaves as:[7]

$$V(r) \approx \frac{m_1 m_2}{M_f^{2+n}} \frac{1}{r^{n+1}}, \quad r \ll R, \tag{1.114}$$

$$V(r) \approx \frac{m_1 m_2}{M_f^{2+n}} \frac{1}{R^n r}, \quad r \gg R, \tag{1.115}$$

where M_f is the fundamental mass scale of gravity in the full $(4+n)$-D space-time. Hence, in the brane the *effective* 4-D Planck scale is given by:

$$M_P^2 = M_f^{2+n} R^n. \tag{1.116}$$

In this way the fundamental scale M_f can be much lower than the Planck mass. If the fundamental scale is comparable to the electroweak scale, $M_f c^2 \sim M_{ew} c^2 \sim$ 1 TeV, then we have that $n \geq 2$.

Randall and Sundrum (1999) suggested that the bulk geometry might be curved and the brane could have a tension. Hence, the brane becomes a gravitating object, interacting dynamically with the bulk. A Randall-Sundrum (RS) universe consists of two branes of torsion σ_1 and σ_2 bounding a slice of an anti-de Sitter space.[8] The two branes are separated by a distance L and the fifth dimension y is periodic with period $2L$. The bulk Einstein's equations read

$$R_{ab} - \frac{1}{2} R g_{ab} = \Lambda_5 g_{ab}, \tag{1.117}$$

where the bulk cosmological constant Λ_5 can be expressed in terms of the curvature length l as

$$\Lambda_5 = \frac{6}{l^2}. \tag{1.118}$$

The metric is

$$ds^2 = a^2(y) \eta_{\mu\nu} dx^\mu dx^\nu + dy^2. \tag{1.119}$$

Using the previous expressions, we can write the metric as:

$$ds^2 = e^{-2|y|/l} \eta_{\mu\nu} dx^\mu dx^\nu + dy^2. \tag{1.120}$$

The term $e^{-2|y|/l}$ is called the *warp* factor. The effective Planck mass becomes

$$M_P^2 = e^{2L/l} M_f^3 l. \tag{1.121}$$

---

[7] Notice that $G = M_P^{-2} \hbar c$ or $G = M_P^{-2}$ in units of $\hbar c = 1$.

[8] An anti-de Sitter space-time has a metric that is a maximally symmetric vacuum solution of Einstein's field equations with an attractive cosmological constant (corresponding to a negative vacuum energy density and positive pressure). This space-time has a constant negative scalar curvature.

According to the ratio L/l, the effective Planck mass can change. If we wish to have $M_f c^2 \sim 1$ TeV, then we need $L/l \sim 50$ in order to generate the observed Planck mass $M_f c^2 \sim 10^{19}$ GeV.

Other RS universes consist of a single, positive tension brane immersed in an infinite (non-compact) extra dimension. The corresponding metric remains the same:

$$ds^2 = e^{-2|y|/l}\eta_{\mu\nu}dx^\mu dx^\nu + dy^2.$$

The 5-D graviton propagates through the bulk, but only the zero (massless) mode moves on the brane (for details see Maartens 2004).

1.12.3 $f(R)$-Gravity

In $f(R)$-gravity, the Lagrangian of the Einstein-Hilbert action

$$S[g] = \int \frac{1}{2\kappa} R \sqrt{-g}\, \mathrm{d}^4 x \tag{1.122}$$

is generalized to

$$S[g] = \int \frac{1}{2\kappa} f(R) \sqrt{-g}\, \mathrm{d}^4 x, \tag{1.123}$$

where g is the determinant of the metric tensor and $f(R)$ is some function of the curvature (Ricci) scalar.

The field equations are obtained by varying with respect to the metric. The variation of the determinant is

$$\delta(\sqrt{-g}) = -\frac{1}{2}\sqrt{-g}\, g_{\mu\nu}\delta g^{\mu\nu}.$$

The Ricci scalar is defined as

$$R = g^{\mu\nu} R_{\mu\nu}.$$

Therefore, its variation with respect to the inverse metric $g^{\mu\nu}$ is given by

$$\begin{aligned}
\delta R &= R_{\mu\nu}\delta g^{\mu\nu} + g^{\mu\nu}\delta R_{\mu\nu} \\
&= R_{\mu\nu}\delta g^{\mu\nu} + g^{\mu\nu}\left(\nabla_\rho \delta \Gamma^\rho_{\nu\mu} - \nabla_\nu \delta \Gamma^\rho_{\rho\mu}\right)
\end{aligned} \tag{1.124}$$

Since $\delta\Gamma^\lambda_{\mu\nu}$ is actually the difference of two connections, it should transform as a tensor. Therefore, it can be written as

$$\delta\Gamma^\lambda_{\mu\nu} = \frac{1}{2}g^{\lambda a}(\nabla_\mu \delta g_{av} + \nabla_\nu \delta g_{a\mu} - \nabla_a \delta g_{\mu\nu}),$$

and substituting in the equation above:

$$\delta R = R_{\mu\nu}\delta g^{\mu\nu} + g_{\mu\nu}\Box \delta g^{\mu\nu} - \nabla_\mu \nabla_\nu \delta g^{\mu\nu}.$$

The variation in the action reads:

$$\delta S[g] = \frac{1}{2\kappa} \int \left(\delta f(R) \sqrt{-g} + f(R) \delta \sqrt{-g} \right) d^4 x$$

$$= \frac{1}{2\kappa} \int \left(F(R) \delta R \sqrt{-g} - \frac{1}{2} \sqrt{-g} \, g_{\mu\nu} \delta g^{\mu\nu} f(R) \right) d^4 x$$

$$= \frac{1}{2\kappa} \int \sqrt{-g} \left[F(R) \left(R_{\mu\nu} \delta g^{\mu\nu} + g_{\mu\nu} \Box \delta g^{\mu\nu} - \nabla_\mu \nabla_\nu \delta g^{\mu\nu} \right) \right.$$

$$\left. - \frac{1}{2} g_{\mu\nu} \delta g^{\mu\nu} f(R) \right] d^4 x,$$

where $F(R) = \frac{\partial f(R)}{\partial R}$. Integrating by parts on the second and third terms we get

$$\delta S[g] = \frac{1}{2\kappa} \int \sqrt{-g} \, \delta g^{\mu\nu} \left[F(R) R_{\mu\nu} - \frac{1}{2} g_{\mu\nu} f(R) + (g_{\mu\nu} \Box - \nabla_\mu \nabla_\nu) F(R) \right] d^4 x.$$

By demanding that the action remains invariant under variations of the metric, i.e. $\delta S[g] = 0$, we obtain the field equations:

$$F(R) R_{\mu\nu} - \frac{1}{2} f(R) g_{\mu\nu} + [g_{\mu\nu} \Box - \nabla_\mu \nabla_\nu] F(R) = \kappa T_{\mu\nu}, \qquad (1.125)$$

where $T_{\mu\nu}$ is the energy-momentum tensor defined as

$$T_{\mu\nu} = -\frac{2}{\sqrt{-g}} \frac{\delta(\sqrt{-g} \, L_{\mathrm{m}})}{\delta g^{\mu\nu}},$$

and L_{m} is the matter Lagrangian. If $F(R) = 1$, i.e. $f(R) = R$, we recover Einstein's theory.

References

H.I. Arcos, J.G. Pereira, Int. J. Mod. Phys. D **13**, 2193 (2004)
N. Arkani-Hamed, S. Dimopoulos, G.R. Dvali, Phys. Lett. B **429**, 263 (1998)
C.H. Brans, R.H. Dicke, Phys. Rev. **124**, 925 (1961)
L. Corry, J. Renn, J. Stachel, Science **278**, 1270 (1997)
A. Einstein, Ann. Phys. **17**, 891 (1905)
A. Einstein, Jahrb. Radioteh. Elektron. **4**, 411 (1907)
A. Einstein, Preuss. Akad. Wiss., p. 844 (1915)
A. Einstein, Preuss. Akad. Wiss., p. 688 (1916)
A. Einstein, Preuss. Akad. Wiss., p. 154 (1918)
A. Einstein, Preuss. Akad. Wiss., p. 217 (1928)
J. Foster, J.D. Nightingale, *A Short Course in General Relativity*, 3rd edn. (Springer, New York, 2006)
A. Grünbaum, *Philosophical Problems of Space and Time*, 2nd edn. (Kluwer, Dordrecht, 1973)
D. Hilbert, Gött. Nachr., p. 395 (1915)

H.P. Hobson, G. Efstathiou, A.N. Lasenby, *General Relativity* (Cambridge University Press, Cambridge, 2007)

T. Kaluza, Preuss. Akad. Wiss., p. 966 (1921)

O. Klein, Z. Phys. **37**, 895 (1926)

L.D. Landau, E.M. Lifshitz, *The Classical Theory of Fields* (Pergamon, Oxford, 1962)

H.A. Lorentz, Proc. K. Akad. Amsterdam **8**, 603 (1900)

J.-P. Luminet, in *Black Holes: Theory and Observation*, ed. by F.W. Hehl, C. Kiefer, R.J.K. Metzler (Springer, Berlin, 1998), p. 3

R. Maartens, Living Rev. Relativ. **7**(7) (2004)

M. Maggiore, *Gravitational Waves*, vol. 1 (Oxford University Press, Oxford, 2008)

H. Minkowski, Lecture delivered before the Math. Ges. Gött. on November 5th, 1907

H. Minkowski, Lecture Raum und Zeit, 80th Versammlung Deutscher Naturforscher (Köln, 1908). Phys. Z. **10**, 75 (1909)

G. Nordstrøm, Z. Phys. **15**, 504 (1914)

A. Pais, *"Subtle Is the Lord..." the Science and Life of Albert Einstein* (Oxford University Press, Oxford, 1982)

R. Penrose, in *General Relativity*, ed. by S.W. Hawking, W. Israel (Cambridge University Press, Cambridge, 1979), p. 581

S.E. Perez-Bergliaffa, G.E. Romero, H. Vucetich, Int. J. Theor. Phys. **37**, 2281 (1998)

S. Perlmutter et al., Astrophys. J. **517**, 565 (1999)

H. Poincaré, C. R. Acad. Sci. Paris **140**, 1504 (1905)

L. Randall, R. Sundrum, Phys. Rev. Lett. **83**, 3370 (1999)

A.G. Reiss et al., Astron. J. **116**, 1009 (1998)

G.E. Romero, Found. Sci. **17**, 291 (2012)

G.E. Romero, Found. Sci. **18**, 139 (2013). arXiv:1205.0804

S. Weinberg, *Gravitation and Cosmology: Principles and Applications of the General Theory of Relativity* (Wiley, New York, 1972)

R. Whisker, Braneworld black holes. PhD thesis, University of Durham, Durham (2006)

Chapter 2
Black Holes

2.1 Dark Stars: A Historical Note

It is usual in textbooks to credit John Michell and Pierre-Simon Laplace for the idea
of black holes, in the XVIII Century. The idea of a body so massive that even light
could not escape was put forward by geologist Rev. John Michell in a letter written
to Henry Cavendish in 1783 to the Royal Society:

> If the semi-diameter of a sphere of the same density as the Sun were to exceed that of the
> Sun in the proportion of 500 to 1, a body falling from an infinite height toward it would
> have acquired at its surface greater velocity than that of light, and consequently supposing
> light to be attracted by the same force in proportion to its inertia, with other bodies, all
> light emitted from such a body would be made to return toward it by its own proper gravity.
> (Michell 1784).

In 1796, the mathematician Pierre-Simon Laplace promoted the same idea in the
first and second editions of his book *Exposition du système du Monde* (it was removed from later editions). Such "dark stars" were largely ignored in the nineteenth
century, since light was then thought to be a massless wave and therefore not influenced by gravity. Unlike the modern concept of black hole, the object behind the
horizon in black stars is assumed to be stable against collapse. Moreover, no equation of state was adopted neither by Michell nor by Laplace. Hence, their dark stars
were Newtonian objects, infinitely rigid, and they have nothing to do with the nature
of space and time, which were considered by them as absolute concepts. Nonetheless, Michel and Laplace could calculate correctly the size of such objects from the
simple device of equating the potential and escape energy from a body of mass M:

$$\frac{1}{2}mv^2 = \frac{GMm}{r^2}.$$

(2.1)

Just setting $v = c$ and assuming that the gravitational and the inertial mass are the
same, we get

$$r_{\text{dark star}} = \sqrt{\frac{2GM}{c^2}}.$$

(2.2)

G.E. Romero, G.S. Vila, *Introduction to Black Hole Astrophysics*,
Lecture Notes in Physics 876, DOI 10.1007/978-3-642-39596-3_2,
© Springer-Verlag Berlin Heidelberg 2014

Dark stars are objects conceivable only within the framework of the Newtonian theory of matter and gravitation. In the context of general relativistic theories of gravitation, collapsed objects have quite different properties. Before exploring particular situations that can be represented by different solutions of Einstein's field equations, it is convenient to introduce a general definition of a collapsed gravitational system in a general space-time framework. This is what we do in the next section.

2.2 A General Definition of Black Hole

We shall now provide a general definition of a black hole, independently of the coordinate system adopted in the description of space-time, and even of the exact form of the field equations. First, we shall introduce some preliminary useful definitions (e.g. Hawking and Ellis 1973; Wald 1984).

Definition A causal curve in a space-time $(M, g_{\mu\nu})$ is a curve that is non space-like, that is, piecewise either time-like or null (light-like).

We say that a given space-time $(M, g_{\mu\nu})$ is *time-orientable* if we can define over M a smooth non-vanishing time-like vector field.

Definition If $(M, g_{\mu\nu})$ is a time-orientable space-time, then $\forall p \in M$, the causal future of p, denoted $J^+(p)$, is defined by:

$$J^+(p) \equiv \left\{q \in M | \exists \text{ a future-directed causal curve from } p \text{ to } q\right\}. \qquad (2.3)$$

Similarly,

Definition If $(M, g_{\mu\nu})$ is a time-orientable space-time, then $\forall p \in M$, the causal past of p, denoted $J^-(p)$, is defined by:

$$J^-(p) \equiv \{q \in M | \exists \text{ a past-directed causal curve from } p \text{ to } q\}. \qquad (2.4)$$

The causal future and past of any set $S \subset M$ are given by:

$$J^+(S) = \bigcup_{p \in S} J^+(P) \qquad (2.5)$$

and,

$$J^-(S) = \bigcup_{p \in S} J^-(P). \qquad (2.6)$$

A set S is said *achronal* if no two points of S are time-like related. A Cauchy surface (Sect. 1.8) is an achronal surface such that every non space-like curve in M

crosses it once, and only once. A space-time $(M, g_{\mu\nu})$ is *globally hyperbolic* if it admits a space-like hypersurface $S \subset M$ which is a Cauchy surface for M.

Causal relations are invariant under conformal transformations of the metric. In this way, the space-times $(M, g_{\mu\nu})$ and $(M, \widetilde{g}_{\mu\nu})$, where $\widetilde{g}_{\mu\nu} = \Omega^2 g_{\mu\nu}$, with Ω a non-zero C^r function, have the same causal structure.

Particle horizons occur whenever a particular system never gets to be influenced by the whole space-time. If a particle crosses the horizon, it will not exert any further action upon the system with respect to which the horizon is defined.

Definition For a causal curve γ the associated future (past) particle horizon is defined as the boundary of the region from which the causal curves can reach some point on γ.

Finding the particle horizons (if one exists at all) requires a knowledge of the global space-time geometry.

Let us now consider a space-time where all null geodesics that start in a region $\mathcal{J}^-$ end at $\mathcal{J}^+$. Then, such a space-time, $(M, g_{\mu\nu})$, is said to contain a *black hole* if M *is not* contained in $J^-(\mathcal{J}^+)$. In other words, there is a region from where no null geodesic can reach the *asymptotic flat*[1] future space-time, or, equivalently, there is a region of M that is causally disconnected from the global future. The *black hole region*, BH, of such space-time is $BH = [M - J^-(\mathcal{J}^+)]$, and the boundary of BH in M, $H = J^-(\mathcal{J}^+) \bigcap M$, is the *event horizon*.

Notice that a black hole is conceived as a space-time *region*, i.e. what characterizes the black hole is its metric and, consequently, its curvature. What is peculiar of this space-time region is that it is causally disconnected from the rest of the space-time: no events in this region can make any influence on events outside the region. Hence the name of the boundary, event horizon: events inside the black hole are separated from events in the global external future of space-time. The events in the black hole, nonetheless, as all events, are causally determined by past events. A black hole does not represent a breakdown of classical causality. As we shall see, even when closed time-like curves are present, *local* causality still holds along with global consistency constrains. And in case of singularities, they do not belong to space-time, so they are not *predicable* (i.e. we cannot attach any predicate to them, nothing can be said about them) in the theory. More on this in Sect. 3.6.

2.3 Schwarzschild Black Holes

The first exact solution of Einstein's field equations was found by Karl Schwarzschild in 1916. This solution describes the geometry of space-time outside a spherically symmetric matter distribution.

---

[1]Asymptotic flatness is a property of the geometry of space-time which means that in appropriate coordinates, the limit of the metric at infinity approaches the metric of the flat (Minkowskian) space-time.

2.3.1 Schwarzschild Solution

The most general spherically symmetric metric is

$$ds^2 = \alpha(r,t)dt^2 - \beta(r,t)dr^2 - \gamma(r,t)d\Omega^2 - \delta(r,t)drdt, \qquad (2.7)$$

where $d\Omega^2 = d\theta^2 + \sin^2\theta d\phi^2$. We are using spherical polar coordinates. The metric (2.7) is invariant under rotations (isotropic).

The invariance group of general relativity is formed by the group of general transformations of coordinates of the form $x'^\mu = f^\mu(x)$. This yields 4 degrees of freedom, two of which have been used when adopting spherical coordinates (the transformations that do not break the central symmetry are $r' = f_1(r,t)$ and $t' = f_2(r,t)$). With the two available degrees of freedom we can freely choose two metric coefficients, whereas the other two are determined by Einstein's equations. Some possibilities are:

- *Standard gauge.*

$$ds^2 = c^2 A(r,t)dt^2 - B(r,t)dr^2 - r^2 d\Omega^2.$$

- *Synchronous gauge.*

$$ds^2 = c^2 dt^2 - F^2(r,t)dr^2 - R^2(r,t)d\Omega^2.$$

- *Isotropic gauge.*

$$ds^2 = c^2 H^2(r,t)dt^2 - K^2(r,t)\left[dr^2 + r^2(r,t)d\Omega^2\right].$$

- *Co-moving gauge.*

$$ds^2 = c^2 W^2(r,t)dt^2 - U(r,t)dr^2 - V(r,t)d\Omega^2.$$

Adopting the standard gauge and a static configuration (no dependence of the metric coefficients on t), we can get equations for the coefficients A and B of the standard metric:

$$ds^2 = c^2 A(r)dt^2 - B(r)dr^2 - r^2 d\Omega^2. \qquad (2.8)$$

Since we are interested in the solution *outside* the spherical mass distribution, we only need to require the Ricci tensor to vanish:

$$R_{\mu\nu} = 0.$$

According to the definition of the curvature tensor and the Ricci tensor, we have:

$$R_{\mu\nu} = \partial_\nu \Gamma^\sigma_{\mu\sigma} - \partial_\sigma \Gamma^\sigma_{\mu\nu} + \Gamma^\rho_{\mu\sigma}\Gamma^\sigma_{\rho\nu} - \Gamma^\rho_{\mu\nu}\Gamma^\sigma_{\rho\sigma} = 0. \qquad (2.9)$$

If we remember that the affine connection depends on the metric as

$$\Gamma^\sigma_{\mu\nu} = \frac{1}{2}g^{\rho\sigma}(\partial_\nu g_{\rho\mu} + \partial_\mu g_{\rho\nu} - \partial_\rho g_{\mu\nu}),$$

we see that we have to solve a set of differential equations for the components of the metric $g_{\mu\nu}$.

The metric coefficients are:

$$g_{00} = A(r),$$

$$g_{11} = -B(r),$$

$$g_{22} = -r^2,$$

$$g_{33} = -r^2 \sin^2\theta,$$

$$g^{00} = 1/A(r),$$

$$g^{11} = -1/B(r),$$

$$g^{22} = -1/r^2,$$

$$g^{33} = -1/r^2 \sin^2\theta.$$

Then, only nine of the 40 independent connection coefficients are different from zero. They are:

$$\Gamma^1_{01} = A'/(2A),$$

$$\Gamma^1_{22} = -r/B,$$

$$\Gamma^2_{33} = -\sin\theta\cos\theta,$$

$$\Gamma^1_{00} = A'/(2B),$$

$$\Gamma^1_{33} = -(r\sin^2/B),$$

$$\Gamma^3_{13} = 1/r,$$

$$\Gamma^1_{11} = B'/(2B),$$

$$\Gamma^2_{12} = 1/r,$$

$$\Gamma^3_{23} = \cot\theta.$$

Replacing in the expression for $R_{\mu\nu}$:

$$R_{00} = -\frac{A''}{2B} + \frac{A'}{4B}\left(\frac{A'}{A} + \frac{B'}{B}\right) - \frac{A'}{rB},$$

$$R_{11} = \frac{A''}{2A} - \frac{A'}{4A}\left(\frac{A'}{A} + \frac{B'}{B}\right) - \frac{B'}{rB},$$

$$R_{22} = \frac{1}{B} - 1 + \frac{r}{2B}\left(\frac{A'}{A} - \frac{B'}{B}\right),$$

$$R_{33} = R_{22}\sin^2\theta.$$

Einstein's field equations for the region of empty space then become:

$$R_{00} = R_{11} = R_{22} = 0$$

(the fourth equation has no additional information). Multiplying the first equation by B/A and adding the result to the second equation, we get:

$$A'B + AB' = 0,$$

from which $AB = \text{constant}$. We can write then $B = \alpha A^{-1}$. Going to the third equation and replacing B we obtain: $A + rA' = \alpha$, or:

$$\frac{d(rA)}{dr} = \alpha.$$

The solution of this equation is:

$$A(r) = \alpha\left(1 + \frac{k}{r}\right),$$

with k another integration constant. For B we get:

$$B = \left(1 + \frac{k}{r}\right)^{-1}.$$

If we now consider the Newtonian limit:

$$\frac{A(r)}{c^2} = 1 + \frac{2\Phi}{c^2},$$

with $\Phi = -GM/r$ the Newtonian gravitational potential, we conclude that

$$k = -\frac{2GM}{c^2}$$

and

$$\alpha = c^2.$$

Therefore, the Schwarzschild solution for a static mass M can be written in spherical coordinates (t, r, θ, ϕ) as

$$ds^2 = \left(1 - \frac{2GM}{rc^2}\right)c^2 dt^2 - \left(1 - \frac{2GM}{rc^2}\right)^{-1} dr^2 - r^2\left(d\theta^2 + \sin^2\theta \, d\phi^2\right). \quad (2.10)$$

As mentioned, this solution corresponds to the vacuum region exterior to the spherical object of mass M. Inside the object, space-time will depend on the peculiarities of the physical object.

The metric given by Eq. (2.10) has some interesting properties. Let's assume that the mass M is concentrated at $r = 0$. There seems to be two singularities at which the metric diverges: one at $r = 0$ and the other at

$$r_{\text{Schw}} = \frac{2GM}{c^2}. \tag{2.11}$$

The length r_{Schw} is know as the *Schwarzschild radius* of the object of mass M. Usually, at normal densities, r_{Schw} is well inside the outer radius of the physical system, and the solution does not apply in the interior but only to the exterior of the object. For instance, for the Sun $r_{\text{Schw}} \sim 3$ km. However, for a point mass, the Schwarzschild radius is in the vacuum region and space-time has the structure given by (2.10). In general, we can write

$$r_{\text{Schw}} \sim 3\left(\frac{M}{M_\odot}\right) \text{ km},$$

where $M_\odot = 1.99 \times 10^{33}$ g is the mass of the Sun.

It is easy to see that strange things occur close to r_{Schw}. For instance, for the proper time we get:

$$d\tau = \left(1 - \frac{2GM}{rc^2}\right)^{1/2} dt, \tag{2.12}$$

or

$$dt = \left(1 - \frac{2GM}{rc^2}\right)^{-1/2} d\tau. \tag{2.13}$$

When $r \to \infty$ both times agree, so t is interpreted as the proper time measured from an infinite distance. As the system with proper time τ approaches to r_{Schw}, dt tends to infinity according to Eq. (2.13). The object never reaches the Schwarzschild surface when seen by an infinitely distant observer. The closer the object is to the Schwarzschild radius, the slower it moves for the external observer.

A direct consequence of the difference introduced by gravity in the local time with respect to the time at infinity is that the radiation that escapes from a given radius $r > r_{\text{Schw}}$ will be redshifted when received by a distant and static observer. Since the frequency (and hence the energy) of the photon depends on the time interval, we can write, from Eq. (2.13):

$$\lambda_\infty = \left(1 - \frac{2GM}{rc^2}\right)^{-1/2} \lambda. \tag{2.14}$$

Since the redshift is:

$$z = \frac{\lambda_\infty - \lambda}{\lambda}, \tag{2.15}$$

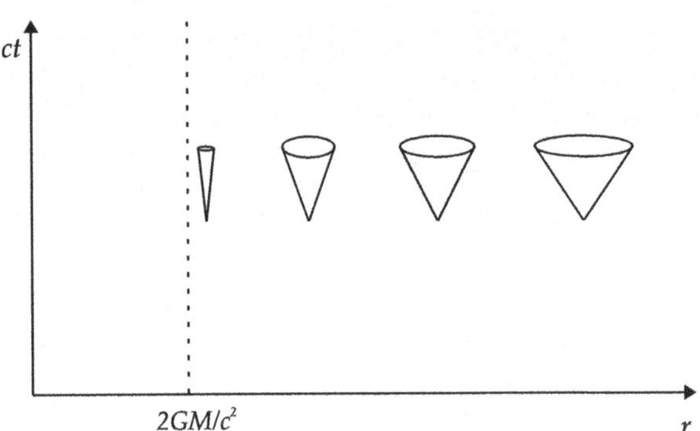

Fig. 2.1 Space-time diagram in Schwarzschild coordinates showing the light cones of events at different distances from the event horizon. Adapted form Carroll (2003)

then

$$1 + z = \left(1 - \frac{2GM}{rc^2}\right)^{-1/2}, \tag{2.16}$$

and we see that when $r \to r_{\text{Schw}}$ the redshift becomes infinite. This means that a photon needs infinite energy to escape from inside the region determined by r_{Schw}. Events that occur at $r < r_{\text{Schw}}$ are disconnected from the rest of the universe. Hence, we call the surface determined by $r = r_{\text{Schw}}$ an *event horizon*. Whatever crosses the event horizon will never return. This is the origin of the expression "black hole", introduced by John A. Wheeler in the mid 1960s. The black hole is the region of space-time inside the event horizon. We can see in Fig. 2.1 what happens with the light cones as an event is closer to the horizon of a Schwarzschild black hole. The shape of the cones can be calculated from the metric (2.10) imposing the null condition $ds^2 = 0$. Then,

$$\frac{dr}{dt} = \pm\left(1 - \frac{2GM}{r}\right), \tag{2.17}$$

where we made $c = 1$. Notice that when $r \to \infty$, $dr/dt \to \pm 1$, as in Minkowski space-time. When $r \to 2GM$, $dr/dt \to 0$, and light moves along the surface $r = 2GM$, which is consequently a null surface. For $r < 2GM$, the sign of the derivative is inverted. The inward region of $r = 2GM$ is time-like for any physical system that has crossed the boundary surface.

What happens to an object when it crosses the event horizon? According to Eq. (2.10), there is a singularity at $r = r_{\text{Schw}}$. The metric coefficients, however, can be made regular by a change of coordinates. For instance we can consider Eddington-Finkelstein coordinates. Let us define a new radial coordinate r_* such that radial null rays satisfy $d(ct \pm r_*) = 0$. Using Eq. (2.10) it can be shown that:

$$r_* = r + \frac{2GM}{c^2} \log\left|\frac{r - 2GM/c^2}{2GM/c^2}\right|.$$

Then, we introduce:

$$v = ct + r_*.$$

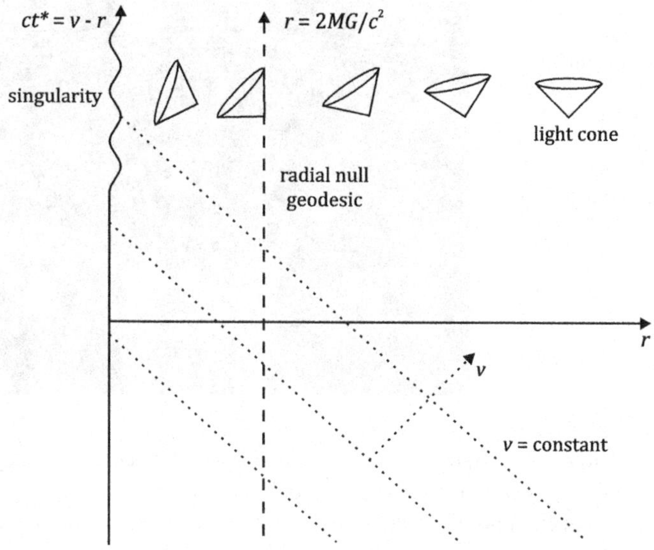

Fig. 2.2 Space-time diagram in Eddington-Finkelstein coordinates showing the light cones close to and inside a black hole. Here, $r = 2GM/c^2 = r_{\mathrm{Schw}}$ is the Schwarzschild radius where the event horizon is located. Adapted form Townsend (1997)

The new coordinate v can be used as a time coordinate replacing t in Eq. (2.10). This yields:

$$ds^2 = \left(1 - \frac{2GM}{rc^2}\right)\left(c^2 dt^2 - dr_*^2\right) - r^2 d\Omega^2$$

or

$$ds^2 = \left(1 - \frac{2GM}{rc^2}\right)dv^2 - 2dr\,dv - r^2 d\Omega^2, \tag{2.18}$$

where

$$d\Omega^2 = d\theta^2 + \sin^2\theta d\phi^2.$$

Notice that in Eq. (2.18) the metric is non-singular at $r = 2GM/c^2$. The only real singularity is at $r = 0$, since there the Riemann tensor diverges. In order to plot the space-time in a (t, r)-plane, we can introduce a new time coordinate $ct_* = v - r$. From the metric (2.18) or from Fig. 2.2 we see that the line $r = r_{\mathrm{Schw}}$, $\theta = \mathrm{constant}$, and $\phi = \mathrm{constant}$ is a null ray, and hence, the surface at $r = r_{\mathrm{Schw}}$ is a null surface. This null surface is an event horizon because inside $r = r_{\mathrm{Schw}}$ all cones have $r = 0$ in their future (see Fig. 2.2). The object in $r = 0$ is the source of the gravitational field and is called the *singularity*. We shall say more about it in Sect. 3.6. For the moment, we only remark that everything that crosses the event horizon will end at the singularity. This is the inescapable fate for everything inside a Schwarzschild black hole. There is no way to avoid it: in the future of every event inside the event

Fig. 2.3 Embedding
space-time diagram in
Eddington-Finkelstein
coordinates showing the light
cones of events at different
distances from a
Schwarzschild black hole.
From http://www.oglethorpe.
edu/faculty/~m_rulison/
ChangingViews/Lecture7.htm

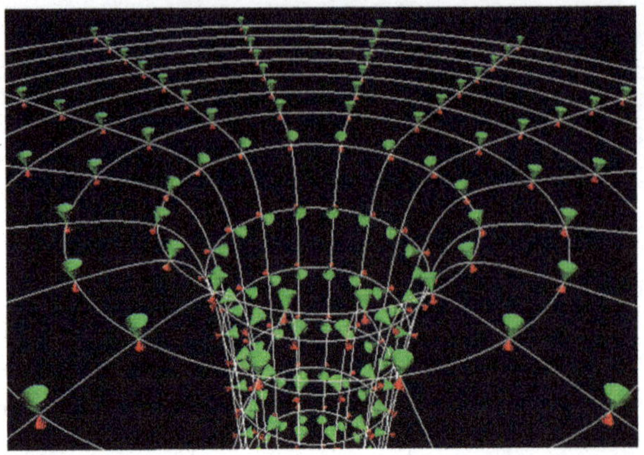

horizon is the singularity. There is no escape, no hope, no freedom, inside the black
hole. There is just the singularity, whatever such a thing might be.

We see now that the name "black hole" is not strictly correct for space-time regions isolated by event horizons. There is no hole to other place. Whatever falls
into the black hole, goes to the singularity. The central object increases its mass and
energy with the accreted bodies and fields, and then the event horizon grows. This
would not happen if what falls into the hole were able to pass through, like through
a hole in a wall. A black hole is more like a space-time precipice, deep, deadly,
and with something unknown at the bottom. A graphic depiction with an embedding diagram of a Schwarzschild black hole is shown in Fig. 2.3. An embedding
is an immersion of a given manifold into a manifold of lower dimensionality that
preserves the metric properties.

2.3.2 Birkhoff's Theorem

If we consider the isotropic but *not static* line element,

$$ds^2 = c^2 A(r,t)dt^2 - B(r,t)dr^2 - r^2 d\Omega^2, \tag{2.19}$$

and substitute it into Einstein's empty-space field equations $R_{\mu\nu} = 0$ to obtain the
functions $A(r,t)$ and $B(r,t)$, the result would be exactly the same:

$$A(r,t) = A(r) = \left(1 - \frac{2GM}{rc^2}\right),$$

and

$$B(r,t) = B(r) = \left(1 - \frac{2GM}{rc^2}\right)^{-1}.$$

This result is general and known as Birkhoff's theorem:

The space-time geometry outside a general spherically symmetric matter distribution is the Schwarzschild geometry.

Birkhoff's theorem implies that strictly radial motions do not perturb the space-time metric. In particular, a pulsating star, if the pulsations are strictly radial, does not produce gravitational waves.

The converse of Birkhoff's theorem is not true, i.e.,

If the region of space-time is described by the metric given by expression (2.10), then the matter distribution that is the source of the metric does not need to be spherically symmetric.

2.3.3 Orbits

Orbits around a Schwarzschild black hole can be easily calculated using the metric and the relevant symmetries (see. e.g. Raine and Thomas 2005; Frolov and Zelnikov 2011). Let us call k^μ a vector in the direction of a given symmetry (i.e. k^μ is a Killing vector). A static situation is symmetric in the time direction, hence we can write $k^\mu = (1, 0, 0, 0)$. The 4-velocity of a particle with trajectory $x^\mu = x^\mu(\tau)$ is $u^\mu = dx^\mu/d\tau$. Then, since $u^0 = E/c$, where E is the energy, we have:

$$g_{\mu\nu}k^\mu u^\nu = g_{00}k^0 u^0 = g_{00}u^0 = \eta_{00}\frac{E}{c} = \frac{E}{c} = \text{constant}. \qquad (2.20)$$

If the particle moves along a geodesic in a Schwarzschild space-time, we obtain from Eq. (2.20):

$$c\left(1 - \frac{2GM}{c^2 r}\right)\frac{dt}{d\tau} = \frac{E}{c}. \qquad (2.21)$$

Similarly, for the symmetry in the azimuthal angle ϕ we have $k^\mu = (0, 0, 0, 1)$, in such a way that:

$$g_{\mu\nu}k^\mu u^\nu = g_{33}k^3 u^3 = g_{33}u^3 = -L = \text{constant}. \qquad (2.22)$$

In the Schwarzschild metric we find, then,

$$r^2\frac{d\phi}{d\tau} = L = \text{constant}. \qquad (2.23)$$

If we now divide the Schwarzschild interval (2.10) by $c^2 d\tau^2$ we get

$$1 = \left(1 - \frac{2GM}{c^2 r}\right)\left(\frac{dt}{d\tau}\right)^2 - c^{-2}\left(1 - \frac{2GM}{c^2 r}\right)^{-1}\left(\frac{dr}{d\tau}\right)^2 - c^{-2}r^2\left(\frac{d\phi}{d\tau}\right)^2, \qquad (2.24)$$

and using the conservation equations (2.21) and (2.23) we obtain:

$$\left(\frac{dr}{d\tau}\right)^2 = \frac{E^2}{c^2} - \left(c^2 + \frac{L^2}{r^2}\right)\left(1 - \frac{2GM}{c^2 r}\right). \qquad (2.25)$$

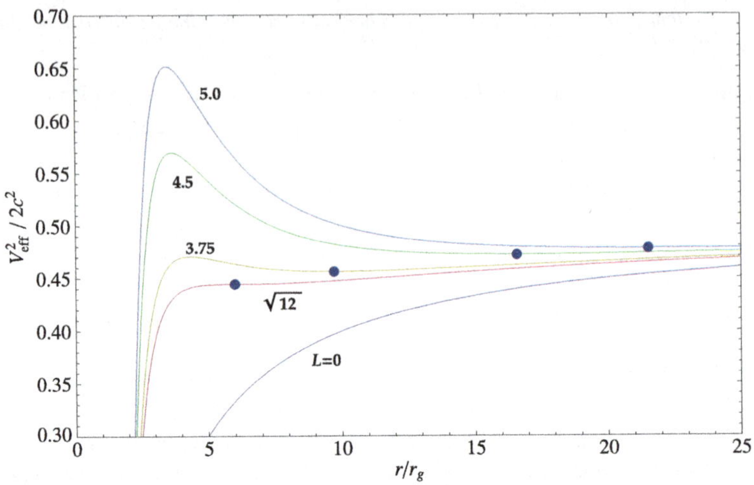

Fig. 2.4 General relativistic effective potential plotted for several values of angular momentum

Then, expressing the energy in units of mc^2 and introducing an effective potential V_{eff},

$$\left(\frac{dr}{d\tau}\right)^2 = \frac{E^2}{c^2} - V^2_{\text{eff}}. \tag{2.26}$$

For circular orbits of a massive particle we have the conditions

$$\frac{dr}{d\tau} = 0 \quad \text{and} \quad \frac{d^2r}{d\tau^2} = 0.$$

The orbits are possible only at the turning points of the effective potential:

$$V_{\text{eff}} = \sqrt{\left(c^2 + \frac{L^2}{r^2}\right)\left(1 - \frac{2r_g}{r}\right)}, \tag{2.27}$$

where L is the angular momentum in units of mc and $r_g = GM/c^2$ is the gravitational radius. Then,

$$r = \frac{L^2}{2cr_g} \pm \frac{1}{2}\sqrt{\frac{L^4}{c^2r_g^2} - 12L^2}. \tag{2.28}$$

The effective potential is shown in Fig. 2.4 for different values of the angular momentum.

For $L^2 > 12c^2r_g^2$ there are two solutions. The negative sign corresponds to a maximum of the potential and is unstable. The positive sign corresponds to a minimum, which is, consequently, stable. At $L^2 = 12c^2r_g^2$ there is a single stable orbit. It is the innermost marginally stable orbit, and it occurs at $r = 6r_g = 3r_{\text{Schw}}$. The specific angular momentum of a particle in a circular orbit at r is:

$$L = c\left(\frac{r_g r}{1 - 3r_g/r}\right)^{1/2}.$$

Its energy (units of mc^2) is:

$$E = \left(1 - \frac{2r_g}{r}\right)\left(1 - \frac{3r_g}{r}\right)^{-1/2}.$$

The proper and observer's periods are:

$$\tau = \frac{2\pi}{c}\left(\frac{r^3}{r_g}\right)^{1/2}\left(1 - \frac{3r_g}{r}\right)^{1/2}$$

and

$$T = \frac{2\pi}{c}\left(\frac{r^3}{r_g}\right)^{1/2}.$$

Notice that when $r \to 3r_g$ both L and E tend to infinity, so only massless particles can orbit at such a radius.

The local velocity at r of an object falling from rest to the black hole is (e.g. Raine and Thomas 2005):

$$v_{\text{loc}} = \frac{\text{proper distance}}{\text{proper time}} = \frac{dr}{(1 - 2GM/c^2r)dt}.$$

Hence, using the expression for dr/dt from the metric (2.10)

$$\frac{dr}{dt} = -c\left(\frac{2GM}{c^2r}\right)^{1/2}\left(1 - \frac{2GM}{c^2r}\right), \tag{2.29}$$

we have,

$$v_{\text{loc}} = \left(\frac{2r_g}{r}\right)^{1/2} \quad \text{(in units of } c\text{)}. \tag{2.30}$$

Then, the differential acceleration the object will experience along an element dr is:[2]

$$dg = \frac{2r_g}{r^3}c^2 dr. \tag{2.31}$$

The tidal acceleration on a body of finite size Δr is simply $(2r_g/r^3)c^2\Delta r$. This acceleration and the corresponding force becomes infinite at the singularity. As the object falls into the black hole, tidal forces act to tear it apart. This painful process is

----

[2]Notice that $dv_{\text{loc}}/d\tau = (dv_{\text{loc}}/dr)(dr/d\tau) = (dv_{\text{loc}}/dr)v_{\text{loc}} = r_g c^2/r^2$.

known as "spaghettification". The process can last significant long before the object crosses the event horizon, depending on the mass of the black hole.

The energy of a particle in the innermost stable orbit can be obtained from the above equation for the energy setting $r = 6r_g$. This yields (units of mc^2):

$$E = \left(1 - \frac{2r_g}{6r_g}\right)\left(1 - \frac{3r_g}{6r_g}\right)^{-1/2} = \frac{2}{3}\sqrt{2}.$$

Since a particle at rest at infinity has $E = 1$, then the energy that the particle should release to fall into the black hole is $1 - (2/3)\sqrt{2} = 0.057$. This means 5.7 % of its rest mass energy, significantly higher than the energy release that can be achieved through nuclear fusion.

An interesting question is what is the gravitational acceleration at the event horizon as seen by an observer from infinity. The acceleration relative to a hovering frame system of a freely falling object at rest at r is (Raine and Thomas 2005):

$$g_r = -c^2\left(\frac{GM/c^2}{r^2}\right)\left(1 - \frac{2GM/c^2}{r}\right)^{-1/2}.$$

So, the energy spent to move the object a distance dl will be $dE_r = mg_r dl$. The energy expended respect to a frame at infinity is $dE_\infty = mg_\infty dl$. Because of the conservation of energy, both quantities should be related by a redshift factor:

$$\frac{E_r}{E_\infty} = \frac{g_r}{g_\infty} = \left(1 - \frac{2GM/c^2}{r}\right)^{-1/2}.$$

Hence, using the expression for g_r we get:

$$g_\infty = c^2\frac{GM/c^2}{r^2}. \tag{2.32}$$

Notice that for an observer at r, $g_r \to \infty$ when $r \to r_{\text{Schw}}$. From infinity, however, the required force to hold the object hovering at the horizon is

$$mg_\infty = c^2\frac{GmM/c^2}{r_{\text{Schw}}^2} = \frac{mc^4}{4GM}.$$

This is the *surface gravity* of the black hole.

2.3.4 Radial Motion of Photons

In the case of photons we have that $ds^2 = 0$. The radial motion, then, satisfies:

$$\left(1 - \frac{2GM}{rc^2}\right)c^2 dt^2 - \left(1 - \frac{2GM}{rc^2}\right)^{-1} dr^2 = 0. \tag{2.33}$$

From here,

$$\frac{dr}{dt} = \pm c\left(1 - \frac{2GM}{rc^2}\right).$$

(2.34)

Integrating, we have:

$$ct = r + \frac{2GM}{c^2}\ln\left|\frac{rc^2}{2GM} - 1\right| + \text{constant outgoing photons,}$$

(2.35)

$$ct = -r - \frac{2GM}{c^2}\ln\left|\frac{rc^2}{2GM} - 1\right| + \text{constant incoming photons.}$$

(2.36)

Notice that in a (ct, r)-diagram the photons have world-lines with slopes ± 1 as $r \to \infty$, indicating that space-time is asymptotically flat. As the events that generate the photons approach to $r = r_{\text{Schw}}$, the slopes tend to $\pm\infty$. This means that the light cones become thinner and thinner for events close to the event horizon. At $r = r_{\text{Schw}}$ the photons cannot escape and they move along the horizon (see Fig. 2.1). An observer in the infinity will never detect them.

2.3.5 Circular Motion of Photons

In this case, fixing $\theta = $ constant due to the symmetry, we have that photons will move in a circle of $r = $ constant and $ds^2 = 0$. Then, from (2.10), we have:

$$\left(1 - \frac{2GM}{rc^2}\right)c^2dt^2 - r^2d\phi^2 = 0.$$

(2.37)

This means that

$$\dot\phi = \frac{c}{r}\sqrt{\left(1 - \frac{2GM}{rc^2}\right)} = \text{constant.}$$

The circular velocity is:

$$v_{\text{circ}} = \frac{r\dot\phi}{\sqrt{g_{00}}} = \frac{\Omega r}{(1 - 2GM/c^2r)^{1/2}}.$$

(2.38)

Setting $v_{\text{circ}} = c$ for photons and using $\Omega = (GM/r^3)^{1/2}$, we get that the only possible radius for a circular photon orbit is:

$$r_{\text{ph}} = \frac{3GM}{c^2}.$$

(2.39)

For a compact object of 1 $M_\odot$, $r_{\text{ph}} \approx 4.5$ km, in comparison with the Schwarzschild radius of 3 km. Photons moving at this distance form the "photosphere" of the black hole. The orbit, however, is unstable, as it can be seen from the effective potential:

$$V_{\text{eff}} = \frac{L_{\text{ph}}^2}{r^2}\left(1 - \frac{2r_{\text{g}}}{r}\right). \tag{2.40}$$

Notice that the four-acceleration for circular motion is $a_\mu = u_\mu u_\nu{}^{;\nu}$. The radial component in the Schwarzschild metric is:

$$a_r = \frac{GM/r^2 - \Omega^2 r}{1 - 2GM/c^2 r - \Omega^2 r^2/c^2}. \tag{2.41}$$

The circular motion along a geodesic line corresponds to the case $a_r = 0$ (free motion). This gives from Eq. (2.41) the usual expression for the Keplerian angular velocity

$$\Omega_{\text{K}} = \left(\frac{GM}{r^3}\right)^{1/2},$$

already used in deriving r_{ph}. The angular velocity, however, can have any value determined by the metric and can be quite different from the corresponding Keplerian value. In general:

$$v = \frac{r\Omega_{\text{K}}}{(1 - 2GM/c^2 r)^{1/2}} = \left(\frac{GM}{r}\right)^{1/2}\left(1 - \frac{2GM}{c^2 r}\right)^{-1/2}. \tag{2.42}$$

From this latter equation and the fact that $v \leq c$ it can be concluded that pure Keplerian motion is only possible for $r \geq 1.5 r_{\text{Schw}}$. At $r \leq 1.5 r_{\text{Schw}}$ any massive particle will find its mass increased by special relativistic effects in such a way that the gravitational attraction will outweigh any centrifugal force.

2.3.6 Gravitational Capture

A particle coming from infinity is captured if its trajectory ends in the black hole. The angular momentum of a non-relativistic particle with velocity v_∞ at infinity is $L = m v_\infty b$, where b is an impact parameter. The condition $L/mcr_{\text{Schw}} = 2$ defines $b_{\text{cr,non-rel}} = 2r_{\text{Schw}}(c/v_\infty)$. Then, the capture cross section is:

$$\sigma_{\text{non-rel}} = \pi b_{\text{cr}}^2 = 4\pi \frac{c^2 r_{\text{Schw}}^2}{v_\infty^2}. \tag{2.43}$$

For an ultra-relativistic particle, $b_{\text{cr}} = 3\sqrt{3}\, r_{\text{Schw}}/2$, and then

$$\sigma_{\text{rel}} = \pi b_{\text{cr}}^2 = \frac{27}{4}\pi r_{\text{Schw}}^2. \tag{2.44}$$

2.3.7 *Other Coordinate Systems*

Other coordinates can be introduced to study additional properties of black holes. We refer the reader to the books of Frolov and Novikov (1998), Raine and Thomas (2005), and Frolov and Zelnikov (2011) for further details. Here we shall only introduce the Kruskal-Szekeres coordinates. These coordinates have the advantage that they cover the entire space-time manifold of the maximally extended Schwarzschild solution and are well-behaved everywhere outside the physical singularity. They allow to remove the non-physical singularity at $r = r_{\text{Schw}}$ and provide new insights on the interior solution, on which we shall return later.

Let us consider the following coordinate transformation:

$$u = \left(\frac{r}{r_{\text{Schw}}} - 1\right)^{1/2} e^{\frac{r}{2r_{\text{Schw}}}} \cosh\left(\frac{ct}{2r_{\text{Schw}}}\right),$$

$$v = \left(\frac{r}{r_{\text{Schw}}} - 1\right)^{1/2} e^{\frac{r}{2r_{\text{Schw}}}} \sinh\left(\frac{ct}{2r_{\text{Schw}}}\right), \tag{2.45}$$

$$\text{if } r > r_{\text{Schw}},$$

and

$$u = \left(1 - \frac{r}{r_{\text{Schw}}}\right)^{1/2} e^{\frac{r}{2r_{\text{Schw}}}} \sinh\left(\frac{ct}{2r_{\text{Schw}}}\right),$$

$$v = \left(1 - \frac{r}{r_{\text{Schw}}}\right)^{1/2} e^{\frac{r}{2r_{\text{Schw}}}} \cosh\left(\frac{ct}{2r_{\text{Schw}}}\right), \tag{2.46}$$

$$\text{if } r < r_{\text{Schw}}.$$

The line element in the Kruskal-Szekeres coordinates is completely regular, except at $r = 0$:

$$ds^2 = \frac{4r_{\text{Schw}}^3}{r} e^{\frac{r}{r_{\text{Schw}}}} \left(dv^2 - du^2\right) - r^2 d\Omega^2. \tag{2.47}$$

The curves at $r = \text{constant}$ are hyperbolic and satisfy:

$$u^2 - v^2 = \left(\frac{r}{r_{\text{Schw}}} - 1\right)^{1/2} e^{\frac{r}{r_{\text{Schw}}}}, \tag{2.48}$$

whereas the curves at $t = \text{constant}$ are straight lines that pass through the origin:

$$\frac{u}{v} = \tanh\frac{ct}{2r_{\text{Schw}}}, \quad r < r_{\text{Schw}},$$

$$\frac{u}{v} = \coth\frac{ct}{2r_{\text{Schw}}}, \quad r > r_{\text{Schw}}. \tag{2.49}$$

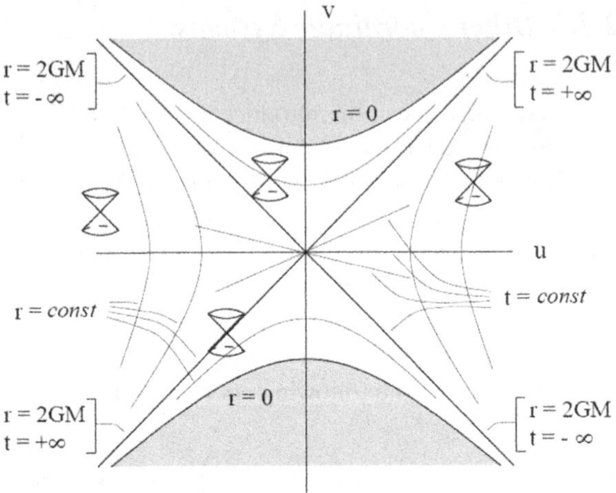

Fig. 2.5 The Schwarzschild metric in Kruskal-Szekeres coordinates ($c = 1$)

In Fig. 2.5 we show the Schwarzschild space-time in Kruskal-Szekeres coordinates. Each hyperbola represents a set of events of constant radius in Schwarzschild coordinates. A radial worldline of a photon in this diagram ($ds = 0$) is represented by a straight line forming an angle of $\pm 45°$ with the u axis. A time-like trajectory has always a slope larger than that of $45°$; and a space-like one, a smaller slope. A particle falling into the black hole crosses the line at $45°$ and reaches the future singularity at $r = 0$. For an external observer this occurs in an infinite time. The Kruskal-Szekeres coordinates have the useful feature that outgoing null geodesics are given by $u = $ constant, whereas ingoing null geodesics are given by $v = $ constant. Furthermore, the (future and past) event horizon(s) are given by the equation $uv = 0$, and the curvature singularity is given by the equation $uv = 1$.

A closely related diagram is the so-called Penrose or Penrose-Carter diagram. This is a two-dimensional diagram that captures the causal relations between different points in space-time. It is an extension of a Minkowski diagram (light cone) where the vertical dimension represents time, and the horizontal dimension represents space, and slanted lines at an angle of $45°$ correspond to light rays. The biggest difference with a Minkowski diagram is that, locally, the metric on a Penrose diagram is conformally equivalent[3] to the actual metric in space-time. The conformal factor is chosen such that the entire infinite space-time is transformed into a Penrose diagram of finite size. For spherically symmetric space-times, every point in the diagram corresponds to a 2-sphere. In Fig. 2.6 we show a Penrose diagram of a Minkowskian space-time.

This type of diagrams can be applied to Schwarzschild black holes. The result is shown in Fig. 2.7. The trajectory represents a particle that goes from some point

---

[3]We remind that two geometries are conformally equivalent if there exists a conformal transformation (an angle-preserving transformation) that maps one geometry to the other. More generally, two (pseudo) Riemannian metrics on a manifold M are conformally equivalent if one is obtained from the other through multiplication by a function on M.

Fig. 2.6 Penrose diagram of
a Minkowskian space-time

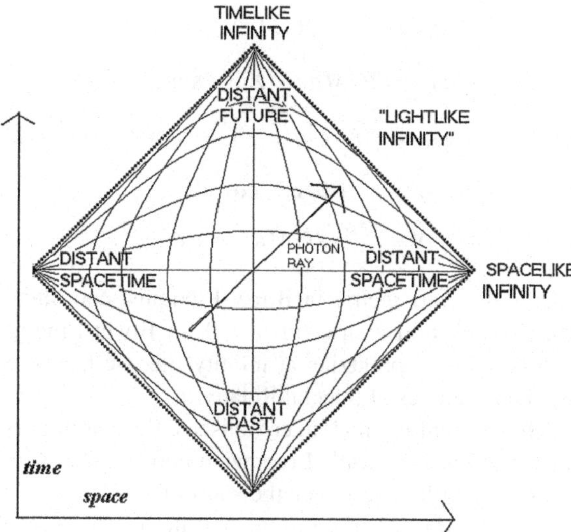

Fig. 2.7 Penrose diagram of
a Schwarzschild black hole

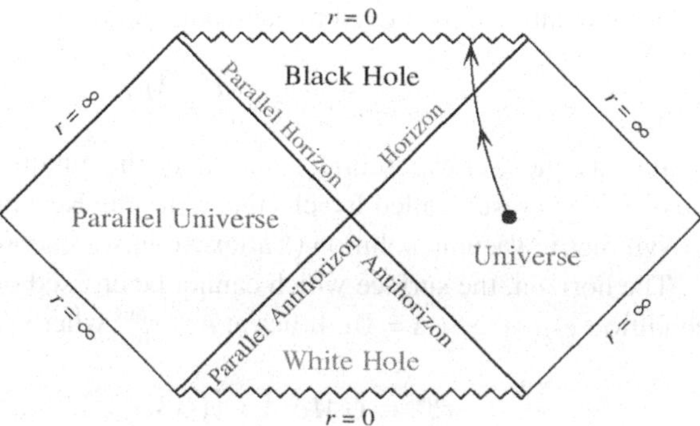

in our universe into the black hole, ending in the singularity. Notice that there is
a mirror extension, also present in the Kruskal-Szekeres diagram, representing a
white hole and a parallel, but inaccessible universe. A white hole presents a naked
singularity. These type of extensions of solutions of Einstein's field equations will
be discussed later.

Now, we turn to axially symmetric (rotating) solutions of the field equations.

2.4 Kerr Black Holes

A Schwarzschild black hole does not rotate. The solution of the field equations
(1.36) for a rotating body of mass M and angular momentum per unit mass a was
found by Roy Kerr (1963):

$$ds^2 = g_{tt}dt^2 + 2g_{t\phi}dtd\phi - g_{\phi\phi}d\phi^2 - \Sigma\Delta^{-1}dr^2 - \Sigma d\theta^2 \qquad (2.50)$$

$$g_{tt} = \left(c^2 - 2GMr\,\Sigma^{-1}\right) \tag{2.51}$$

$$g_{t\phi} = 2GMac^{-2}\Sigma^{-1}r\sin^2\theta \tag{2.52}$$

$$g_{\phi\phi} = \left[\left(r^2 + a^2c^{-2}\right)^2 - a^2c^{-2}\Delta\sin^2\theta\right]\Sigma^{-1}\sin^2\theta \tag{2.53}$$

$$\Sigma \equiv r^2 + a^2c^{-2}\cos^2\theta \tag{2.54}$$

$$\Delta \equiv r^2 - 2GMc^{-2}r + a^2c^{-2}. \tag{2.55}$$

This is the Kerr metric in Boyer-Lindquist coordinates (t, r, θ, ϕ), which reduces to Schwarzschild metric for $a = 0$. In Boyer-Lindquist coordinates the metric is approximately Lorentzian at infinity (i.e. we have a Minkowski space-time in the usual coordinates of Special Relativity).

The element $g_{t\phi}$ no longer vanishes. Even at infinity this element remains (hence we wrote *approximately* Lorentzian above). The Kerr parameter ac^{-1} has dimensions of length. The larger the ratio of this scale to GMc^{-2} (the *spin parameter* $a_* \equiv ac/GM$), the more aspherical the metric. Schwarzschild's black hole is the special case of Kerr's for $a = 0$. Notice that, with the adopted conventions, the angular momentum J is related to the parameter a by:

$$J = Ma. \tag{2.56}$$

Just as the Schwarzschild solution is the unique static vacuum solution of Eqs. (1.36) (a result called Israel's theorem), the Kerr metric is the unique stationary axisymmetric vacuum solution (Carter-Robinson theorem).

The horizon, the surface which cannot be crossed outward, is determined by the condition $g_{rr} \to \infty$ ($\Delta = 0$). It lies at $r = r_h^{out}$ where

$$r_h^{out} \equiv GMc^{-2} + \left[\left(GMc^{-2}\right)^2 - a^2c^{-2}\right]^{1/2}. \tag{2.57}$$

Indeed, the track $r = r_h^{out}$, $\theta = $ constant with $d\phi/d\tau = a(r_h^2 + a^2)^{-1}dt/d\tau$ has $ds = 0$ (it represents a photon circling azimuthally *on* the horizon, as opposed to hovering at it). Hence the surface $r = r_h^{out}$ is tangent to the local light cone. Because of the square root in Eq. (2.57), the horizon is well defined only for $a_* = ac/GM \leq 1$. An *extreme* (i.e. maximally rotating) Kerr black hole has a spin parameter $a_* = 1$. Notice that for $(GMc^{-2})^2 - a^2c^{-2} > 0$ we have actually two horizons. The second, the *inner* horizon, is located at:

$$r_h^{inn} \equiv GMc^{-2} - \left[\left(GMc^{-2}\right)^2 - a^2c^{-2}\right]^{1/2}. \tag{2.58}$$

This horizon is not seen by an external observer, but it hides the singularity to any observer that has already crossed r_h and is separated from the rest of the universe. For $a = 0$, $r_h^{inn} = 0$ and $r_h^{out} = r_{Schw}$. The case $(GMc^{-2})^2 - a^2c^{-2} < 0$ corresponds to no horizons and it is thought to be unphysical.

A study of the orbits around a Kerr black hole is beyond the limits of the present text (the reader is referred to Frolov and Novikov 1998; Pérez et al. 2013), but we shall mention several interesting features. One is that if a particle initially falls radially with no angular momentum from infinity to the black hole, it gains angular motion during the infall. The angular velocity as seen from a distant observer is:

$$\Omega(r,\theta) = \frac{d\phi}{dt} = \frac{(2GM/c^2)ar}{(r^2 + a^2c^{-2})^2 - a^2c^{-2}\Delta\sin^2\theta}. \tag{2.59}$$

The particle will acquire angular velocity in the direction of the spin of the black hole. As the black hole is approached, the particle will find an increasing tendency to get carried away in the same sense in which the black hole is rotating. To keep the particle stationary with respect to the distant stars, it will be necessary to apply a force against this tendency. The closer the particle will be to the black hole, the stronger the force. At a point r_e it becomes impossible to counteract the rotational sweeping force. The particle is in a kind of space-time maelstrom. The surface determined by r_e is the *static limit*: from there in, you cannot avoid rotating. Space-time is rotating here in such a way that you cannot do anything in order to not co-rotate with it. You can still escape from the black hole, since the outer event horizon has not been crossed, but rotation is inescapable. The region between the static limit and the event horizon is called the *ergosphere*. The ergosphere is not spherical but its shape changes with the latitude θ. It can be determined through the condition $g_{tt} = 0$. Consider a stationary particle, $r = $ constant, $\theta = $ constant, and $\phi = $ constant. Then:

$$c^2 = g_{tt}\left(\frac{dt}{d\tau}\right)^2. \tag{2.60}$$

When $g_{tt} \leq 0$ this condition cannot be fulfilled, and hence a massive particle cannot be stationary inside the surface defined by $g_{tt} = 0$. For photons, since $ds = cd\tau = 0$, the condition is satisfied at the surface. Solving $g_{tt} = 0$ we obtain the shape of the ergosphere:

$$r_e = \frac{GM}{c^2} + \frac{1}{c^2}\left(G^2M^2 - a^2c^2\cos^2\theta\right)^{1/2}. \tag{2.61}$$

The static limit lies outside the horizon except at the poles where both surfaces coincide. The phenomenon of "frame dragging" is common to all axially symmetric metrics with $g_{t\phi} \neq 0$.

Roger Penrose (1969) suggested that a projectile thrown from outside into the ergosphere begins to rotate acquiring more rotational energy than it originally had. Then the projectile can break up into two pieces, one of which will fall into the black hole, whereas the other can go out of the ergosphere. The piece coming out will then have more energy than the original projectile. In this way, we can extract energy from a rotating black hole. In Fig. 2.8 we illustrate the situation and show the static limit, the ergosphere and the outer/inner horizons of a Kerr black hole.

The innermost marginally stable circular orbit r_{ms} around an extreme rotating black hole ($ac^{-1} = GM/c^2$) is given by Raine and Thomas (2005):

$$\left(\frac{r_{ms}}{GM/c^2}\right)^2 - 6\left(\frac{r_{ms}}{GM/c^2}\right) \pm 8\left(\frac{r_{ms}}{GM/c^2}\right)^{1/2} - 3 = 0. \tag{2.62}$$

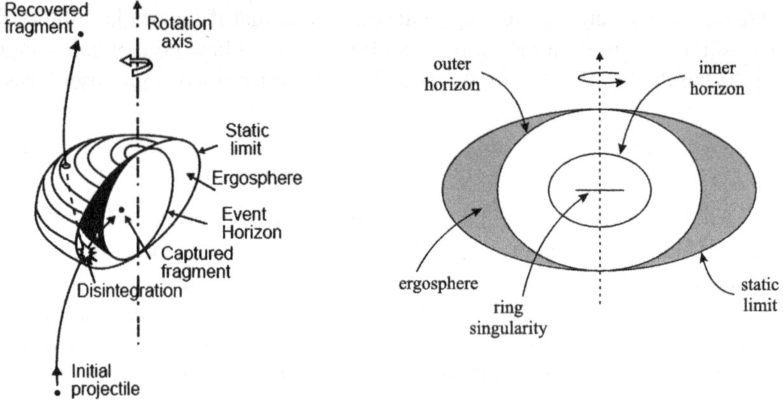

Fig. 2.8 *Left*: a rotating black hole and the Penrose process. From Luminet (1998). *Right*: sketch of the interior of a Kerr black hole

For the "+" sign this is satisfied by $r_{\mathrm{ms}} = GM/c^2$, whereas for the "−" sign the solution is $r_{\mathrm{ms}} = 9GM/c^2$. The first case corresponds to a co-rotating particle and the second one to a counter-rotating particle. The energy of the co-rotating particle in the innermost orbit is $1/\sqrt{3}$ (units of mc^2). The binding energy of a particle in an orbit is the difference between the orbital energy and its energy at infinity. This means a binding energy of 42 % of the rest energy at infinity! For the counter-rotating particle, the binding energy is 3.8 %, smaller than for a Schwarzschild black hole.

An essential singularity occurs when $g_{tt} \to \infty$; this happens if $\Sigma = 0$. This condition implies:

$$r^2 + a^2 c^{-2} \cos^2 \theta = 0. \tag{2.63}$$

Such a condition is fulfilled only by $r = 0$ and $\theta = \frac{\pi}{2}$. This translates in Cartesian coordinates to:[4]

$$x^2 + y^2 = a^2 c^{-2} \quad \text{and} \quad z = 0. \tag{2.64}$$

The singularity is a ring of radius ac^{-1} on the equatorial plane. If $a = 0$, then Schwarzschild's point-like singularity is recovered. If $a \neq 0$ the singularity is not necessarily in the future of all events at $r < r_{\mathrm{h}}^{\mathrm{inn}}$: the singularity can be avoided by some geodesics.

---

[4]The relation with Boyer-Lindquist coordinates is $z = r \cos \theta$, $x = \sqrt{r^2 + a^2 c^{-2}} \sin \theta \cos \phi$, $y = \sqrt{r^2 + a^2 c^{-2}} \sin \theta \sin \phi$.

2.4.1 Pseudo-Newtonian Potentials for Black Holes

The full effective general relativistic potential for particle orbits around a Kerr black hole is quite complex. Instead, pseudo-Newtonian potentials can be used. The first of such potentials, derived by Bohdam Paczyński and used by first time by Paczyński and Wiita (1980), for a non-rotating black hole with mass M, is:

$$\Phi = -\frac{GM}{r - 2r_{\mathrm{g}}},$$
(2.65)

where as before $r_{\mathrm{g}} = GM/c^2$ is the gravitational radius. With this potential one can use Newtonian theory and obtain the same behavior of the Keplerian circular orbits of free particles as in the exact theory: orbits with $r < 9r_{\mathrm{g}}$ are unstable, and orbits with $r < 6r_{\mathrm{g}}$ are unbound. However, velocities of material particles obtained with the potential (2.65) are not accurate, since special relativistic effects are not included (Abramowicz et al. 1996). The velocity $v_{\mathrm{p-N}}$ calculated with the pseudo-Newtonian potential should be replaced by the corrected velocity $v_{\mathrm{p-N}}^{\mathrm{corr}}$ such that

$$v_{\mathrm{p-N}} = v_{\mathrm{p-N}}^{\mathrm{corr}} \gamma_{\mathrm{p-N}}^{\mathrm{corr}}, \qquad \gamma_{\mathrm{p-N}}^{\mathrm{corr}} = \frac{1}{\sqrt{1 - \left(\frac{v_{\mathrm{p-N}}^{\mathrm{corr}}}{c}\right)^2}}.$$
(2.66)

This re-scaling works amazingly well (see Abramowicz et al. 1996) compared with the actual velocities. The agreement with General Relativity is better than 5 %.

For the Kerr black hole, a pseudo-Newtonian potential was found by Semerák and Karas (1999). It can be found in the expression (19) of their paper. However, the use of this potential is almost as complicated as dealing with the full effective potential of the Kerr metric in General Relativity.

2.5 Reissner-Nordström Black Holes

The Reissner-Nordström metric is a spherically symmetric solution of Eqs. (1.36). However, it is not a vacuum solution, since the source has an electric charge Q, and hence there is an electromagnetic field. The energy-momentum tensor of this field is:

$$T_{\mu\nu} = -\mu_0^{-1}\left(F_{\mu\rho}F_{\nu}^{\rho} - \frac{1}{4}g_{\mu\nu}F_{\rho\sigma}F^{\rho\sigma}\right),$$
(2.67)

where $F_{\mu\nu} = \partial_\mu A_\nu - \partial_\nu A_\mu$ is the electromagnetic field strength tensor and A_μ is the electromagnetic 4-potential. Outside the charged object the 4-current j^μ is zero, so Maxwell's equations are:

$$F^{\mu\nu}_{\;;\mu} = 0,$$
(2.68)

$$F_{\mu\nu;\sigma} + F_{\sigma\mu;\nu} + F_{\nu\sigma;\mu} = 0.$$
(2.69)

Einstein's and Maxwell's equations are coupled since $F^{\mu\nu}$ enters into the gravitational field equations through the energy-momentum tensor and the metric $g_{\mu\nu}$ enters into the electromagnetic equations through the covariant derivative. Because of the symmetry constraints we can write:

$$[A^{\mu}] = \left(\frac{\varphi(r)}{c^2}, a(r), 0, 0\right), \tag{2.70}$$

where $\varphi(r)$ is the electrostatic potential, and $a(r)$ is the radial component of the 3-vector potential as $r \to \infty$.

The solution for the metric is given by

$$ds^2 = \Delta c^2 dt^2 - \Delta^{-1} dr^2 - r^2 d\Omega^2, \tag{2.71}$$

where

$$\Delta = 1 - \frac{2GM/c^2}{r} + \frac{q^2}{r^2}. \tag{2.72}$$

In this expression, M is once again interpreted as the mass of the hole and

$$q = \frac{GQ^2}{4\pi\epsilon_0 c^4} \tag{2.73}$$

is related to the total electric charge Q.

The metric has a coordinate singularity at $\Delta = 0$, in such a way that:

$$r_{\pm} = r_{\mathrm{g}} \pm \left(r_{\mathrm{g}}^2 - q^2\right)^{1/2}. \tag{2.74}$$

Here, $r_{\mathrm{g}} = GM/c^2$ is the gravitational radius. For $r_{\mathrm{g}} = q$, we have an *extreme* Reissner-Nordström black hole with a unique horizon at $r = r_{\mathrm{g}}$. Notice that a Reissner-Nordström black hole can be more compact than a Schwarzschild black hole of the same mass. For the case $r_{\mathrm{g}}^2 > q^2$, both $r_{\pm}$ are real and there are two horizons as in the Kerr solution. Finally, in the case $r_{\mathrm{g}}^2 < q^2$ both $r_{\pm}$ are imaginary there is no coordinate singularities, no horizon hides the intrinsic singularity at $r = 0$. It is thought, however, that naked singularities do not exist in Nature (see Sect. 3.6 below).

2.6 Kerr-Newman Black Holes

The Kerr-Newman metric of a charged spinning black hole is the most general black hole solution. It was found by Ezra "Ted" Newman in 1965 (Newman et al. 1965). This metric can be obtained from the Kerr metric (2.50) in Boyer-Lindquist coordinates with the replacement:

$$\frac{2GM}{c^2}r \longrightarrow \frac{2GM}{c^2}r - q^2,$$

where q is related to the charge Q by Eq. (2.73).

The full expression reads:

$$ds^2 = g_{tt}dt^2 + 2g_{t\phi}dtd\phi - g_{\phi\phi}d\phi^2 - \Sigma\Delta^{-1}dr^2 - \Sigma d\theta^2 \tag{2.75}$$

$$g_{tt} = c^2\left[1 - \left(2GMrc^{-2} - q^2\right)\Sigma^{-1}\right] \tag{2.76}$$

$$g_{t\phi} = a\sin^2\theta\,\Sigma^{-1}\left(2GMrc^{-2} - q^2\right) \tag{2.77}$$

$$g_{\phi\phi} = \left[\left(r^2 + a^2c^{-2}\right)^2 - a^2c^{-2}\Delta\sin^2\theta\right]\Sigma^{-1}\sin^2\theta \tag{2.78}$$

$$\Sigma \equiv r^2 + a^2c^{-2}\cos^2\theta \tag{2.79}$$

$$\Delta \equiv r^2 - 2GMc^{-2}r + a^2c^{-2} + q^2 \equiv \left(r - r_{\mathrm{h}}^{\mathrm{out}}\right)\left(r - r_{\mathrm{h}}^{\mathrm{inn}}\right), \tag{2.80}$$

where all symbols have the same meaning as in the Kerr metric and the outer horizon is located at

$$r_{\mathrm{h}}^{\mathrm{out}} = GMc^{-2} + \left[\left(GMc^{-2}\right)^2 - a^2c^{-2} - q^2\right]^{1/2}. \tag{2.81}$$

The inner horizon is located at:

$$r_{\mathrm{h}}^{\mathrm{inn}} = GMc^{-2} - \left[\left(GMc^{-2}\right)^2 - a^2c^{-2} - q^2\right]^{1/2}. \tag{2.82}$$

The Kerr-Newman solution is a non-vacuum solution either. It shares with the Kerr and Reissner-Nordström solutions the existence of two horizons, and as the Kerr solution it presents an ergosphere. At a latitude θ, the radial coordinate for the ergosphere is:

$$r_{\mathrm{e}} = GMc^{-2} + \left[\left(GMc^{-2}\right)^2 - a^2c^{-2}\cos^2\theta - q^2\right]^{1/2}. \tag{2.83}$$

Like the Kerr metric for an uncharged rotating mass, the Kerr-Newman interior solution exists mathematically but is probably not representative of the actual metric of a physically realistic rotating black hole due to stability problems (see next chapter). The surface area of the horizon is:

$$A = 4\pi\left(r_{\mathrm{h}}^{\mathrm{out}2} + a^2c^{-2}\right). \tag{2.84}$$

The Kerr-Newman metric represents the simplest stationary, axisymmetric, asymptotically flat solution of Einstein's equations in the presence of an electromagnetic field in four dimensions. It is sometimes referred to as an "electrovaccuum" solution of Einstein's equations. Any Kerr-Newman source has its rotation axis aligned with its magnetic axis (Punsly 1998a). Thus, a Kerr-Newman source is different from commonly observed astronomical bodies, for which there might be a substantial angle between the rotation axis and the magnetic moment.

Since the electric field cannot remain static in the ergosphere, a magnetic field is generated as seen by an observer outside the static limit. This is illustrated in Fig. 2.9. You can find the expressions for the components of the fields in Punsly (2001).

Pekeris and Frankowski (1987) have calculated the interior electromagnetic field of the Kerr-Newman source, i.e., the ring singularity. The electric and magnetic

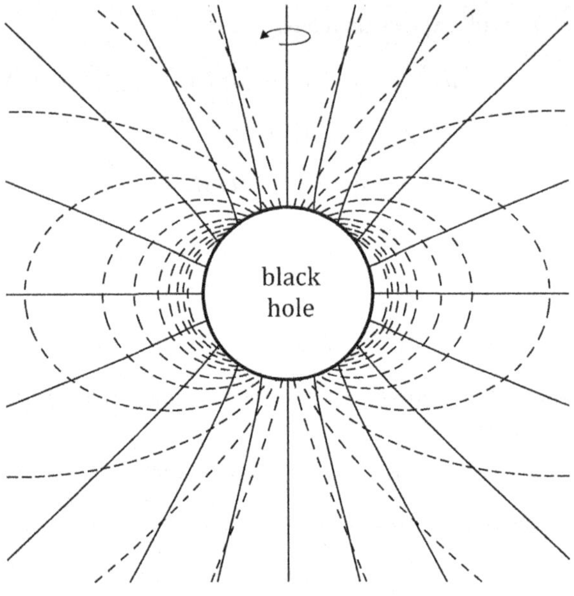

Fig. 2.9 The electric (*solid*) and magnetic (*dashed*) field lines of a Kerr-Newman black hole. The rotation axis of the hole is indicated

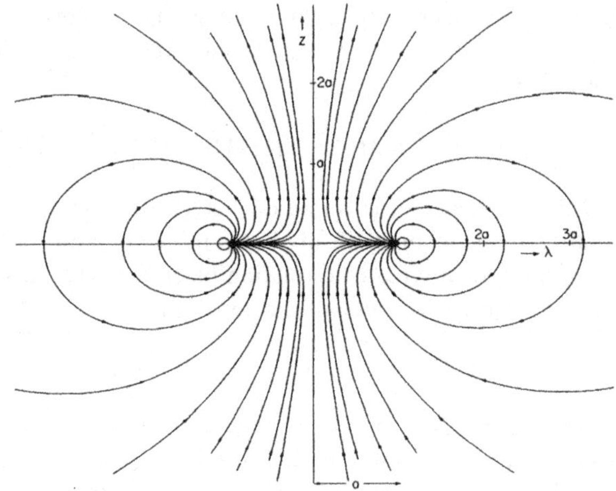

Fig. 2.10 Magnetic field of a Kerr-Newman source. See text for units. Reprinted figure with permission from Pekeris and Frankowski (1987). Copyright (1987) by the American Physical Society

fields are shown in Figs. 2.10 and 2.11, in a (λ, z)-plane, with $\lambda = (x^2 + y^2)^{1/2}$. The general features of the magnetic field are that at distances much larger than ac^{-1} it resembles closely a dipole field, with a dipolar magnetic moment $\mu_{\rm d} = Qac^{-1}$. On the disc of radius ac^{-1} the z-component of the field vanishes, in contrast with the interior of Minkowskian ring-current models. The electric field for a positive charge distribution is attractive for positive charges toward the interior disc. At the ring there is a charge singularity and at large distances the field corresponds to that of a point-like charge Q.

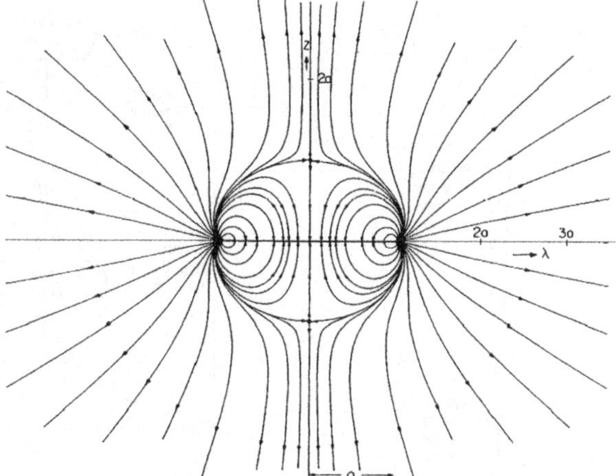

Fig. 2.11 Electric field of a Kerr-Newman source. See text for units. Reprinted figure with permission from Pekeris and Frankowski (1987). Copyright (1987) by the American Physical Society

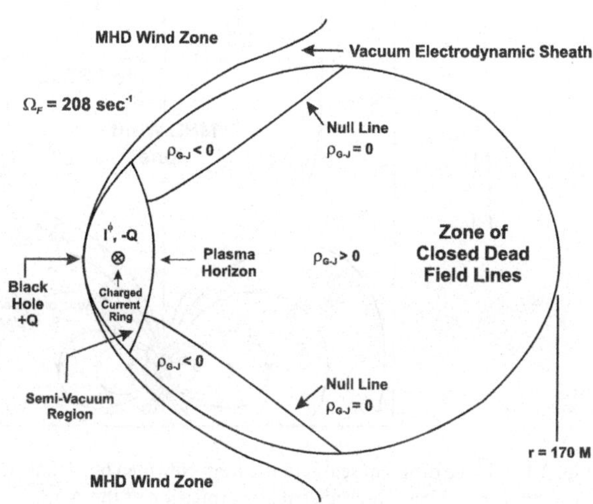

Fig. 2.12 Charged and rotating black hole magnetosphere. The black hole has charge $+Q$ whereas the current ring circulating around it has opposite charge. The figure shows (units $G = c = 1$) the region of closed lines determined by the light cylinder, the open lines that drive a magnetohydrodynamical wind, and the vacuum region in between. Adapted from Punsly (1998a). Reproduced by permission of the AAS

Charged black holes might be a natural result from charge separation during the gravitational collapse of a star. It is thought that an astrophysical charged object would discharge quickly by accretion of charges of opposite sign. There remains the possibility, however, that the charge separation could lead to a configuration where the black hole has a charge and a superconducting ring around it would have the same but opposite charge, in such a way the whole system seen from infinity is neutral. In such a case a Kerr-Newman black hole might survive for some time, depending on the environment. For further details, the reader is referred to the highly technical book by Brian Punsly (2001) and related articles (Punsly 1998a, 1998b, and Punsly et al. 2000). In Figs. 2.12 and 2.13 the magnetic field around a Kerr-Newman black hole surrounded by a charged current ring is shown. The opposite charged black hole and ring are the minimum energy configuration for the system black hole plus magnetosphere. Since the system is neutral from the infinity, it dis-

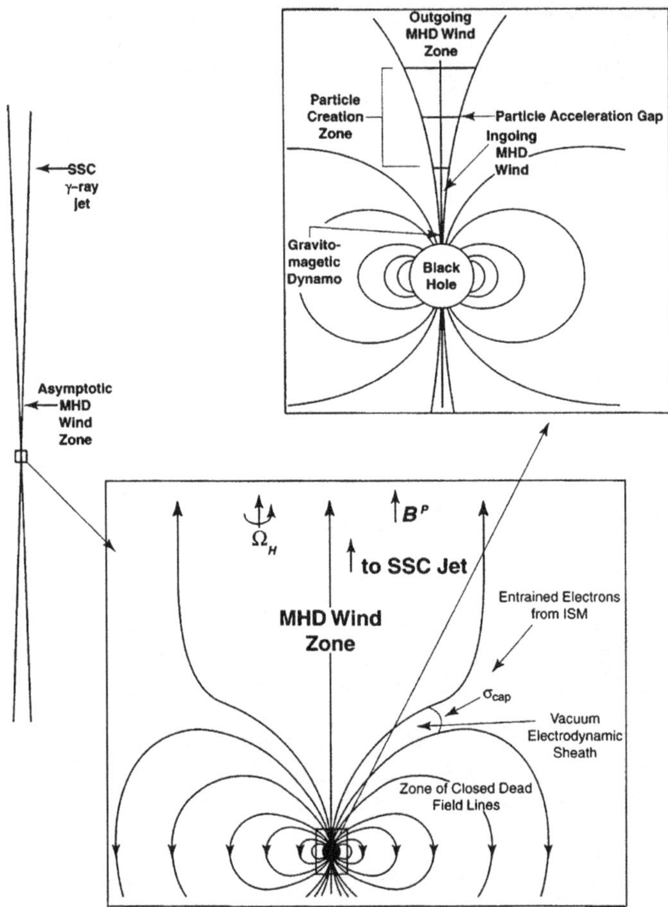

Fig. 2.13 Three different scales of the Kerr-Newman black hole model developed by Brian Punsly. From Punsly (1998a). Reproduced by permission of the AAS

charges slowly and can survive for a few thousand years. During this period, the source can be active through the capture of free electrons from the environment and the production of gamma rays by inverse Compton up-scattering of synchrotron photons produced by electrons accelerated in the polar gap of the hole. In Fig. 2.14 we show the corresponding spectral energy distribution obtained by Punsly et al. (2000) for such a configuration of Kerr-Newman black hole magnetosphere.

2.6.1 Einstein-Maxwell Equations

In order to determine the gravitational and electromagnetic fields over a region of a space-time we have to solve the Einstein-Maxwell equations:

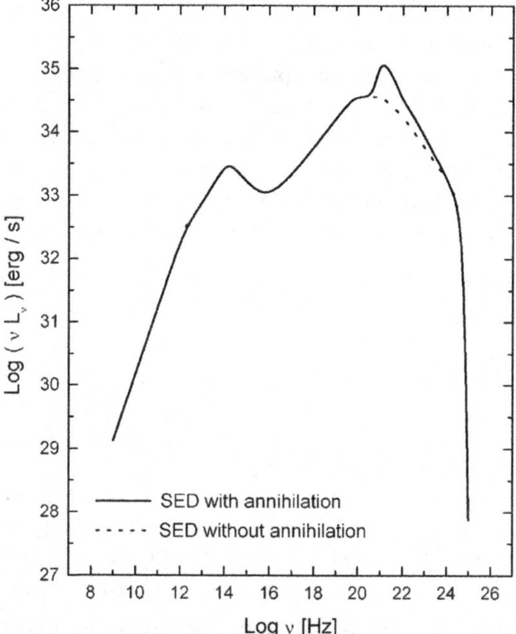

Fig. 2.14 The spectral energy distribution resulting from a Kerr-Newman black hole slowly accreting from the interstellar medium. From Punsly et al. (2000), reproduced with permission ©ESO

$$R_{\mu\nu} - \frac{1}{2}Rg_{\mu\nu} + \Lambda g_{\mu\nu} = -\frac{8\pi G}{c^4}(T_{\mu\nu} + E_{\mu\nu}), \tag{2.85}$$

$$\frac{4\pi}{c}E_\nu^\mu = -F^{\mu\rho}F_{\rho\nu} + \frac{1}{4}\delta_\nu^\mu F^{\sigma\lambda}F_{\sigma\lambda}, \tag{2.86}$$

$$F_{\mu\nu} = A_{\mu;\nu} - A_{\nu;\mu}, \tag{2.87}$$

$$F_\mu^{\nu;\nu} = \frac{4\pi}{c}J_\mu. \tag{2.88}$$

Here $T_{\mu\nu}$ and $E_{\mu\nu}$ are the energy-momentum tensors of matter and electromagnetic fields, $F_{\mu\nu}$ and J_μ are the electromagnetic field and current density, A_μ is the 4-dimensional potential, and Λ is the cosmological constant.

The solution of this system of equations is non-trivial since they are coupled. The electromagnetic field is a source of the gravitational field and this field enters into the electromagnetic equations through the covariant derivatives indicated by the semi-colons. For an exact and relevant solution of the problem see Manko and Sibgatullin (1992).

2.7 Other Black Holes

2.7.1 Born-Infeld Black Holes

Born and Infeld (1934) to avoid the singularities associated with charged point particles in Maxwell theory. Almost immediately, Hoffmann (1935) coupled General

Relativity with Born-Infeld electrodynamics to obtain a spherically symmetric solution representing the gravitational field of a charged object. This solution, forgotten during decades, can represent a charged black hole in nonlinear electrodynamics. In Born-Infeld electrodynamics the trajectories of photons in curved space-times are not null geodesics of the background metric. Instead, they follow null geodesics of an effective geometry determined by the nonlinearities of the electromagnetic field.

The action of Einstein gravity coupled to Born-Infeld electrodynamics has the form (in this section we adopt, for simplicity, $c = G = 4\pi\epsilon_0 = (4\pi)^{-1}\mu_0 = 1$):

$$S = \int dx^4 \sqrt{-g}\left(\frac{R}{16\pi} + L_{\mathrm{BI}}\right),\qquad(2.89)$$

with

$$L_{\mathrm{BI}} = \frac{1}{4\pi b^2}\left(1 - \sqrt{1 + \frac{1}{2}F_{\sigma\nu}F^{\sigma\nu}b^2 - \frac{1}{4}\tilde{F}_{\sigma\nu}F^{\sigma\nu}b^4}\right),\qquad(2.90)$$

where g is the determinant of the metric tensor, R is the scalar of curvature, $F_{\sigma\nu} = \partial_\sigma A_\nu - \partial_\nu A_\sigma$ is the electromagnetic tensor, $\tilde{F}_{\sigma\nu} = \frac{1}{2}\sqrt{-g}\,\varepsilon_{\alpha\beta\sigma\nu}F^{\alpha\beta}$ is the dual of $F_{\sigma\nu}$ (with $\varepsilon_{\alpha\beta\sigma\nu}$ the Levi-Civita symbol), and b is a parameter that indicates how much Born-Infeld and Maxwell electrodynamics differ. For $b \to 0$ the Einstein-Maxwell action is recovered. The maximal possible value of the electric field in this theory is b, and the self-energy of point charges is finite. The field equations can be obtained by varying the action with respect to the metric $g_{\sigma\nu}$ and the electromagnetic potential A_ν.

We can write L_{BI} in terms of the electric and magnetic fields:

$$L_{\mathrm{BI}} = \frac{b^2}{4\pi}\left[1 - \sqrt{1 - \frac{B^2 - E^2}{b^2} - \frac{(\mathbf{E}\cdot\mathbf{B})^2}{b^4}}\right].\qquad(2.91)$$

The Lagrangian depends non-linearly of the electromagnetic invariants:

$$F = \frac{1}{4}F_{\alpha\beta}F^{\alpha\beta} = \frac{1}{2}\left(B^2 - E^2\right),\qquad(2.92)$$

$$\tilde{G} = \frac{1}{4}F_{\alpha\beta}\tilde{F}^{\alpha\beta} = -\mathbf{B}\cdot\mathbf{E}.\qquad(2.93)$$

Introducing the Hamiltonian formalism,

$$P^{\alpha\beta} = 2\frac{\partial L}{\partial F_{\alpha\beta}} = \frac{\partial L}{\partial F}F^{\alpha\beta} + \frac{\partial L}{\partial \tilde{G}}\tilde{F}^{\alpha\beta},\qquad(2.94)$$

$$H = \frac{1}{2}P^{\alpha\beta}F_{\alpha\beta} - L\left(F, \tilde{G}^2\right),\qquad(2.95)$$

and adopting the notation

$$P = \frac{1}{4}P_{\alpha\beta}P^{\alpha\beta},\qquad(2.96)$$

$$\tilde{Q} = \frac{1}{4} P_{\alpha\beta} \tilde{P}^{\alpha\beta}, \tag{2.97}$$

we can express $F^{\alpha\beta}$ as a function of $P^{\alpha\beta}$, P, and $\tilde{Q}$:

$$F^{\alpha\beta} = 2\frac{\partial H}{\partial P_{\alpha\beta}} = \frac{\partial H}{\partial P} P^{\alpha\beta} + \frac{\partial H}{\partial \tilde{Q}} \tilde{P}^{\alpha\beta}. \tag{2.98}$$

The Hamiltonian equations in the P and $\tilde{Q}$ formalism can be written as:

$$\left(\frac{\partial H}{\partial P} \tilde{P}^{\alpha\beta} + \frac{\partial H}{\partial \tilde{Q}} P^{\alpha\beta} \right)_{,\beta} = 0. \tag{2.99}$$

The coupled Einstein-Born-Infeld equations are:

$$4\pi T_{\mu\nu} = \frac{\partial H}{\partial P} P_{\mu\alpha} P^{\alpha}_{\nu} - g_{\mu\nu}\left(2P\frac{\partial H}{\partial P} + \tilde{Q}\frac{\partial H}{\partial \tilde{Q}} - H \right), \tag{2.100}$$

$$R = 8\left(P\frac{\partial H}{\partial P} + \tilde{Q}\frac{\partial H}{\partial \tilde{Q}} - H \right). \tag{2.101}$$

The field equations have spherically symmetric black hole solutions given by

$$ds^2 = \psi(r)dt^2 - \psi(r)^{-1}dr^2 - r^2 d\Omega^2, \tag{2.102}$$

with

$$\psi(r) = 1 - \frac{2M}{r} + \frac{2}{b^2 r}\int_r^{\infty} (\sqrt{x^4 + b^2 Q^2} - x^2)dx, \tag{2.103}$$

$$D(r) = \frac{Q_E}{r^2}, \tag{2.104}$$

$$B(r) = Q_M \sin\theta, \tag{2.105}$$

where M is the mass, $Q^2 = Q_E^2 + Q_M^2$ is the sum of the squares of the electric Q_E and magnetic Q_M charges, $B(r)$ and $D(r)$ are the magnetic and the electric inductions in the local orthonormal frame. In the limit $b \to 0$, the Reissner-Nordström metric is obtained. The metric (2.102) is also asymptotically Reissner-Nordström for large values of r. With the units adopted above, M, Q and b have dimensions of length. The metric function $\psi(r)$ can be expressed in the form

$$\psi(r) = 1 - \frac{2M}{r} + \frac{2}{3b^2}\left\{ r^2 - \sqrt{r^4 + b^2 Q^2} \right.$$

$$\left. + \frac{\sqrt{|bQ|^3}}{r} F\left[\arccos\left(\frac{r^2 - |bQ|}{r^2 + |bQ|} \right), \frac{\sqrt{2}}{2} \right] \right\}, \tag{2.106}$$

where $F(\gamma, k)$ is the elliptic integral of the first kind.[5] As in Schwarzschild and Reissner-Nordström cases, the metric (2.102) has a singularity at $r = 0$.

The zeros of $\psi(r)$ determine the position of the horizons, which have to be obtained numerically. For a given value of b, when the charge is small, $0 \leq |Q|/M \leq \nu_1$, the function $\psi(r)$ has one zero and there is a regular event horizon. For intermediate values of charge, $\nu_1 < |Q|/M < \nu_2$, $\psi(r)$ has two zeros, so there are, as in the Reissner-Nordström geometry, an inner horizon and an outer regular event horizon. When $|Q|/M = \nu_2$, there is one degenerate horizon. Finally, if the values of charge are large, $|Q|/M > \nu_2$, the function $\psi(r)$ has no zeros and a naked singularity is obtained. The values of $|Q|/M$ where the number of horizons change, $\nu_1 = (9|b|/M)^{1/3}[F(\pi, \sqrt{2}/2)]^{-2/3}$ and ν_2, which should be calculated numerically from the condition $\psi(r_h) = \psi'(r_h) = 0$, are increasing functions of $|b|/M$. In the Reissner-Nordström limit ($b \to 0$) it is easy to see that $\nu_1 = 0$ and $\nu_2 = 1$.

The paths of photons in nonlinear electrodynamics are not null geodesics of the background geometry. Instead, they follow null geodesics of an effective metric generated by the self-interaction of the electromagnetic field, which depends on the particular nonlinear theory considered. In Einstein gravity coupled to Born-Infeld electrodynamics the effective geometry for photons is given by Bretón (2002):

$$ds_{\text{eff}}^2 = \omega(r)^{1/2}\psi(r)dt^2 - \omega(r)^{1/2}\psi(r)^{-1}dr^2 - \omega(r)^{-1/2}r^2 d\Omega^2, \tag{2.107}$$

where

$$\omega(r) = 1 + \frac{Q^2 b^2}{r^4}. \tag{2.108}$$

Then, to calculate the deflection angle for photons passing near the black holes, it is necessary to use the effective metric (2.107) instead of the background metric (2.102). The horizon structure of the effective metric is the same as that of metric (2.102), but the trajectories of photons are different.

2.7.2 Regular Black Holes

Solutions of Einstein's field equations representing black holes where the metric is always regular (i.e. free of intrinsic singularities where $R^{\mu\nu\rho\sigma} R_{\mu\nu\rho\sigma}$ diverges) can be found for some choices of the equation of state. For instance, Mbonye and Kazanas (2005) have suggested the following equation:

$$p_r(\rho) = \left[\alpha - (\alpha + 1)\left(\frac{\rho}{\rho_{\text{max}}}\right)^m\right]\left(\frac{\rho}{\rho_{\text{max}}}\right)^{1/n}\rho. \tag{2.109}$$

---

[5] $F(\gamma, k) = \int_0^\gamma (1 - k^2 \sin^2\phi)^{-1/2}d\phi = \int_0^{\sin\gamma}[(1 - z^2)(1 - k^2 z^2)]^{-1/2}dz.$

The maximum limiting density $\rho_{\max}$ is concentrated in a region of radius

$$r_0 = \sqrt{\frac{1}{G\rho_{\max}}}. \tag{2.110}$$

At low densities $p_r \propto \rho^{1+1/n}$ and the equation reduces to that of a polytrope gas. At high densities close to $\rho_{\max}$ the equation becomes $p_r = -\rho$ and the system behaves as a gravitational field dominated by a cosmological term in the field equations. The exact values of m, n, and α determine the sound speed in the system. Imposing that the maximum sound speed $c_s = (dp/d\rho)^{1/2}$ be finite everywhere, it is possible to constrain the free parameters. Adopting $m = 2$ and $n = 1$ Eq. (2.109) becomes:

$$p_r(\rho) = \left[\alpha - (\alpha + 1)\left(\frac{\rho}{\rho_{\max}}\right)^2\right]\left(\frac{\rho}{\rho_{\max}}\right)\rho. \tag{2.111}$$

The model introduced by Mbonye and Kazanas represents a regular static black hole, with a matter source that smoothly goes from a de Sitter behavior near the origin to Schwarzschild's spacetime outside the object. A space-time metric well-adapted to examine the properties of this system is (Mbonye et al. 2011):

$$ds^2 = -B(r)dt^2 + \left(1 - \frac{2m(r)}{r}\right)^{-1} dr^2 + r^2(d\theta^2 + \sin^2\theta d\phi^2), \tag{2.112}$$

where

$$B(r) = \exp\int_{r_0}^{r} \frac{2}{r'^2}\left[m(r') + 4\pi r'^3 p_{sfr'}\right]$$

$$\times \left[\frac{1}{(1 - \frac{2m(r')}{r'})}\right]dr', \tag{2.113}$$

and

$$m(r) = 4\pi \int_0^{r} \rho(r')r'^2 dr'. \tag{2.114}$$

Outside the body $\rho \to 0$, and Eq. (2.112) becomes Schwarzschild solution for $R_{\mu\nu} = 0$. When $r \to 0$, $\rho = \rho_{\max}$ and the metric becomes of de Sitter type:

$$ds^2 = \left(1 - \frac{r^2}{r_0^2}\right)c^2 dt^2 - \left(1 - \frac{r^2}{r_0^2}\right)^{-1} dr^2 - r^2(d\theta^2 + \sin^2\theta d\phi^2), \tag{2.115}$$

with

$$r_0 = \sqrt{\frac{3}{8\pi G\rho_{\max}}}. \tag{2.116}$$

There is no singularity at $r = 0$ and the black hole is regular. For $0 \leq r < 1$ it has constant positive density $\rho_{\max}$ and negative pressure $p_r = -\rho_{\max}$ and space-time becomes asymptotically de Sitter in the innermost region. It might be speculated that the transition in the equation of state occurs because at very high densities the matter field couples with a scalar field that provides the negative pressure.

Other assumptions for the equation of state can lead to different (but still regular) behavior, like a bouncing close to $r = 0$ and the development of an expanding closed universe inside the black hole (Frolov et al. 1990). Unstable behavior, both dynamic and thermodynamic, seems to be a characteristic of these type of black hole solutions (Pérez et al. 2011).

Regular black holes might also be found in $f(R)$ gravity for some suitable function of the curvature scalar.

2.7.3 $f(R)$ Black Holes

2.7.3.1 Rewriting the Field Equations of $f(R)$ Gravity

In order to study the possible black hole solutions obtained from any $f(R)$ theory, we rewrite the action (1.122) in the form

$$S = S_g + S_m, \tag{2.117}$$

where S_g is the gravitational action, with the scalar function now written as $R + f(R)$

$$S_g = \frac{1}{16\pi G} \int d^4x \sqrt{|g|} \big(R + f(R)\big). \tag{2.118}$$

From the matter term S_m, we define the energy momentum tensor as

$$T^{\mu\nu} = -\frac{2}{\sqrt{|g|}} \frac{\delta S_m}{\delta g_{\mu\nu}}. \tag{2.119}$$

By performing variations of (2.117) with respect to the metric tensor, we obtain the field equations in metric formalism in a more convenient way:

$$R_{\mu\nu}\big(1 + f'(R)\big) - \frac{1}{2}g_{\mu\nu}\big(R + f(R)\big)$$
$$+ (\nabla_\mu \nabla_\nu - g_{\mu\nu}\Box)f'(R) + 8\pi G T_{\mu\nu} = 0, \tag{2.120}$$

with $\Box = \nabla_\beta \nabla^\beta$ as before and $f'(R) = df(R)/dR$. Clearly, for $f(R) = 0$ standard General Relativity is recovered. Taking the trace of this equation yields:

$$R\big(1 + f'(R)\big) - 2\big(R + f(R)\big) - 3\Box f'(R) + 8\pi G T = 0, \tag{2.121}$$

where $T = T^\mu_\mu$. Unlike the case of General Relativity, vacuum solutions ($T = 0$) do not necessarily imply a null curvature $R = 0$. From Eq. (2.120) we obtain the condition for vacuum constant scalar curvature $R = R_0$ solutions:

$$R_{\mu\nu}\left(1 + f'(R_0)\right) - \frac{1}{2}g_{\mu\nu}\left(R_0 + f(R_0)\right) = 0. \tag{2.122}$$

and the Ricci tensor becomes proportional to the metric,

$$R_{\mu\nu} = \frac{R_0 + f(R_0)}{2(1 + f'(R_0))}g_{\mu\nu}, \tag{2.123}$$

with $1 + f'(R_0) \neq 0$. Taking the trace on the previous equation:

$$R_0\left(1 + f'(R_0)\right) - 2\left(R_0 + f(R_0)\right) = 0, \tag{2.124}$$

and therefore

$$R_0 = \frac{2f(R_0)}{f'(R_0) - 1}. \tag{2.125}$$

If we want to find viable black hole solutions of Eq. (2.120), some conditions must be imposed in order to make $f(R)$ theories consistent with known gravitational and cosmological facts. These conditions are (Cembranos et al. 2011):

1. $f''(R) \geq 0$ for $R \gg f''(R)$. This is the stability requirement for a high-curvature classical regime and that of the existence of a matter dominated era in cosmological evolution. A simple physical interpretation can be given to this condition: if an effective gravitational constant $G_{\text{eff}} \equiv G/(1 + f'(R))$ is defined, then the sign of its variation with respect to R, dG_{eff}/dR, is uniquely determined by the sign of $f''(R)$, so in case $f''(R) < 0$, G_{eff} would grow as R does, because R generates more and more curvature. This mechanism would destabilize the metric field, since it would not have a fundamental state because any small curvature would grow to infinity. Instead, if $f''(R) \geq 0$, a counter reaction mechanism operates to compensate this R growth and stabilize the system.
2. $1 + f'(R) > 0$. This conditions ensures that the effective gravitational constant is positive, as it can be checked from the previous definition of G_{eff}.
3. $f'(R) < 0$. Keeping in mind the strong restrictions of Big Bang nucleosynthesis and cosmic microwave background, this condition ensures that the expected behavior be recovered at early times, that is, $f(R)/R \to 0$ and $f'(R) \to 0$ as $R \to \infty$. Conditions 1 and 2 together demand $f(R)$ to be a monotonously increasing function between the values $-1 < f'(R) < 0$.
4. $f'(R)$ must be small in recent epochs. This condition is mandatory in order to satisfy imposed restrictions by local (solar and galactic) gravity tests.

In looking for constant curvature R_0 vacuum solutions for fields generated by massive charged objects we follow Cembranos et al. (2011). The action (in units of $G = c = \hbar = k = 1$) is:

$$S = \frac{1}{16\pi}\int d^4x \sqrt{|g|}\left(R + f(R) - F_{\mu\nu}F^{\mu\nu}\right), \tag{2.126}$$

where $F_{\mu\nu} = \partial_\mu A_\nu - \partial_\nu A_\mu$ and A_μ the electromagnetic potential. This action leads to the field equations:

$$R_{\mu\nu}\left(1 + f'(R_0)\right) - \frac{1}{2}g_{\mu\nu}\left(R_0 + f(R_0)\right)$$

$$- 2\left(F_{\mu\alpha}F^\alpha_{\ \nu} - \frac{1}{4}g_{\mu\nu}F_{\alpha\beta}F^{\alpha\beta}\right) = 0. \tag{2.127}$$

If we take the trace of the previous equation, then (2.124) is recovered due to the fact that $F^\mu_{\ \mu} = 0$.

The axisymmetric, stationary and constant curvature R_0 solution that describes a black hole with mass, electric charge, and angular momentum was originally found by Carter (1973). In Boyer-Lindquist coordinates, the metric describing with no coordinate singularities the spacetime exterior to the black hole and interior to the cosmological horizon (provided it exists), is:

$$ds^2 = \frac{\rho^2}{\Delta_r}dr^2 + \frac{\rho^2}{\Delta_\theta}d\theta^2 + \frac{\Delta_\theta \sin^2\theta}{\rho^2}\left[a\frac{dt}{\Xi} - (r^2+a^2)\frac{d\phi}{\Xi}\right]^2$$

$$\frac{\Delta_r}{\rho^2}\left(\frac{dt}{\Xi} - a\sin^2\theta\frac{d\phi}{\Xi}\right)^2, \tag{2.128}$$

with

$$a\Delta_r \equiv \left(r^2+a^2\right)\left(1 - \frac{R_0}{12}r^2\right) - 2Mr + \frac{q^2}{(1+f'(R_0))},$$

$$\rho^2 \equiv r^2 + a^2\cos^2\theta,$$

$$\Delta_\theta \equiv 1 + \frac{R_0}{12}a^2\cos^2\theta, \tag{2.129}$$

$$\Xi \equiv 1 + \frac{R_0}{12}a^2,$$

where M, a and q, as before, denote the mass, spin, and electric charge parameters, respectively.

The potential vector and electromagnetic field tensor in Eq. (2.127) for metric (2.128) are:

$$A = -\frac{qr}{\rho^2}\left(\frac{dt}{\Xi} - a\sin^2\theta\frac{d\phi}{\Xi}\right),$$

$$F = -\frac{q(r^2 - a^2\cos^2\theta)}{\rho^4}\left(\frac{dt}{\Xi} - a\sin^2\theta\frac{d\phi}{\Xi}\right)\wedge dr \tag{2.130}$$

$$-\frac{2qra\cos\theta\sin\theta}{\rho^4}d\theta\wedge\left[a\frac{dt}{\Xi} - (r^2+a^2)\frac{d\phi}{\Xi}\right].$$

We adopt $Q^2 \equiv q^2/(1 + f'(R_0))$ in what follows to refer to the electric charge parameter of the black hole.

If we take the limits $M \to 0$, $Q \to 0$, $a \to 0$ to the metric, we obtain a constant curvature spacetime metric:

$$ds^2 = -\left(1 - \frac{R_0 r^2}{12}\right)dt^2 + \frac{1}{(1 - \frac{R_0 r^2}{12})}dr^2 + r^2 d\Omega^2, \qquad (2.131)$$

that corresponds to either a de Sitter or an anti-de Sitter space-time depending on the sign of R_0. Clearly, when $a \to 0$ and $Q \to 0$, Schwarzschild black hole is recovered.

Calculating $R^{\mu\nu\sigma\rho}R_{\mu\nu\sigma\rho}$, only $\rho = 0$ happens to be an intrinsic singularity, and considering the definition of ρ in (2.129), such singularity is given by:

$$r = 0 \quad \text{and} \quad \theta = \pi/2. \qquad (2.132)$$

Keeping in mind that we are working with Boyer-Lindquist coordinates, the set of points given by $r = 0$ and $\theta = \pi/2$ represent a ring in the equatorial plane of radius a centered on the rotation axis of the black hole, just as in Kerr black holes.

The horizons are found from $g^{rr} = 0$, i.e. their location is given by the roots of the equation $\Delta_r = 0$:

$$r^4 + \left(a^2 - \frac{12}{R_0}\right)r^2 + \frac{24M}{R_0}r - \frac{12}{R_0}\left(a^2 + Q^2\right) = 0. \qquad (2.133)$$

This is a fourth order equation that can be rewritten as:

$$(r - r_-)(r - r_{\text{int}})(r - r_{\text{ext}})(r - r_{\text{cosm}}) = 0, \qquad (2.134)$$

where r_- is always a negative solution with no physical meaning, r_{int} and r_{ext} are the interior and exterior horizons respectively, and r_{cosm} represents the cosmological event horizon for observers between r_{ext} and r_{cosm}. This horizon divides the region that the observer could see from the region she/he could never see if she/he waited long enough. The existence of real solutions for this equation is given by a factor h, called *horizon parameter* (Cembranos et al. 2011):

$$h \equiv \left[\frac{4}{R_0}\left(1 - \frac{R_0}{12}a^2\right)^2 - 4\left(a^2 + Q^2\right)\right]^3$$

$$+ \frac{4}{R_0}\left\{\left(1 - \frac{R_0}{12}a^2\right)\left[\frac{4}{R_0}\left(1 - \frac{R_0}{12}a^2\right)^2\right.\right.$$

$$\left.\left. + 12\left(a^2 + Q^2\right)\right] - 18M^2\right\}^2. \qquad (2.135)$$

For a negative scalar curvature R_0, three options may be considered: (i) $h > 0$: there are only two real solutions, r_{int} and r_{ext}, lacking this configuration a cosmological horizon, as it is expected for an anti-de Sitter like Universe. (ii) $h = 0$: there is only a degenerated root, particular case of an extremal black hole, whose interior and exterior horizons have merged into one single horizon with a null surface gravity. (iii) $h < 0$: there is no real solution to (2.135), which means absence of horizons and then a naked singularity.

For a positive curvature R_0, there are also several configurations depending on the value of h: (i) $h < 0$: both r_{int}, r_{ext}, and r_{cosm} are positive and real, thus the black hole possesses a well-defined horizon structure in an Universe with a cosmological horizon. (ii) $h = 0$: two different cases may be described, either r_{int} and r_{ext} become degenerated solutions, or r_{ext} and r_{cosm} do so. The first case represents an extremal black hole. The second case can be understood as the cosmological limit for which a black hole preserves its exterior horizon without being "torn apart" by the relative recession speed between two radially separated points induced by the cosmic expansion in an Universe described by a constant positive curvature. (iii) $h > 0$: there is only one positive root, that may be either r_{int} or r_{cosm}. In the first case, the mass of the black hole has exceeded the limit imposed by the cosmology ($h = 0$), and there are neither exterior nor cosmological horizon. This situation just leaves the interior horizon to cover the singularity (*marginal naked singularity case*). If the root corresponds to r_{cosm}, there is a naked singularity with a cosmological horizon. From a certain positive value of the curvature R_0^{crit} onward, the h factor goes to zero for two values of a, i.e., apart from the usual a_{max} for which the black hole turns extremal, there is now a spin lower bound a_{min}, below which the black hole turns into a *marginally extremal* black hole. Therefore

$$h\left(a_{max}, M, |R_0| \geq 0, Q\right) = 0$$

$$\Rightarrow \quad a_{max} \equiv a_{max}\left(M, |R_0| \geq 0, Q\right), \tag{2.136}$$

$$h\left(a_{min}, M, R_0 \geq R_0^{crit} > 0, Q\right) = 0$$

$$\Rightarrow \quad a_{min} \equiv a_{min}\left(M, R_0 \geq R_0^{crit} > 0, Q\right). \tag{2.137}$$

Another interesting feature of Kerr-Newman black holes is the presence of an ergosphere bounded by a Stationary Limit Surface (SLS), given by $g_{tt} = 0$. In Boyer-Lindquist coordinates:

$$\frac{\Delta_\theta \sin^2\theta a^2}{\rho^2 \Xi^2} - \frac{\Delta_r}{\rho^2 \Xi^2} = 0, \tag{2.138}$$

that leads to the fourth order equation

$$r^4 + \left(a^2 - \frac{12}{R_0}\right)r^2 + \frac{24M}{R_0}r - \left(a^2\cos^2\theta + \frac{12}{R_0}\right)a^2\sin^2\theta$$

$$-\frac{12}{R_0}\left(a^2 + Q^2\right) = 0, \tag{2.139}$$

which can be rewritten as:

$$(r - r_{S-})(r - r_{S\,int})(r - r_{S\,ext})(r - r_{S\,cosm}) = 0. \tag{2.140}$$

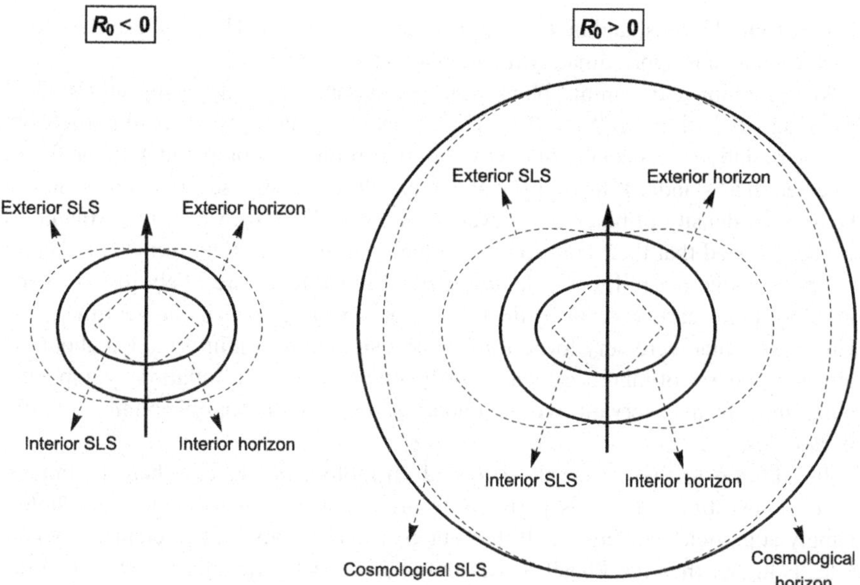

Fig. 2.15 *Left*: Diagram of a Kerr-Newman black hole structure with negative curvature solution $R_0 = -0.4 < 0$, $M = 1$, $a = 0.85$ and $Q = 0.35$ ($h > 0$). *Right*: Black hole structure with positive curvature solution $R_0 = 0.4 > 0$, $M = 1$, $a = 0.9$ and $Q = 0.4$ ($h < 0$). *Dotted surfaces* represent the static limit surfaces (SLS) whereas horizons are shown with *continuous lines*. The rotation axis of the black hole is indicated by the *vertical arrow*. In both types of black hole, the region between the exterior SLS $r_{\text{S ext}}$ and its associated exterior horizon r_{ext} is known as *ergoregion*. From Cembranos et al. (2011). Reproduced by permission of the authors

From this equation it follows that each horizon has an "associated" SLS. Both hypersurfaces coincide at $\theta = 0, \pi$ as seen when comparing (2.139) with Eq.(2.133). A scheme of black hole horizons and the corresponding ergospheres is shown in Fig. 2.15 for both signs of R_0.

For some general properties of static and spherically symmetric black holes in $f(R)$-theories see Perez-Bergliaffa and Chifarelli de Oliveira Nunes (2011). For Kerr-$f(R)$ black holes and disks around them see Pérez et al. (2013).

2.7.4 Mini Black Holes

In principle, a black hole can have any mass above the Planck mass. A black hole is formed always that there is an energy density large enough as to curve the space-time forming a null closed surface. A lower possible mass for a black hole is imposed by the Compton wavelength, $\lambda_C = h/Mc$, which represents a limit on the minimum size of the region in which a mass M at rest can be localized. For a sufficiently small M, the reduced Compton wavelength ($\bar{\lambda}_C = \hbar/Mc$) exceeds half the

Schwarzschild radius, and no black hole description exists. The smallest mass for a black hole is thus approximately the Planck mass.

Very low-mass or "mini" black holes evaporate very quickly by emission of Hawking's radiation (see Sect. 3.3). The absence of notable excesses of particles in cosmic radiation—especially in the form of anti-protons—compared with the fluxes expected in a "standard" astrophysical context allows to impose strict constraints on the number density of black holes evaporating in today's Universe. In particular, it can be deduced that their contribution to the total mass of the universe is today no higher than one ten millionth. As these small black holes are likely to have been produced in the early cosmos by the fluctuations in energy density at that time—and with masses that were very low—it is possible to obtain vital information about the universe's degree of inhomogeneity shortly after the period of inflation by imposing upper limits from observations of secondary particles to the possible number of mini black holes.

In addition to these astrophysical and cosmological aspects, there is another route of investigation that is particularly promising for microscopic black holes, namely at particle accelerators. If the center-of-mass energy of two elementary particles is higher than the Planck scale, and their impact parameter is lower than the Schwarzschild radius, a black hole must be produced. If the Planck scale is thus in the TeV range, the 14 TeV center-of-mass energy of the Large Hadron Collider might allow it to become a black-hole factory with a production rate as high as about one per second.

The possible presence of extra compact dimensions (Sect. 1.12) would be beneficial for the production of black holes. The key point is that it allows the Planck scale to be reduced to accessible values, but it also allows the Schwarzschild radius to be significantly increased, thus making the condition for the impact parameter to be smaller than the Schwarzschild radius easier to satisfy. It is important to note that the resulting mini black holes have radii that are much smaller than the size of extra dimensions, and that they can therefore be considered as totally immersed in a D-dimensional space, which has, to a good approximation, a time dimension and $D-1$ non-compact space dimensions. The black hole thus acts like a quasi-selective source of S waves and sees our brane in the same way as the "bulk" associated with the extra dimensions. As the particles residing in the brane greatly outnumber those living in the bulk (essentially gravitons), the black hole evaporates into particles of the Standard Model. Its lifetime is very short (of the order of 10^{-26} s) and its temperature (typically about 100 GeV here) is much lower than it would be with the same mass in a four-dimensional space. The black hole nevertheless retains its characteristic spectrum in the form of a quasi-thermal law peaked around its temperature. From the point of view of detection, it is not too difficult to find a signature for such events: they have a high multiplicity, a large transverse energy, a "democratic" coupling to all particles, and a rapid increase in the production cross-section with energy.

First of all the reconstruction of temperature (determined by the energy spectrum of the particles emitted when the black hole evaporates) as a function of mass (determined by the total energy deposited) allows information to be gained about the dimensionality of space-time. In the case of Planck scales close to the TeV mark, the number of extra dimensions could thus be revealed quite easily by the characteristics of the emitted particles. One can go even further. In particular, quantum gravity effects could be revealed, as behavior during evaporation in the Planck region is sensitive to the details of the gravitational theory used (for more on mini black holes see Frolov and Zelnikov 2011, and references therein).

References

M.A. Abramowicz, A.M. Beloborodov, X.-M. Chen, I.V. Igumenshchev, Astron. Astrophys. **313**, 334 (1996)

M. Born, L. Infeld, Proc. R. Soc. Lond. A **144**, 425 (1934)

N. Bretón, Class. Quantum Gravity **19**, 601 (2002)

S. Carroll, *Space-Time and Geometry: An Introduction to General Relativity* (Addison-Wesley, New York, 2003)

B. Carter, in *Les Astres Occlus*, ed. by C.M. DeWitt (Gordon and Breach, New York, 1973)

J.A.R. Cembranos, A. de la Cruz-Dombriz, P. Jimeno Romero, arXiv:1109.4519 (2011)

V.P. Frolov, I.D. Novikov, *Black Hole Physics* (Kluwer, Dordrecht, 1998)

V.P. Frolov, A. Zelnikov, *Introduction to Back Hole Physics* (Oxford University Press, Oxford, 2011)

V.P. Frolov, M.A. Markov, V.F. Mukhanov, Phys. Rev. D **41**, 383 (1990)

S.W. Hawking, G.F.R. Ellis, *The Large-Scale Structure of Space-Time* (Cambridge University Press, Cambridge, 1973)

B. Hoffmann, Phys. Rev. **47**, 877 (1935)

R.P. Kerr, Phys. Rev. Lett. **11**, 237 (1963)

P.-S. Laplace, Exposition du Système du Monde, Paris. (first edition) (1796)

J.-P. Luminet, in *Black Holes: Theory and Observation*, ed. by F.W. Hehl, C. Kiefer, R.J.K. Metzler (Springer, Berlin, 1998), p. 3

V.S. Manko, N.R. Sibgatullin, Phys. Rev. D **46**, R4122 (1992)

J. Michell, Philos. Trans. R. Soc. Lond. **74**, 35 (1784)

M.R. Mbonye, D. Kazanas, Phys. Rev. D **72**, 024016 (2005)

M.R. Mbonye, N. Battista, R. Farr, Int. J. Mod. Phys. D **20**, 1 (2011)

E.T. Newman, E. Couch, K. Chinnapared, A. Exton, A. Prakash, R. Torrence, J. Math. Phys. **6**, 918 (1965)

B. Paczyński, P. Wiita, Astron. Astrophys. **88**, 23 (1980)

C.L. Pekeris, K. Frankowski, Phys. Rev. A **36**, 5118 (1987)

R. Penrose, Riv. Nuovo Cimento **I**, 252 (1969)

D. Pérez, G.E. Romero, C.A. Correa, S.E. Perez-Bergliaffa, Int. J. Mod. Phys. Conf. Ser. **3**, 396 (2011)

D. Pérez, G.E. Romero, S.E. Perez-Bergliaffa, Astron. Astrophys. **551**, A4 (2013)

S.E. Perez-Bergliaffa, Y.E. Chifarelli de Oliveira Nunes, Phys. Rev. D **84**, 084006 (2011)

B. Punsly, Astrophys. J. **498**, 640 (1998a)

B. Punsly, Astrophys. J. **498**, 660 (1998b)

B. Punsly, *Black Hole Gravitohydromagnetics* (Springer, Berlin, 2001)

B. Punsly, G.E. Romero, D.F. Torres, J.A. Combi, Astron. Astrophys. **364**, 552 (2000)

D. Raine, E. Thomas, *Black Holes: An Introduction* (Imperial College Press, London, 2005)

O. Semerák, V. Karas, Astron. Astrophys. **343**, 325 (1999)

P.K. Townsend, *Black Holes (Lecture Notes)* (University of Cambridge, Cambridge, 1997). arXiv: gr-qc/9707012

R.M. Wald, *General Relativity* (The University of Chicago Press, Chicago, 1984)

Chapter 3
Black Hole Physics

3.1 Black Hole Formation

3.1.1 Stellar Structure

The idea that stars are self-gravitating gaseous bodies was introduced in the XIX Century by Lane, Kelvin, and Helmholtz. They suggested that stars should be understood in terms of the equation of hydrostatic equilibrium:

$$\frac{dp(r)}{dr} = -\frac{GM(r)\rho(r)}{r^2},$$

(3.1)

where the pressure P is given by

$$P = \frac{\rho kT}{\mu m_p}.$$

(3.2)

Here, k is Boltzmann's constant, μ is the mean molecular weight, T the temperature, ρ the mass density, and m_p the mass of the proton. Kelvin and Helmholtz suggested that the source of heat was the gravitational contraction of the gas. However, if the luminosity of a star like the Sun is taken into account, the total energy available would be released in 10^7 yr, which is in contradiction with the geological evidence that can be found on Earth.

Arthur S. Eddington made two fundamental contributions to the theory of stellar structure proposing that (i) the source of energy was of thermonuclear nature and (ii) the outward pressure of radiation should be included in Eq. (3.1). Then, the basic equations for stellar equilibrium become (Eddington 1926):

$$\frac{d}{dr}\left[\frac{\rho kT}{\mu m_p} + \frac{1}{3}aT^4\right] = -\frac{GM(r)\rho(r)}{r^2},$$

(3.3)

$$\frac{dp_{\mathrm{rad}}(r)}{dr} = -\left(\frac{L(r)}{4\pi r^2 c}\right)\frac{1}{l},$$

(3.4)

G.E. Romero, G.S. Vila, *Introduction to Black Hole Astrophysics*,
Lecture Notes in Physics 876, DOI 10.1007/978-3-642-39596-3_3,
© Springer-Verlag Berlin Heidelberg 2014

$$\frac{dL(r)}{dr} = 4\pi r^2 \varepsilon \rho, \tag{3.5}$$

where l is the mean free path of the photons, L the luminosity, and ε the energy generated per gram of material per unit time.

Once the nuclear power of the star is exhausted, the contribution from the radiation pressure decreases dramatically when the temperature diminishes. The star then contracts until a new source of pressure helps to balance gravity's attraction: the degeneracy pressure of the electrons. The equation of state for a degenerate gas of electrons is:

$$p_{\mathrm{rel}} = K\rho^{4/3}. \tag{3.6}$$

Then, using Eq. (3.1),

$$\frac{M^{4/3}}{r^5} \propto \frac{GM^2}{r^5}. \tag{3.7}$$

Since the radius cancels out, this relation can be satisfied by a unique mass:

$$M = 0.197\left[\left(\frac{hc}{G}\right)^3 \frac{1}{m_p^2}\right]\frac{1}{\mu_e^2} = 1.4M_\odot, \tag{3.8}$$

where μ_e is the mean molecular weight of the electrons. The result implies that a completely degenerated star has this and only this mass. This limit was found by Chandrasekhar (1931) and is known as the *Chandrasekhar limit*.

In 1939 Chandrasekhar conjectured that massive stars could develop a degenerate core. If the degenerate core attains sufficiently high densities the protons and electrons will combine to form neutrons. "This would cause a sudden diminution of pressure resulting in the collapse of the star to the neutron core giving rise to an enormous liberation of gravitational energy. This may be the origin of the supernova phenomenon." (Chandrasekhar 1939). An implication of this prediction is that the masses of neutron stars (objects supported by the degeneracy pressure of nucleons) should be close to $1.4M_\odot$, the maximum mass for white dwarfs. Not long before, Baade and Zwicky commented: "With all reserve we advance the view that supernovae represent the transitions from ordinary stars into *neutron stars* which in their final stages consist of extremely closely packed neutrons". In a single paper Baade and Zwicky not only invented neutron stars and provided a theory for supernova explosions, but also proposed that these explosion were the origin of cosmic rays (Baade and Zwicky 1934).

In the 1930s, neutron stars were not taken as a serious physical possibility. Oppenheimer and Volkoff (1939) concluded that if the neutron core was massive enough, then "either the Fermi equation of state must fail at very high densities, or the star will continue to contract indefinitely never reaching equilibrium". In a subsequent paper, Oppenheimer and Snyder (1939) chose between these two possibilities: "when all thermonuclear sources of energy are exhausted a sufficiently heavy star will collapse. This contraction will continue indefinitely till the radius of the star approaches asymptotically its gravitational radius. Light from the surface of the star will be progressively reddened and can escape over a progressively narrower range of angles till eventually the star tends to close itself off from any communication with a distant observer". What we now understand as a black hole was then conceived. The scientific community paid no attention to these results, and Oppenheimer and many other scientists turned their efforts to win a war.

3.1.2 Stellar Collapse

Black holes will form every time that matter and fields are compressed beyond the corresponding Schwarzschild radius. This can occur in a variety of forms, from particle collisions to the implosion of stars or the collapse of dark matter in the early universe. The most common black hole formation mechanism in our Galaxy seems to be gravitational collapse. A normal star is stable as long as the nuclear reactions occurring in its interior provide thermal pressure to support it against gravity. Nuclear burning gradually transforms the stellar core from H to He and in the case of massive stars ($M > 5M_\odot$) then to C and finally to Fe. The core contracts in the process, in order to achieve the ignition of each phase of thermonuclear burning.

Finally, the endothermic disintegration of iron-group nuclei, which are those with the tightest bound, precipitates the collapse of the core to a stellar-mass black hole. Stars with masses in the range 20–$30M_\odot$ produce black holes with $M > 1.8M_\odot$. Low-mass black holes ($1.5M_\odot < M < 1.8M_\odot$) can result from the collapse of stars of 18–$20M_\odot$ along with the ejection of the outer layers of the star by a shock wave in an event known as Type II supernova. A similar event, occurring in stars with 10–$18M_\odot$ leaves behind a neutron star. Very massive stars with high spin likely end producing a gamma-ray burst and a very massive ($M > 10M_\odot$) black hole. The binary stellar systems have a different evolution. The interested reader can find a comprehensive review in Brown et al. (2000).

In Fig. 3.1 we show the Eddington-Finkelstein diagram of the gravitational collapse of a star. Once the null surface of the light cones points along the time axis the black hole is formed: light rays will never be again able to escape to the outer universe. The different paths that can lead to a stellar-mass black hole are illustrated in Fig. 3.2.

If the collapse is not perfectly symmetric, any asymmetry in the resulting black hole is radiated away as gravitational waves, in such a way that the final result is a black hole that is completely characterized by the three parameters M, J, and Q. The black hole, once formed, has no hints about the details of the formation process and its previous history.

Gravitational collapse can also be the result of inhomogeneities in the original metric giving rise to mini-black holes as proposed by Hawking (1971), although the number of microscopic black holes is strongly constrained by observations of cosmic gamma-ray background emission.

Let us restrict ourselves to the homogeneous collapse of a spherically symmetric star. The dynamical interior of the collapsing star will depend on the details of the

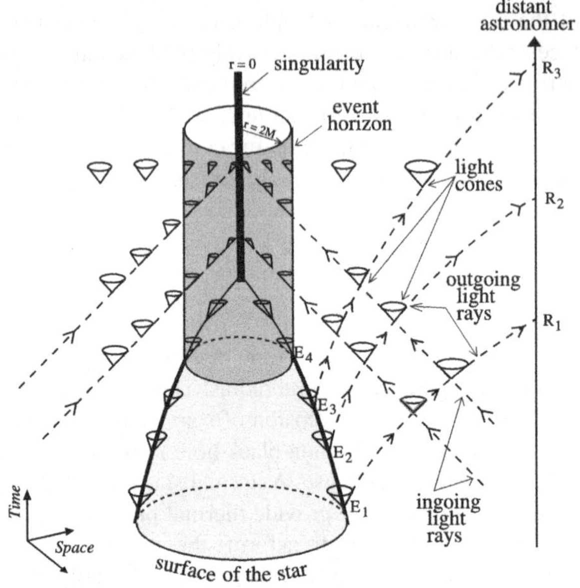

Fig. 3.1 An Eddington-Finkelstein diagram of a collapsing star with the subsequent black hole formation. From Luminet (1998)

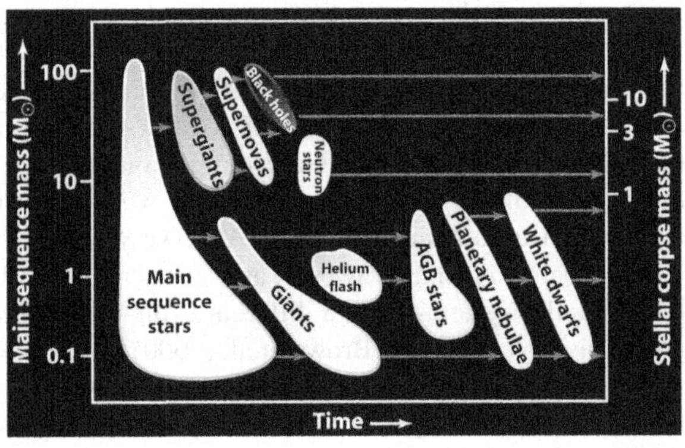

Fig. 3.2 Life cycle of stars and channels for black hole formation

stellar matter, described by the equation of state. If as a first approximation, as done by Oppenheimer and Snyder (1939), the star is modeled as a spherical cloud of dust, the interior is pressure-less: $P = 0$ and the energy-momentum tensor is $T^{\mu\nu} = \rho u^\mu u^\nu$, with ρ the energy density and u^μ the 4-velocity field of the fluid. Solving Einstein's field equations determines the metric coefficients completely, yielding a line element that is identical to that of the close homogeneous isotropic Friedmann model:

$$ds^2 = dt^2 - R^2(r,t)\left[\frac{dr^2}{1-r^2} + r^2 d\Omega^2\right], \tag{3.9}$$

where $R(r,t)$ is a time-dependent scale factor, $d\Omega^2 = d\theta^2 + \sin^2\theta d\phi^2$ is the metric on a 2-sphere, and we adopted units of $c = 1$. Since cold dust does not radiate, the exterior solution is Schwarzschild space-time in accordance with Birkhoff's theorem. The interior solution and the vacuum exterior must match at the surface of the collapsing cloud. When the collapse is total, the final space-time becomes Schwarzschild's.

Misner and Sharp (1964) developed a general formalism for spherically symmetric gravitational collapse including pressure. The energy-momentum tensor now is $T^{\mu\nu} = (\rho + P)u^\mu u^\nu + P g^{\mu\nu}$ and the line element in co-moving coordinates is (Misner and Sharp 1964; Joshi 2007):

$$ds^2 = e^{2\varphi}dt^2 - e^\lambda dr^2 - R^2(r,t)d\Omega^2, \tag{3.10}$$

where φ and λ are functions of r and t. The components of the 4-velocity are: $u^0 = e^{-\varphi}$, $u^i = 0$, for $i = r, \theta, \phi$.

We can now introduce a function $m(r, t)$ by defining:

$$e^\lambda = \left(1 + \dot{R}^2 - \frac{2m}{r}\right)^{-1}\left(\frac{\partial R}{\partial r}\right)^2, \tag{3.11}$$

where a dot means differentiation with respect to t and multiplication by $e^{-\varphi}$,

$$\dot{f} = u^\mu \frac{\partial f}{\partial x^\mu} = e^{-\varphi}\left(\frac{\partial f}{\partial t}\right). \tag{3.12}$$

This is the co-moving proper time derivative.

Integrating the conservation equation $T^{\mu\nu}_{\ \ ;\nu} = 0$ and solving Einstein's field equations, we can find the Misner-Sharp equations for spherically symmetric collapse:

$$\dot{m} = -4\pi R^2 P \dot{R}, \tag{3.13}$$

$$\ddot{R} = \left(\frac{1 + \dot{R}^2 - 2m/r}{\rho + P}\right)\left(\frac{\partial P}{\partial R}\right) - \frac{m + 4\pi R^3 P}{R^2}, \tag{3.14}$$

$$\frac{\partial m}{\partial R} = 4\pi R^2 \rho. \tag{3.15}$$

These equations, along with the equation of state relating P and ρ, determine the dynamical evolution of the spherical homogeneous collapse of the star. If $P = 0$ the results of Oppenheimer and Snyder (1939) are recovered. When $P \neq 0$ the solution requires numerical integration.[1] Any solution demands the specification of initial values for $R(r, 0)$, $m(r, 0)$, and $U(r, 0)$, with $U = e^{-\varphi}\dot{R}$. It is also required that at $r = 0$ the functions R, m, and U all vanish. If r_b defines the outer boundary of the distribution of matter, then $m(r_b, t) = M$ is constant and the interior metric can be smoothly joined at the surface $r = r_b$ to an exterior Schwarzschild metric of mass M.

---

[1]The reader is referred for numerical calculations to the book by Baumgarte and Shapiro (2010) and references therein.

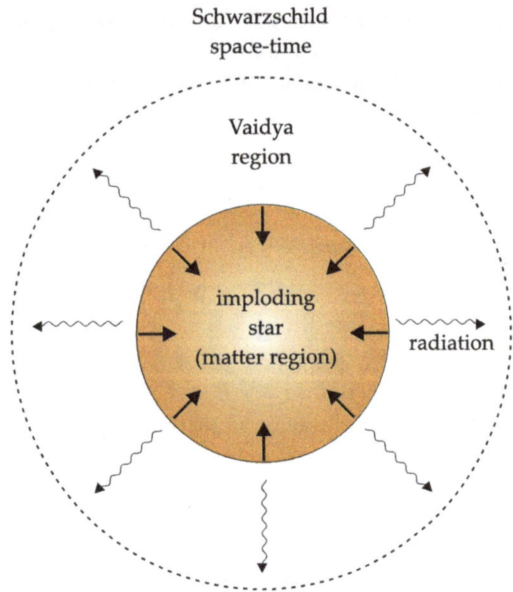

Fig. 3.3 A radiating collapsing star and the different regions associated with the collapsing matter, the radiation field, and the vacuum exterior

The analysis outlined above does not consider the effects of the radiation field of the collapsing star. If radiation is being produced by the star, the exterior solution is not a vacuum solution such as Schwarzschild's. Although the effects of radiation of a normal star on the space-time metric are negligible, in the case of the later stages of gravitational collapse they can become important. A collapsing radiating star would be surrounded by an expanding zone of radiation. Only far from the radiation zone space-time can be described by the Schwarzschild solution. The system, then, comprises three regions: the Misner-Sharp collapsing space-time, the radiation zone, and an exterior Schwarzschild space-time, which asymptotically tends to Minskowski space-time (see Fig. 3.3).

To find the metric in the radiation region, Einstein's field equations must be solved for an energy-momentum tensor of the form:

$$T_{\mu\nu} = \sigma k_\mu k_\nu, \tag{3.16}$$

where k_μ is a null radial vector pointing outwards, and σ is the energy density of radiation in a local frame with velocity u_μ.

A solution of this kind was found by Vaidya (1943, 1951). In null coordinates (u, r, θ, ϕ) the metric is given by the line element:

$$ds^2 = \left(1 - \frac{2m(u)}{r}\right)du^2 + 2dudr - r^2d\Omega^2, \tag{3.17}$$

where $m(u)$ is an arbitrary non-increasing function of the retarded time coordinate u. The latter is related with Schwarzschild's time t by:

$$u = t - r - 2m\log(r - 2m). \tag{3.18}$$

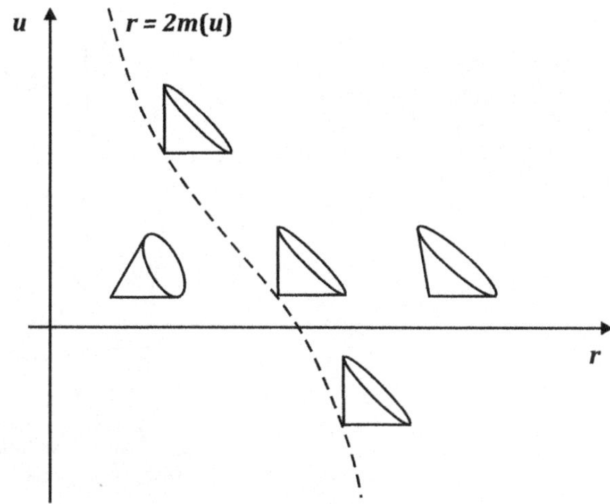

Fig. 3.4 Light cones for Vaidya's metric. The surface $r = 2m(u)$ is a space-like surface. Adapted from Lindquist et al. (1965)

From (3.16), (3.17), and Einstein's equations:

$$\sigma = -\frac{1}{4\pi r^2}\frac{dm(u)}{du}. \tag{3.19}$$

The total luminosity of the star at infinity is $L = -dm(u)/du$. Hence, m is interpreted as the mass of the system and the energy flux is its negative rate of change.

The properties of Vaidya space-time have been discussed by Lindquist et al. (1965). In particular, they have shown that the surface $r = 2m$, unlike the Schwarzschild case, is a space-like hypersurface; no time-like or null trajectories can cross it from the outside to reach the inner region. This is illustrated in Fig. 3.4.

An interesting aspect of the gravitational collapse of a radiating star is that for some prescriptions of the mass function (e.g. $m(u) = \lambda u$, where the rate of collapse is controlled by the value of λ), it is possible to have a slow collapse (adopting λ small), in such a way that the onset of the event horizon is delayed beyond the formation of the singularity, allowing, in principle, outgoing null geodesics originated close to the non-regular region to reach the exterior. Such results can also be obtained with inhomogeneous collapse (see Fig. 3.5 for a sketch of the situation). Whether singularities formed by gravitational collapse can be glimpsed by external observers is an open topic (for more on this see Joshi 2007 and Joshi and Malafarina 2011).

3.1.3 Supermassive Black Holes

Supermassive black holes can result from a variety of processes occurring at the center of galaxies as discussed by Rees (1984); see Fig. 3.6 for a sketch of some possible formation paths. Some current views, however, suggest that galaxies were

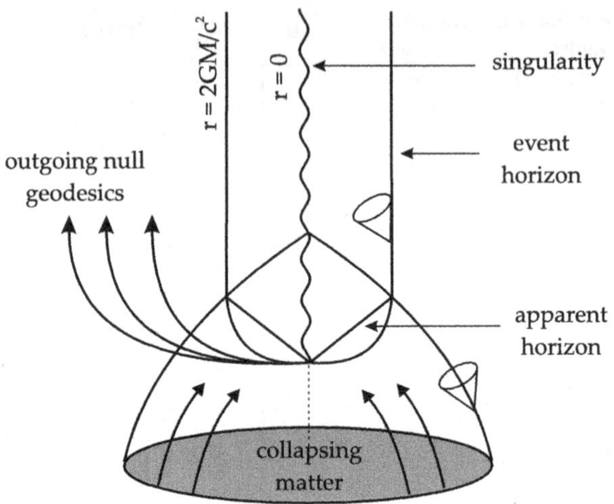

Fig. 3.5 Formation of a black holes with visible singularity on the onset. Compare with Fig. 3.1 where the horizon always covers the singularity. Adapted from Joshi and Malafarina (2011)

formed around seed massive black holes which were the result of the gravitational collapse of dark matter.

Dynamical evidence supports the existence of supermassive black holes in the centers of most nearby galaxies. The masses of these black holes ranges from $\sim 10^6$ to $\sim 10^9 M_\odot$. In the case of our own galaxy the black hole has a mass $\sim 4 \times 10^6 M_\odot$.

The existence of black holes with masses up to $\sim 10^9 M_\odot$ has been inferred in active galactic nuclei (AGN) with cosmological redshift $z > 6$. Less massive but anyway significant black holes might have existed in large numbers during the Dark Ages of the Universe, after the first combination of protons and electrons (Volonteri 2010). Such an early black hole formation might be the result of one or several of the following processes: (1) gravitational collapse of the first generation of stars—the so-called Population III stars—, (2) gas-dynamical instabilities, (3) stellar-dynamical stabilities, (4) dark matter collapse, and (5) primordial fluctuations. We shall briefly discuss these possibilities.

Population III stars are formed out of the collapse of zero-metallicity gas after the first combination. These stars are expected to be very massive ($M_* > 100 M_\odot$, e.g. Gao et al. 2007). Massive stars evolve fast and undergo at the end of their lifetimes total gravitational collapse leaving behind black holes with masses between $40 M_\odot$ and perhaps up to $1000 M_\odot$, except in the range ~ 140–$260 M_\odot$, where the pair instability supernovae leave no compact remnant (e.g. Fryer et al. 2001). Galaxies then might be formed around these early black holes which would group following some of the paths indicated in Fig. 3.6.

If fragmentation can be inhibited in early massive clouds (e.g. by turbulence), and cooling proceeds gradually, the gas will contract until rotation can stop the collapse. In such a case, global dynamical instabilities, like bar-instabilities (e.g. Begelman et al. 2006), can transport angular momentum outwards, allowing the core collapse to continue. Then, the gas accumulated in the center can give rise to a very massive central object. Eventually, total collapse might yield a black hole with a mass of $\sim 10^4 M_\odot$ or even higher (Volonteri 2010, and references therein).

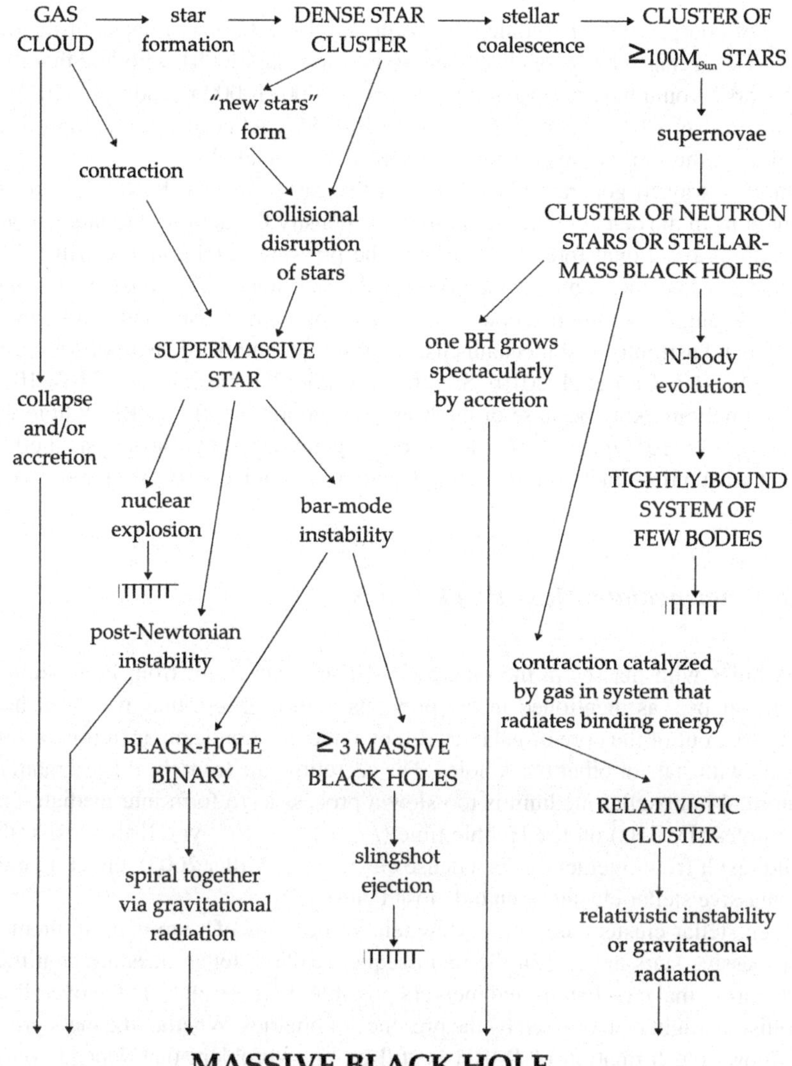

Fig. 3.6 Massive and supermassive black hole formation channels. Adapted from Rees (1984)

On the other hand, fragmentation and star formation in a collapsing cloud of gas can form a very compact nuclear stellar cluster. Frequent stellar collisions can lead to the formation of a black hole of $\sim 10^2$–$10^4 M_\odot$ (Devecchi and Volonteri 2009). The process has been originally predicted by Begelman and Rees (1978). For details see Spitzer (1987).

Direct collapse of dark matter proceeds as dust, without any heating, until the annihilation rate of dark particles (say WIMPs) becomes significant. Then, a "dark-matter star" might be supported by dark matter annihilation radiation. What is supported, of course, is the baryonic content pulled down by the collapse of the dark matter. The scenario has been discussed by Freese et al. (2008), who find that these "dark stars" would have surface temperatures of 4000–10000 K, radii of $\sim 10^{14}$ cm, luminosities of $\sim 10^6 L_\odot$,[2] and masses of $\sim 1000 M_\odot$. After the annihilation of the dark matter, the star would collapse into a massive black hole.

Another way to generate black holes in the early universe is from primordial fluctuations in the density field. Anywhere the density fluctuations are large enough as for the gravitational force to overcome the pressure, total collapse will follow, with the consequent formation of primordial black holes. The mass of the black holes can range from microscopic to thousands of solar masses. However, several physical and astrophysical mechanisms constrain the number density of primordial black holes (e.g. Carr et al. 2010). Significant primordial black holes of 10^9–10^{17} g are extremely unlikely because of unobserved evaporation effects. Black hole with masses up to $\sim 10^{40}$ g ($\sim 10^7 M_\odot$) are strongly constrained by astrophysical effects to be at most a tiny fraction of the total density of the universe (Carr et al. 2010).

3.1.4 Intermediate-Mass Black Holes

Black holes with masses in the range 10^2–$10^3 M_\odot$ can result from Population III stellar collapse, as mentioned in the previous section. More massive black holes might exist out of the core of galaxies if they undergo significant accretion or often mergers with stars or other black holes. The accretion rate from the diffuse matter in the normal interstellar medium is too slow a process as to form intermediate-mass black holes (IMBHs) on the Hubble time ($t \sim H_0^{-1} \times 10^{12}$ yr). If they exist, they should result from interactions in a dense cluster (e.g. Miller 2003), either a young, very massive stellar cluster or an old closed cluster.

Open stellar clusters are young (few tens of millions of years) and in them the most massive stars are still in the main sequence. The stellar sizes are significant enough as to make collisions and mergers possible. In these clusters the overall rate of collision might be increased by the presence of binaries. Whether the mass growth rate allows the formation of IMBH is still an open problem that depends on the delicate balance between the frequency of collisions, and hence the mass accretion onto a seed black hole, and the mass lost by the cluster in winds, supernovae, and other expansive processes. So far, there is no conclusive evidence for the existence of IMBH in young clusters.

Globular clusters are old and most of the stars in them are compact remnants, such as neutron stars or white dwarfs. The cross section for collisions of such objects is negligible, so seed black holes are not expected to grow because of mergers. Nonetheless, large numbers of binaries are expected to exist in old stellar clusters. The binaries lose energy by gravitational radiation, and finally merge into a black

---

[2] The solar luminosity is $L_\odot \approx 3.8 \times 10^{33}$ erg s^{-1}.

hole. If a black hole has a mass above $\sim 50 M_\odot$, its inertia keeps it at the center of the cluster, where repeated mergers can rise its mass up to $\sim 10^3 M_\odot$ or more on the Hubble time (Miller and Hamilton 2002). Additionally, interactions of single black holes and binaries inside the cluster can result in recoil kicks able to expel black holes at speeds of more than 50 km s^{-1} (e.g. Webbink 1985). This might produce black hole mergers well outside the globular cluster when binaries are kicked out.

Ultra-luminous X-ray sources have been found in a number of galaxies. These sources display luminosities in the X-ray band of more than $10^{40} \text{ erg s}^{-1}$, reaching in one case even $3 \times 10^{41} \text{ s}^{-1}$. It has been suggested that these and similar sources might be IMBH accreting at sub-Eddington luminosities, but to find definitive evidence has proven to be highly elusive so far. We shall say more on this in Chap. 6.

3.1.5 Mini Black Holes

Black holes with masses well below the solar mass might be formed out of primordial fluctuations or through particle collisions if the Planck scale is lowered by extra dimensions (Sect. 1.12.2). In general, if within some region of space density fluctuations are large enough as to enforce the gravitational field to overcome pressure, the whole region will collapse and form a primordial black hole. The mass of the primordial black hole due to a cosmological density fluctuation at a time t after the Big Bang will have a mass (Carr et al. 2010):

$$M \sim \frac{c^3 t}{G} \sim 10^{15} \left(\frac{t}{10^{-23} \text{ s}} \right) \text{ g}. \tag{3.20}$$

Black holes formed by the Planck time would have a Planck mass of $\sim 10^{-5}$ g, and black holes formed just $\sim 10^{-23}$ s after the Big Bang would have $\sim 10^{15}$ g. According to Eq. (3.29) below (Sect. 3.3), black holes of this mass or less should have evaporated by now. Since these black holes should produce photons with energy around 100 MeV at the present epoch, the observational limit on the γ-ray background intensity at this energy implies that their density could not exceed about 10^{-8} of the critical energy. This completely rules out small primordial black holes as dark matter candidates (Carr et al. 2010). Current data, however, cannot exclude black holes of sub-lunar mass (10^{20}–10^{26} g) and of intermediate mass (10^2–$10^4 M_\odot$) from having a significant contribution to the critical density of the Universe (e.g. Blais et al. 2002; Saito et al. 2008). Black holes evaporating after the so-called recombination time (approximately 3.8×10^5 yr after the Big Bang) have been even suggested as a source of re-ionization for the Universe (e.g. He and Fang 2002).

3.2 Black Hole Thermodynamics

The area of a Schwarzschild black hole is

$$A_{\text{Schw}} = 4\pi r_{\text{Schw}}^2 = \frac{16\pi G^2 M^2}{c^4}. \tag{3.21}$$

In the case of a Kerr-Newman black hole, the area is

$$A_{\text{KN}} = 4\pi \left(r_+^2 + \frac{a^2}{c^2} \right)$$

$$= 4\pi \left[\left(\frac{GM}{c^2} + \frac{1}{c^2}\sqrt{G^2 M^2 - GQ^2 - a^2} \right)^2 + \frac{a^2}{c^2} \right]. \tag{3.22}$$

Notice that expression (3.22) reduces to (3.21) for $a = Q = 0$.

When a black hole absorbs a mass δM, its mass increases to $M + \delta M$, and hence, the area increases as well. Since the horizon can be crossed in just one direction, the area of a black hole can only increase. This suggests an analogy with entropy. A variation in the entropy of the black hole will be related to the heat (δQ) absorbed through the following equation:

$$\delta S = \frac{\delta Q}{T_{\text{BH}}} = \frac{\delta M c^2}{T_{\text{BH}}}. \tag{3.23}$$

Particles trapped in the black hole will have a wavelength:

$$\lambda = \frac{\hbar c}{kT} \propto r_{\text{Schw}}, \tag{3.24}$$

where k is the Boltzmann constant. Then,

$$\xi \frac{\hbar c}{kT} = \frac{2GM}{c^2},$$

where ξ is a numerical constant. Hence, we can associate a temperature to the black hole:

$$T_{\text{BH}} = \xi \frac{\hbar c^3}{2GkM},$$

and

$$S = \frac{c^6}{32\pi G^2 M} \int \frac{dA_{\text{Schw}}}{T_{\text{BH}}} = \frac{c^3 k}{16\pi \hbar G \xi} A_{\text{Schw}} + \text{constant}.$$

A quantum mechanical calculation of the horizon temperature in the Schwarzschild case leads to $\xi = (4\pi)^{-1}$. So,

$$T_{\text{BH}} = \frac{\hbar c^3}{8GMk} \cong 10^{-7}\,\text{K}\left(\frac{M_\odot}{M} \right). \tag{3.25}$$

And we can write the entropy of the black hole as:

$$S = \frac{kc^3}{4\pi \hbar G} A_{\text{Schw}} + \text{constant} \sim 10^{77} \left(\frac{M}{M_\odot} \right)^2 k \, \text{J K}^{-1}. \tag{3.26}$$

The formation of a black hole implies a huge increase of entropy. Just to compare, a star has an entropy $\sim$20 orders of magnitude lower than the corresponding black hole. This tremendous increase of entropy is related to the loss of all the structure of the original system (e.g. a star) once the black hole is formed.

The analogy between area and entropy allows to state a set of laws for black holes thermodynamics (Bardeen et al. 1973):

- First law (energy conservation): $dM = T_{\text{BH}} dS + \Omega_+ dJ + \Phi dQ + \delta M$. Here, Ω_+ is the angular velocity, J the angular momentum, Q the electric charge, Φ the electrostatic potential, and δM is the contribution to the change in the black hole mass due to the change in the external stationary matter distribution.
- Second law (entropy never decreases): in all physical processes involving black holes the total surface area of all the participating black holes can never decrease.
- Third law (Nernst's law): the temperature (surface gravity) of a black hole cannot be zero. Since $T_{\text{BH}} = 0$ with $A \neq 0$ for extremal charged and extremal Kerr black holes, these are thought to be limit cases that cannot be reached in Nature.
- Zeroth law (thermal equilibrium): the surface gravity (temperature) is constant over the event horizon of a stationary axially symmetric black hole.

3.3 Quantum Effects in Black Holes

If a temperature can be associated with black holes, then they should radiate as any other body. The luminosity of a Schwarzschild black hole is:

$$L_{\text{BH}} = 4\pi r_{\text{Schw}}^2 \sigma T_{\text{BH}}^4 \sim \frac{16\pi \sigma_{\text{SB}} \hbar^4 c^6}{(8\pi)^4 G^2 M^2 k^4}. \tag{3.27}$$

Here σ_{SB} is the Stefan-Boltzmann constant. This expression can be written as:

$$L_{\text{BH}} = 10^{-17} \left(\frac{M_\odot}{M} \right)^2 \text{erg s}^{-1}. \tag{3.28}$$

The lifetime of a black hole is:

$$\tau \cong \frac{M}{dM/dt} \sim 2.5 \times 10^{63} \left(\frac{M}{M_\odot} \right)^3 \text{yr}. \tag{3.29}$$

Notice that the black hole heats up as it radiates! This occurs because when the hole radiates, its mass decreases and then according to Eq. (3.25) the temperature must rise.

If nothing can escape from black holes because of the existence of the event horizon, what is the origin of this radiation? The answer, found by Hawking (1974), is related to quantum effects close to the horizon. According to the Heisenberg relation $\Delta t \Delta E \geq \hbar/2$ particles can be created out of the ground state of a quantum field as far as the relation is not violated. Particles must be created in pairs, along a tiny time, in order to satisfy conservation laws other than energy. If a pair is created close to the horizon and one particle crosses it, then the other particle can escape provided its momentum is in the outward direction. The virtual particle is then transformed into a real particle, at expense of the black hole energy. The black hole then will lose energy and its area will decrease slowly, violating the second law of thermodynamics. However, there is no violation if we consider a *generalized second law*, that always holds: *in any process, the total generalized entropy $S + S_{BH}$ never decreases* (Bekenstein 1973).

3.4 Black Hole Magnetospheres

In the real universe black holes are not expected to be isolated, hence the ergosphere should be populated by charged particles. This plasma rotates in the same sense as the black hole due to the effects of the frame dragging. A magnetic field will develop and will rotate too, generating a potential drop that might accelerate particles up to relativistic speeds and produce a wind along the rotation axis of the hole. Such a picture has been consistently developed by Punsly and Coroniti (1990a, 1990b) and Punsly (2001).

In Figs. 3.7 and 3.8 we show the behavior of fields and currents in the ergosphere. Since the whole region is rotating, an ergospheric wind arises along the direction of the large scale field.

Blandford and Znajek (1977) developed a general theory of force-free steady state axisymmetric magnetosphere of a rotating black hole. In an accreting black hole, a magnetic field can be sustained by external currents, but as such currents move along the horizon, the field lines are usually representing as originating from the horizon and then being torqued by rotation. The result is an outgoing electromagnetic flux of energy and momentum. The power of the outflow is:

$$L \approx 10^{39} \; \mathrm{erg\, s}^{-1} \left(\frac{M}{10^6 M_\odot} \right)^2 \left(\frac{a}{a_{\max}} \right)^2 \left(\frac{B}{10^4 \, \mathrm{G}} \right)^2. \qquad (3.30)$$

This picture stimulated the development of the so-called "Membrane Paradigm" by Thorne et al. (1986) where the event horizon is attributed with a set of physical properties. This model of black hole has been subjected to strong criticism by Punsly (2001) since General Relativity implies that the horizon is causally disconnected from the outgoing wind. We shall look at this closer in Chap. 5

Recent numerical simulations (e.g. Komissarov 2004) show that the key role in the electrodynamic mechanisms of rotating black holes is played by the ergosphere and not by the horizon. Globally, however, the Blandford-Znajek solution seems to

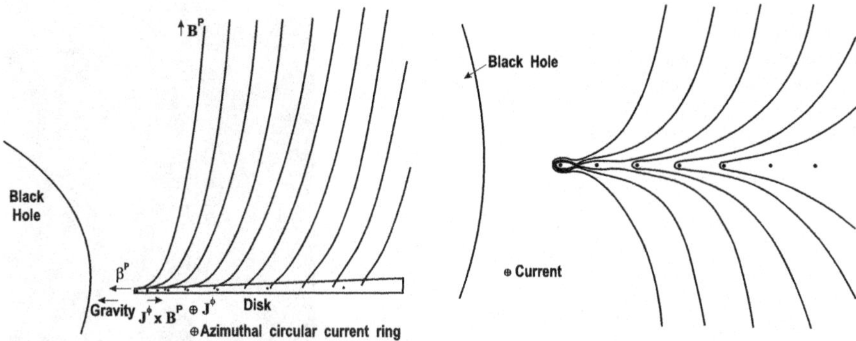

Fig. 3.7 Currents in the infalling matter supports a radial magnetic field. As the inner part of the current sheet approaches the black hole the sources are redshifted to observers at infinity and their contribution to the poloidal magnetic field diminish. At some point X, the field reconnects. Adapted from Punsly and Coroniti (1990b). Reproduced by permission of the AAS

Fig. 3.8 As reconnection proceeds, the magnetic field around the innermost currents is disconnected from the large scale field allowing the destruction of the magnetic flux by the black hole. From Punsly and Coroniti (1990b). Reproduced by permission of the AAS

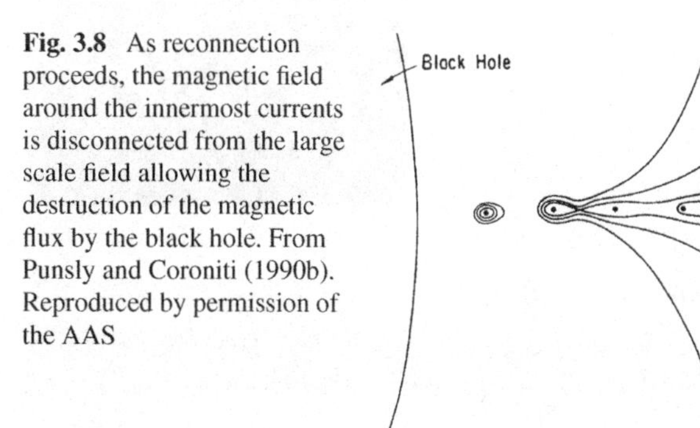

be asymptotically stable. The twisted magnetic fields in the ergosphere of a Kerr black hole are shown in Figs. 3.9, 3.10 and 3.11. The controversy still goes on and a whole bunch of new simulations are exploring the different aspects of relativistic magnetohydrodynamic (MHD) outflows from black hole magnetospheres. We shall say more about these outflows when discussing astrophysical jets.

3.5 Back Hole Interiors

The most relevant feature of a Schwarzschild black hole interior is that the roles of space and time are exchanged: the space radial direction becomes time, and time becomes a space direction. Inside a spherical black hole, the radial coordinate becomes *time-like*: changes occur in a prefer direction, i.e. toward the space-time singularity. This means that the black hole interior is essentially dynamic. In order to see this,

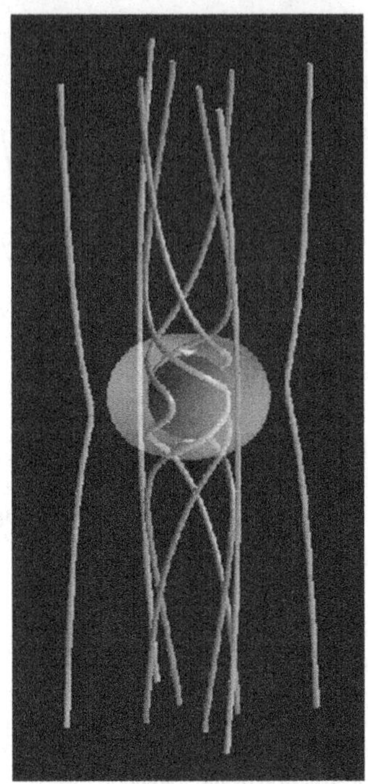

Fig. 3.9 Effects of the ergosphere of a Kerr black hole on external magnetic field lines. Credit: NASA Jet Propulsion Laboratory (NASA-JPL)

let us recall the Schwarzschild metric (2.10):

$$ds^2 = \left(1 - \frac{2GM}{rc^2}\right)c^2dt^2 - \left(1 - \frac{2GM}{rc^2}\right)^{-1}dr^2 - r^2\left(d\theta^2 + \sin^2\theta d\phi^2\right).$$

If we consider a radially infalling test particle:

$$ds^2 = \left(1 - \frac{2GM}{rc^2}\right)c^2dt^2 - \left(1 - \frac{2GM}{rc^2}\right)^{-1}dr^2.$$

The structure of the light cones is defined by the condition $ds = 0$. Writing r_{Schw} once again for the Schwarzschild radius, we get:

$$\left(1 - \frac{r_{\text{Schw}}}{r}\right)c^2dt^2 - \left(1 - \frac{r_{\text{Schw}}}{r}\right)^{-1}dr^2 = 0. \tag{3.31}$$

If we consider the interior of the black hole, $r < r_{\text{Schw}}$. Then,

$$\left(1 - \frac{r_{\text{Schw}}}{r}\right)^{-1}dr^2 - \left(1 - \frac{r_{\text{Schw}}}{r}\right)c^2dt^2 = 0. \tag{3.32}$$

The signs of space and time are now exchanged. The light cones, that in Schwarzschild coordinates are shown in Fig. 2.1, are now oriented with the time axis per-

Fig. 3.10 Three-dimensional graphic of magnetic field lines and plasma flow around the Kerr black hole. The *black sphere* at the center depicts the black hole horizon. The transparent (*gray*) surface around the black hole is that of the ergosphere. The *arrows* show the plasma flow velocity. The *tubes* in the shape of propellers show the magnetic field lines. From Koide (2004). Reproduced by permission of the AAS

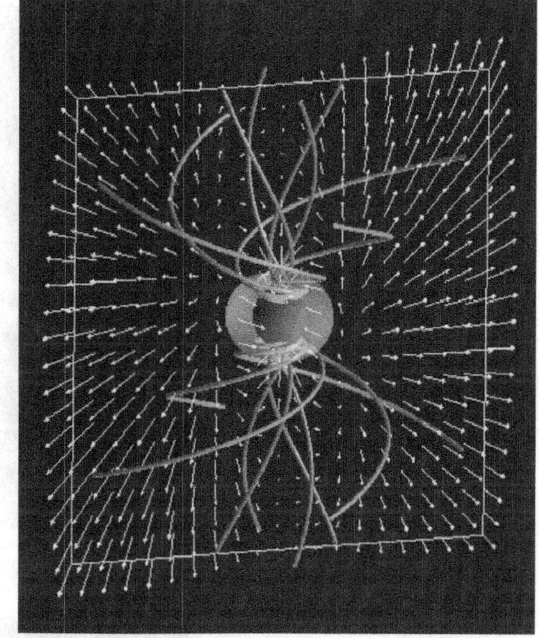

pendicular to the event horizon. The trajectory of photons is given by:

$$\frac{dr}{dt} = \mp c \left| 1 - \frac{r_{\text{Schw}}}{r} \right|, \tag{3.33}$$

with r always decreasing. The light cones are thinner and thinner as r gets closer to the singularity at $r = 0$. In addition to infalling particles, there is a small flux of gravitational radiation into the black hole through the horizon because of small perturbations outside it. This radiation, as the material particles and photons, ends at the singularity.

In the case of a Kerr black hole, between the two horizons space and time also exchange roles as it happens with the Schwarzschild interior black hole space-time. Instead of time always moving inexorably onward, the radial dimension of space moves inexorably inward to the second horizon, that it is also a Cauchy horizon, i.e. a null hyper-surface beyond which predictability breaks down. After that, the Kerr solution predicts a second reversal so that one can avoid the ring singularity and achieve to orbit safely. In this strange region inside the Cauchy horizon the observer can, by selecting a particular orbit around the ring singularity, travel backwards in time and meet himself, i.e. there are closed time-like curves. Another possibility admitted by the equations for the observer in the central region is to plunge through the hole in the ring to emerge in an anti-gravity universe, whose physical laws would be even most peculiar. Or she/he can travel through two further horizons, (or more properly anti-horizons), to emerge at coordinate time $t = -\infty$ into some other uni-

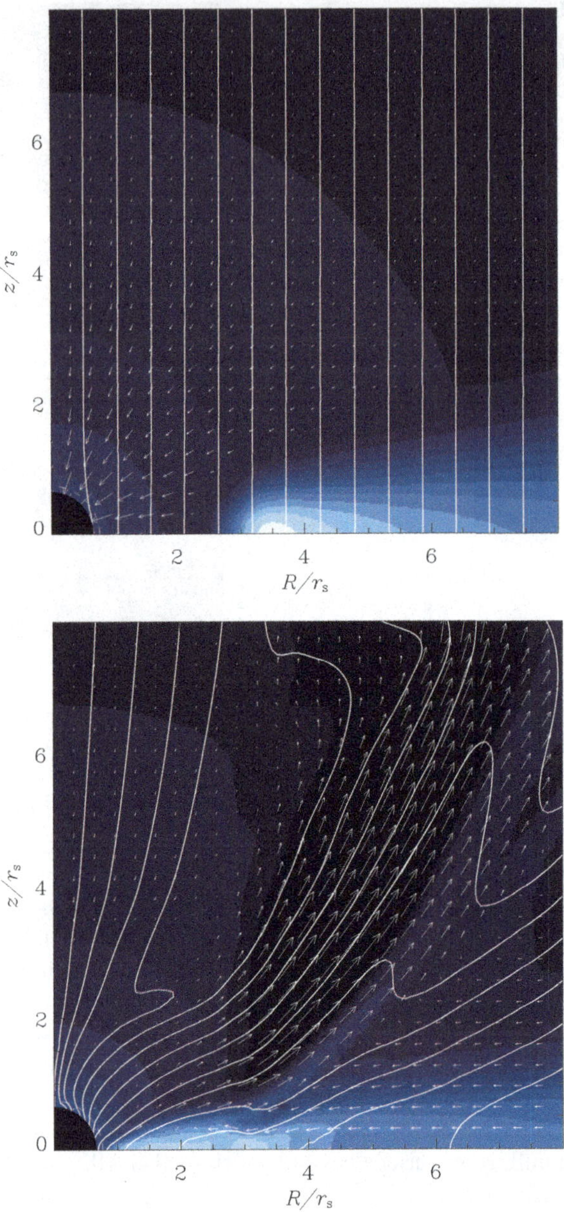

Fig. 3.11 Black hole magnetosphere (corona) and resulting outflows. This article was published in Koide et al. (2000). Copyright Elsevier (2000)

verse. All this can be represented in a Penrose-Carter diagram for a Kerr black hole (see Fig. 3.12).

The above discussion on Kerr black hole interiors is rather academic, since in real black holes the inner horizon is likely unstable. Poisson and Israel (1990) have shown that when the space-time is perturbed by a fully non-linear, ingoing,

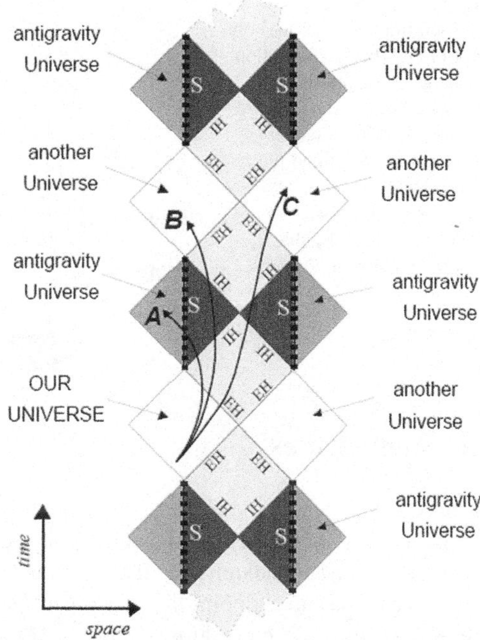

Fig. 3.12 Penrose-Carter diagram of a non-extreme Kerr solution. The figure is repeated infinitely in both directions. One trajectory ends in the singularity (A), the other two (B and C) escape. IH stands for "inner horizon", EH for "external horizon", and S for "singularity". From Luminet (1998)

spherically-symmetric null shell, a null curvature singularity develops at the inner horizon. This singularity is "weak" in the sense that none of the scalar curvature invariants is divergent there. The singularity development at the Cauchy horizon would deal off the "Kerr tunnel" that would lead to other asymptotically flat universes. The key factor producing the instability is the infinite concentration of energy density close to the Cauchy horizon as seen by a free falling observer. The infinite energy density is due to the ingoing gravitational radiation, which is partially backscattered by the inner space-time curvature. The non-linear interaction of the infalling and outgoing gravitational fluxes results in the weak curvature singularity on the Cauchy horizon, where a tremendous inflation of the mass parameter takes place (see Fig. 3.13). This changes the conception of the Kerr black hole interior, since instead of a Cauchy horizon acting as a curtain beyond which predictability breaks down we have a microscopically thin region near the inner horizon where the curvature is extremely high (Poisson and Israel 1990). Other analysis based on plane-symmetric space-time analysis seem to suggest that instead of a null, weak singularity a space-like strong singularity is formed under generic non-linear perturbations (Yurtsever 1993). This is the same result that can be obtained through a linear perturbation analysis of the inner horizon. More recent numerical investigations using regular initial data find a mass inflation-type null singularity (Droz 1997).

The issue of realistic black hole interiors is still an open one, with active research ongoing.

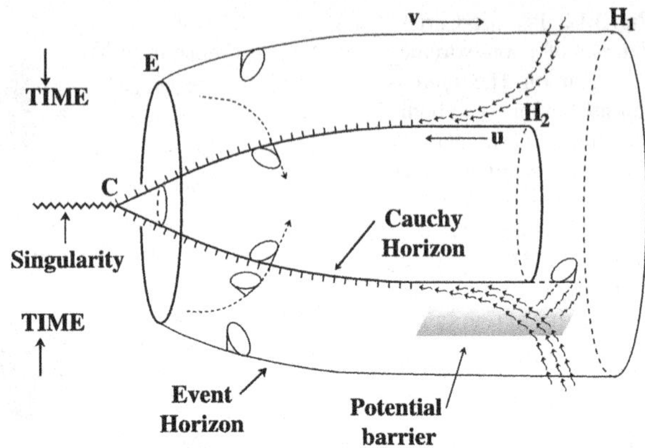

Fig. 3.13 Diagram representing a Kerr black hole interior and the accumulation of energy-momentum at the inner horizon. H_1 and H_2 are the outer and inner horizons, respectively. Reprinted figure with permission from Bonanno et al. (1994). Copyright (1994) by the American Physical Society

3.6 Singularities

A space-time is said to be *singular* if the manifold M that represents it is *incomplete*. A manifold is incomplete if it contains at least one *inextendible* curve. A curve $\gamma : [0, a) \longrightarrow M$ is inextendible if there is no point p in M such that $\gamma(s) \to p$ as $a \to s$, i.e. γ has no endpoint in M. A given space-time (M, g_{ab}) has an *extension* if there is an isometric embedding $\theta : M \to M'$, where (M', g'_{ab}) is a space-time and θ is onto a proper subset of M'. A space-time is *singular* if it contains a curve γ that is inextendible in the sense given above. Singular space-times are said to contain singularities, but this is an abuse of language: singularities are not "things" in space-time, but a pathological feature of the theory. Actually, "singularities" cannot exist in space-time by definition.

A so-called *coordinate singularity* is not a real feature of a singular space-time. The space-time seems to be singular in some representation but the pathologies (divergences) can be removed by a coordinate change, like the "Schwarzschild singularity" at $r_{\mathrm{Scwh}} = 2GM/c^2$ in a Schwarzschild space-time. We can change the description of the space-time to Eddington-Finkelstein coordinates, for instance, and then see that geodesic lines can go through the "singular" point of the manifold. Essential singularities cannot be removed in this way. This occurs, for instance, with the singularity at $r = 0$ in the Schwarzschild space-time or with the ring singularity at $r = 0$ and $\theta = \pi/2$ in the Kerr metric written in Boyer-Lindquist coordinates.[3] In such cases, the curvature scalar $R^{\mu\nu\rho\sigma} R_{\mu\nu\rho\sigma}$ diverges. There is no metric there, and Einstein's equations cannot be defined.

An essential or true singularity should not be interpreted as a representation of a physical object of infinite density, infinite pressure, etc. Since the singularity does not belong to the manifold that represents space-time in General Relativity, it simply cannot be described or represented in the framework of such a theory. General Relativity is incomplete in the sense that it cannot provide a full description of the gravitational behavior of any physical system. True singularities are not within the

---

[3]In Cartesian coordinates the Kerr singularity occurs at $x^2 + y^2 = a^2 c^{-2}$ and $z = 0$.

range of values of the bound variables of the theory: they do not belong to the ontology of a world that can be described with 4-dimensional differential manifolds (for more on this see Romero 2012).

An essential singularity in the solutions of Einstein's field equations is one of two things:

1. A situation where matter is forced to be compressed to a point (a space-like singularity).
2. A situation where certain light rays come from a region with infinite curvature (time-like singularity).

Space-like singularities are a feature of non-rotating uncharged black-holes, whereas time-like singularities are those that occur in charged or rotating black hole exact solutions, where time-like or null curves can always avoid hitting the singularities.

What is referred to as a singularity does not belong to classical space-time. Matter compressed to such a point that its effects on space-time cannot be described by General Relativity is usually designated as a "singularity". At such small scales and high densities, relations among things should be described in a quantum mechanical way. If space-time is formed by the events that occur to things, it should be represented through a quantum theory when the quantum things have effect upon the space-time structure. Since even in the standard quantum theory time appears as a continuum variable, a new approach is necessary (see, for instance, Rovelli 2004; Oriti 2009, and Gambini and Pullin 2011).

Space-time singularities are expected to be covered by horizons. Although formation mechanisms for naked singularities have been proposed, the following conjecture is usually considered valid:

- Cosmic Censorship Conjecture (Roger Penrose): singularities are always hidden behind event horizons.

We emphasize that this conjecture is not proved in General Relativity and hence it has not the strength of a theorem of the theory.

The classical references on singular space-times are Hawking and Ellis (1973) and Clarke (1993).

3.6.1 Singularity Theorems

Several singularity theorems can be proved from pure geometrical properties of the space-time model (Clarke 1993). The most important one is due to Hawking and Penrose (1970):

Theorem *Let (M, g_{ab}) a time-oriented space-time satisfying the following conditions*:

1. $R_{ab}V^a V^b \geq 0$ *for any non space-like V^a.*

2. *Time-like and null generic conditions are fulfilled.*
3. *There are no closed time-like curves.*
4. *At least one of the following conditions holds*

   - *There exists a compact[4] achronal set[5] without edge.*
   - *There exists a trapped surface.*
   - *There is a $p \in M$ such that the expansion of the future (or past) directed null geodesics through p becomes negative along each of the geodesics.*

Then, (M, g_{ab}) contains at least one incomplete time-like or null geodesic.

If the theorem has to be applied to the physical world, the hypothesis must be supported by empirical evidence. Condition 1 will be satisfied if the energy-momentum T^{ab} satisfies the so-called *strong energy condition*: $T_{ab}V^a V^b \geq -(1/2)T^a_a$, for any time-like vector V^a. If the energy-momentum is diagonal:

$$T_{\mu\mu} = (\rho, -P, -P, -P)$$

the strong energy condition can be written as $\rho + 3P \geq 0$ and $\rho + P \geq 0$. Condition 2 requires that any time-like or null geodesic experiences a tidal force at some point in its history. Condition 4a requires that, at least at one time, the Universe is closed and the compact slice that corresponds to such a time is not intersected more than once by a future directed time-like curve. The trapped surfaces mentioned in 4b refers to horizons due to gravitational collapse. Condition 4c requires that the Universe is collapsing in the past or the future.

The theorem is purely geometric, no physical law is invoked. Theorems of this type are a consequence of the gravitational focusing of congruences. A congruence is a family of curves such that exactly one, and only one, time-like geodesic trajectory passes through each point $p \in M$. If the curves are smooth, a congruence defines a smooth time-like vector field on the space-time model. If V^a is the time-like tangent vector to the congruence, we can write the *spatial part* of the metric tensor as:

$$h_{ab} = g_{ab} + V_a V_b. \tag{3.34}$$

---

[4]A space is said to be compact if whenever one takes an infinite number of "steps" in the space, eventually one must get arbitrarily close to some other point of the space. Thus, whereas disks and spheres are compact, infinite lines and planes are not, nor is a disk or a sphere with a missing point. In the case of an infinite line or plane, one can set off making equal steps in any direction without approaching any point, so that neither space is compact. In the case of a disk or sphere with a missing point, one can move toward the missing point without approaching any point within the space. More formally, a topological space is compact if, whenever a collection of open sets covers the space, some sub-collection consisting only of finitely many open sets also covers the space. A topological space is called compact if each of its open covers has a finite sub-cover. Otherwise it is called non-compact. Compactness, when defined in this manner, often allows one to take information that is known locally—in a neighborhood of each point of the space—and to extend it to information that holds globally throughout the space.

[5]A set of points in a space-time with no two points of the set having time-like separation.

For a given congruence of time-like geodesic we can define the *expansion, shear,* and *torsion* tensors as:

$$\theta_{ab} = V_{(i;l)} h^i_a h^l_b,$$
(3.35)

$$\sigma_{ab} = \theta_{ab} - \frac{1}{3} h_{ab}\theta,$$
(3.36)

$$\omega_{ab} = h^i_a h^l_b V_{[i;l]}.$$
(3.37)

Here, the *volume expansion* θ is defined as:

$$\theta = \theta_{ab} h^{ab} = \nabla_a V^a = V^a_{;a}.$$
(3.38)

The rate of change of the volume expansion as the time-like geodesic curves in the congruence are moved along is given by the Raychaudhuri (1955) equation:

$$\frac{d\theta}{d\tau} = -R_{ab} V^a V^b - \frac{1}{3}\theta^2 - \sigma_{ab}\sigma^{ab} + \omega_{ab}\omega^{ab},$$

or

$$\frac{d\theta}{d\tau} = -R_{ab} V^a V^b - \frac{1}{3}\theta^2 - 2\sigma^2 + 2\omega^2.$$
(3.39)

We can use now Einstein's field equations to relate the congruence with the space-time curvature:

$$R_{ab} V^a V^b = \kappa \left[T_{ab} V^a V^b + \frac{1}{2} T \right].$$
(3.40)

The term $T_{ab} V^a V^b$ represents the energy density measured by a time-like observer with unit tangent for velocity V^a. The weak energy condition then states that:

$$T_{ab} V^a V^b \geq 0. \quad \text{WEC}$$
(3.41)

A stronger condition is:

$$T_{ab} V^a V^b + \frac{1}{2} T \geq 0. \quad \text{SEC}$$
(3.42)

Notice that this condition implies, according to Eq. (3.40),

$$R_{ab} V^a V^b \geq 0.$$
(3.43)

We see then that the conditions of the Hawking-Penrose theorem imply that the focusing of the congruence yields:

$$\frac{d\theta}{d\tau} \leq -\frac{\theta^2}{3},$$
(3.44)

where we have used that both the shear and the rotation vanishes. Equation (3.44) indicates that the volume expansion of the congruence must be necessarily decreasing along the time-like geodesic. Integrating, we get:

$$\frac{1}{\theta} \geq \frac{1}{\theta_0} + \frac{\tau}{3},$$ (3.45)

where θ_0 is the initial value of the expansion. Then, $\theta \to -\infty$ in a finite proper time $\tau \leq 3/|\theta_0|$. This means that once a convergence occurs in a congruence of time-like geodesics, a caustic must develop in the space-time model. The non space-like geodesics are in such a case inextendible.

A closely related theorem is due to Hawking (1967):

Theorem *Let* (M, g_{ab}) *a time-oriented space-time satisfying the following conditions*:

1. $R_{ab} V^a V^b \geq 0$ *for any non space-like* V^a.
2. *There exists a compact space-like hypersurface* $\Sigma \subset M$ *without edge.*
3. *The unit normals to* Σ *are everywhere converging* (*or diverging*).

Then, (M, g_{ab}) *is time-like geodesically incomplete.*

Singular space-time models can be classified in accordance with the kind of extension they admit. A C^k-singular space-time model does not have C^k-extensions of the manifold that allow incomplete curves to be extended. The index k measures the *strength* of the singular character of the space-time model. The smaller k, the stronger the singular feature (Clarke 1993).

Singularity theorems do not seem to apply to the Universe as a whole, since there is increasing evidence that the energy conditions are violated on large scales (Reiss et al. 1998; Perlmutter et al. 1999).

References

W. Baade, F. Zwicky, Phys. Rev. **45**, 138 (1934)

J.M. Bardeen, B. Carter, S.W. Hawking, Commun. Math. Phys. **31**, 161 (1973)

T.W. Baumgarte, S.L. Shapiro, *Numerical Relativity* (Cambridge University Press, Cambridge, 2010)

M.C. Begelman, M.J. Rees, Mon. Not. R. Astron. Soc. **185**, 847 (1978)

M.C. Begelman, M.M. Volonteri, M.J. Rees, Mon. Not. R. Astron. Soc. **370**, 289 (2006)

J.D. Bekenstein, Phys. Rev. D **7**, 2333 (1973)

D. Blais, C. Kiefer, D. Polarski, Phys. Lett. B **535**, 11 (2002)

R.D. Blandford, R.L. Znajek, Mon. Not. R. Astron. Soc. **179**, 433 (1977)

A. Bonanno, S. Droz, W. Israel, S.M. Morsink, Phys. Rev. D **50**, 7372 (1994)

G.B. Brown, C.-H. Lee, R.A.M.J. Wijers, H.A. Bethe, Phys. Rep. **333**, 471 (2000)

B.J. Carr, K. Kohri, Y. Sendouda, J. Yokoyama, Phys. Rev. D **81**, 104019 (2010)

S. Chandrasekhar, Astrophys. J. **74**, 81 (1931)

S. Chandrasekhar, in *Novae and White Dwarfs*, ed. by H. Cie (1939)

C.J.S. Clarke, *The Analysis of Space-Time Singularities* (Cambridge University Press, Cambridge, 1993)

B. Devecchi, M. Volonteri, Astrophys. J. **694**, 302 (2009)

S. Droz, Phys. Rev. D **55**, 3575 (1997)

A.S. Eddington, *The Internal Constitution of Stars* (Cambridge University Press, Cambridge, 1926)

K. Freese, P. Bodenheimer, D. Spolyar, P. Gondolo, Astrophys. J. **685**, L101 (2008)

C.L. Fryer, S.E. Woosley, A. Heger, Astrophys. J. **550**, 372 (2001)

R. Gambini, J. Pullin, *A First Course in Loop Quantum Gravity* (Oxford University Press, Oxford, 2011)

L. Gao, N. Yoshida, T. Abel, C.S. Frenk, A. Jenkins, V. Springel, Mon. Not. R. Astron. Soc. **378**, 449 (2007)

S.W. Hawking, Proc. R. Soc. Lond. A **300**, 187 (1967)

S.W. Hawking, Mon. Not. R. Astron. Soc. **152**, 75 (1971)

S.W. Hawking, Nature **248**, 30 (1974)

S.W. Hawking, G.F.R. Ellis, *The Large-Scale Structure of Space-Time* (Cambridge University Press, Cambridge, 1973)

S.W. Hawking, R. Penrose, Proc. R. Soc. Lond. A **314**, 529 (1970)

P. He, L.-Z. Fang, Astrophys. J. **568**, L1 (2002)

P.S. Joshi, *Gravitational Collapse and Spacetime Singularities* (Cambridge University Press, Cambridge, 2007)

P.S. Joshi, D. Malafarina, Int. J. Mod. Phys. D **20**, 2641 (2011)

S. Koide, Astrophys. J. **606**, L45 (2004)

S. Koide, D.L. Meier, K. Shibata, T. Kudoh, Nucl. Phys. B Proc. Suppl. **80**, 01 (2000)

S.S. Komissarov, Mon. Not. R. Astron. Soc. **350**, 427 (2004)

R.W. Lindquist, R.A. Schwartz, C.W. Misner, Phys. Rev. **137**, 1364 (1965)

J.-P. Luminet, in *Black Holes: Theory and Observation*, ed. by F.W. Hehl, C. Kiefer, R.J.K. Metzler (Springer, Berlin, 1998)

M.C. Miller, D.P. Hamilton, Mon. Not. R. Astron. Soc. **330**, 232 (2002)

M.C. Miller, AIP Conf. Proc. **686**, 125 (2003)

C.W. Misner, D.H. Sharp, Phys. Rev. **136**, 571 (1964)

J.R. Oppenheimer, H. Snyder, Phys. Rev. **56**, 455 (1939)

J.R. Oppenheimer, G.M. Volkoff, Phys. Rev. **55**, 374 (1939)

D. Oriti (ed.), *Approaches to Quantum Gravity* (Cambridge University Press, Cambridge, 2009)

S. Perlmutter et al., Astrophys. J. **517**, 565 (1999)

E. Poisson, W. Israel, Phys. Rev. D **41**, 1796 (1990)

B. Punsly, *Black Hole Gravitohydromagnetics* (Springer, Berlin, 2001)

B. Punsly, F.V. Coroniti, Astrophys. J. **350**, 518 (1990a)

B. Punsly, F.V. Coroniti, Astrophys. J. **354**, 583 (1990b)

A.K. Raychaudhuri, Phys. Rev. **98**, 1123 (1955)

M. Rees, Annu. Rev. Astron. Astrophys. **22**, 471 (1984)

A.G. Reiss et al., Astron. J. **116**, 1009 (1998)

G.E. Romero, The ontology of space-time singularities, in *Mario Novello's 70th Anniversary Symposium*, Livraria da Física, Rio de Janeiro, ed. by N. Pinto-Neto, S.E. Perez Bergliaffa (2012), p. 341

C. Rovelli, *Quantum Gravity* (Cambridge University Press, Cambridge, 2004)

R. Saito, J. Yokoyama, R. Nagata, J. Cosmol. Astropart. Phys. **6**, 24 (2008)

L. Spitzer, *Dynamical Evolution of Globular Clusters* (Princeton University Press, Princeton, 1987)

K.S. Thorne, R.H. Price, D.A. Macdonald, *The Membrane Paradigm* (Yale University Press, New Haven, 1986)

P.C. Vaidya, Curr. Sci. **12**, 183 (1943)

P.C. Vaidya, Proc. Indian Acad. Sci. A **33**, 264 (1951)

M. Volonteri, Astron. Astrophys. Rev. **18**, 279 (2010)

R.F. Webbink, IAU Symp. **113**, 541 (1985)

Y. Yurtsever, Class. Quantum Gravity **10**, L17 (1993)

Chapter 4
Accretion onto Black Holes

4.1 Introduction

Accretion is the process of matter falling into the potential well of a gravitating object. The accretion of matter with no angular momentum is basically determined by the relation between the speed of sound a_s in the matter and the relative velocity v_{rel} between the accretor and the medium. The accretion of matter with angular momentum can lead to the formation of an accretion disk around the compact object.

We can distinguish four basic accretion regimes:

- Spherically symmetric accretion. It occurs when $v_{rel} \ll a_s$ and the matter in accretion does not have any significant angular momentum.
- Cylindrical accretion. The angular momentum of the medium remains small but $v_{rel} \gtrsim a_s$.
- Disk accretion. The total angular momentum of matter is enough as to form an accretion disk around the accretor.
- Two-stream accretion. A quasi-spherically symmetric inflow of matter coexists with an accretion disk.

The hydrodynamic description of accretion (or any other physical process) is valid if the mean free path of the particles in the medium is shorter than the typical size scale of the system. In the case of accretion the self-gravitation of the fluid is usually negligible, so the characteristic length scale is, as we shall see, the *gravitational capture radius* or *accretion radius* R_{accr}. This quantity is roughly equal to the distance to the accretor at which the kinetic energy of an element of matter is of the order of its gravitational energy,

$$\frac{1}{2}\left(a_s^2 + v_{rel}^2\right) = \frac{GM}{R_{accr}}. \tag{4.1}$$

Here M is the mass of the accretor. Hence,

$$R_{accr} \approx \frac{2GM}{a_s^2 + v_{rel}^2}. \tag{4.2}$$

G.E. Romero, G.S. Vila, *Introduction to Black Hole Astrophysics*,
Lecture Notes in Physics 876, DOI 10.1007/978-3-642-39596-3_4,
© Springer-Verlag Berlin Heidelberg 2014

If the hydrodynamic description of the flow is appropriate, the basic equations that govern the accretion process are the equations for the conservation of mass (or continuity equation) and the conservation of momentum. In the non-relativistic regime these read

$$\frac{\partial \rho}{\partial t} + \nabla \cdot (\rho \mathbf{v}) = 0 \tag{4.3}$$

and

$$\rho \left[\frac{\partial \mathbf{v}}{\partial t} + (\mathbf{v} \cdot \nabla)\mathbf{v} \right] = -\nabla P + \rho \mathbf{f} + \nabla \cdot \sigma, \tag{4.4}$$

respectively, where $\mathbf{v}$ is the velocity field of the flow, ρ is the mass density, P is the pressure, and $\mathbf{f}$ is the sum of the forces per unit mass. The last term is the divergence of the viscosity stress tensor[1]

$$\sigma_{ij} = 2\eta \tau_{ij}, \quad \tau_{ij} = \frac{1}{2}\left(\frac{\partial v_i}{\partial x_j} + \frac{\partial v_j}{\partial x_i} - \frac{2}{3}\frac{\partial v_k}{\partial x_k}\delta_{ij} \right). \tag{4.5}$$

The coefficient η is the *dynamic viscosity*; it is related to the coefficient of *kinematic viscosity* v as $\eta = \rho v$. In the particular case of constant η the momentum conservation equation becomes

$$\rho \left[\frac{\partial \mathbf{v}}{\partial t} + (\mathbf{v} \cdot \nabla)\mathbf{v} \right] = -\nabla P + \rho \mathbf{f} + \eta \nabla^2 \mathbf{v} + \frac{1}{3}\eta \nabla(\nabla \cdot \mathbf{v}). \tag{4.6}$$

The momentum equation for an ideal fluid (with no viscosity) is generally called Euler equation, whereas for a viscous fluid we speak of the Navier-Stokes equation.

In the most basic description, and provided that the self-gravity of the fluid can be neglected, the only external force is the gravitational attraction of the accretor. Then

$$\mathbf{f} = -\nabla \Phi = \nabla \left(\frac{GM}{r} \right), \tag{4.7}$$

where Φ is the gravitational potential and r is the distance to M.

The continuity equation and the equation of motion must be complemented with an equation of state, that in a very general form we can write as

$$P = P(\rho). \tag{4.8}$$

Finally, we need an equation for the energy balance in the flow. This equation must consider the change in the kinetic plus the internal energy of an element of fluid because of the action of the volume and surface forces acting on it, as well as the possible flux of heat across the surface of the element. Using the equations for the conservation of mass and momentum and the expression for the stress tensor

---

[1] There is a second term $\eta_B(\nabla \cdot \mathbf{v})\delta_{ij}$ in the stress tensor. The coefficient η_B is the *bulk viscosity*, that we shall assume to be zero.

given above, it can be shown that the total time derivative[2] of the internal energy e per unit mass of the fluid is given by

$$\rho \frac{de}{dt} = -P\nabla \cdot \mathbf{v} + 2\eta \left[S_{ij}S_{ij} - \frac{1}{3}(\nabla \cdot \mathbf{v})^2 \right] + \mathcal{Q}, \tag{4.9}$$

where $\mathcal{Q}$ is the net heat exchanged by the element of fluid per unit time per unit volume, and

$$S_{ij} = \frac{1}{2}\left(\frac{\partial v_i}{\partial x_j} + \frac{\partial v_j}{\partial x_i} \right) = \tau_{ij} - \frac{1}{3}(\nabla \cdot \mathbf{v})\delta_{ij}. \tag{4.10}$$

The quantity

$$q^{+} = 2\eta \left[S_{ij}S_{ij} - \frac{1}{3}(\nabla \cdot \mathbf{v})^2 \right] \tag{4.11}$$

represents, then, the rate of energy dissipation per unit volume due to the work done by the viscous forces.

A key parameter in the study of accretion flows is the mass accretion rate $\dot{M}$, defined as the mass per unit time captured by the gravitating center. In a general manner, we can write it as

$$\dot{M} = \sigma_{\mathrm{G}}\rho v_{\mathrm{rel}} \tag{4.12}$$

where σ_{G} is the cross section of gravitational capture. This cross section depends strongly on the nature of the gas. If the gas is made of collisionless non-relativistic particles, the gravitational capture cross section in a Schwarzschild black hole is (e.g. Shapiro and Teukolsky 1983; Frolov and Novikov 1998)

$$\sigma_{\mathrm{G(collisionless)}} = 4\pi \left(\frac{c}{v_{\infty}} \right)^2 R_{\mathrm{Schw}}^2, \tag{4.13}$$

where $v_{\infty} \ll c$ is the velocity of the particles relative to M at infinity. If, on the other hand, the medium can be modeled as a fluid, we shall later see that

$$\sigma_{\mathrm{G(fluid)}} \approx \pi R_{\mathrm{grav}}^2. \tag{4.14}$$

Then

$$\frac{\sigma_{\mathrm{G(collisionless)}}}{\sigma_{\mathrm{G(fluid)}}} \propto \left(\frac{c}{v_{\infty}} \right)^2 \left(\frac{R_{\mathrm{Schw}}}{R_{\mathrm{grav}}} \right)^2 \ll 1. \tag{4.15}$$

Under typical conditions in the interstellar medium the capture cross section for accretion of a fluid is about a million times that for the accretion of collisionless particles.

---

[2]It is a common practice in the study of fluid dynamics to work with the total or material derivative of a function $f(\mathbf{x}, t)$, defined as $df/dt = \partial f/\partial t + (\mathbf{v} \cdot \nabla)f$. The second term is called the convective derivative.

4.2 Spherically Symmetric Accretion

4.2.1 The Bondi Solution

Bondi (1952) was the first to obtain the steady, spherically symmetric solution for accretion onto a mass M at rest with respect to the surrounding medium.

In spherical coordinates (taking the position of M at $r = 0$) Eqs. (4.3) and (4.4) for an ideal fluid (with $\eta = 0$) read

$$\frac{1}{r^2}\frac{d}{dr}\left(r^2 \rho v\right) = 0 \tag{4.16}$$

and

$$v\frac{dv}{dr} + \frac{1}{\rho}\frac{dP}{dr} + \frac{GM}{r^2} = 0, \tag{4.17}$$

where $\mathbf{v} = v\hat{r}$ is the radial velocity of the flow considered positive inwards.

Let ρ_∞, P_∞, and $v_\infty = 0$ be the density, pressure, and velocity of the medium at infinity, respectively. We shall assume that the gas satisfies a polytropic equation of state

$$P = P_\infty\left(\frac{\rho}{\rho_\infty}\right)^\gamma, \tag{4.18}$$

where $1 \le \gamma \le 5/3$ is a constant. In particular, $\gamma = 1$ and $\gamma = 5/3$ represent an isothermic and an adiabatic flow, respectively. From Eq. (4.18) we obtain an expression for the speed of sound,

$$a_s^2 = \gamma\frac{P}{\rho} = a_\infty^2\left(\frac{\rho}{\rho_\infty}\right)^{\gamma-1}, \tag{4.19}$$

where a_∞ is the speed of sound at infinity.

Equation (4.16) can be immediately integrated. Since ρv is the flux of matter, we can write

$$\dot{M} = 4\pi r^2 v\rho, \tag{4.20}$$

where $\dot{M}$ is a constant that we identify with the mass accretion rate.

With the help of Eq. (4.16) and the definition of the speed of sound, Eq. (4.17) can be cast in a convenient form

$$\frac{1}{2}\left(1 - \frac{a_s^2}{v^2}\right)\frac{dv^2}{dr} = -\frac{GM}{r^2}\left(1 - \frac{2a_s^2 r}{GM}\right). \tag{4.21}$$

As a_s remains finite for $r \to \infty$, the right-hand side of Eq. (4.21) is positive for large radii and increases as $r \to 0$. Unless $a_s^2 > GM/2r$ for all r, it vanishes at

$$r_s = \frac{GM}{2a_s^2(r_s)}. \tag{4.22}$$

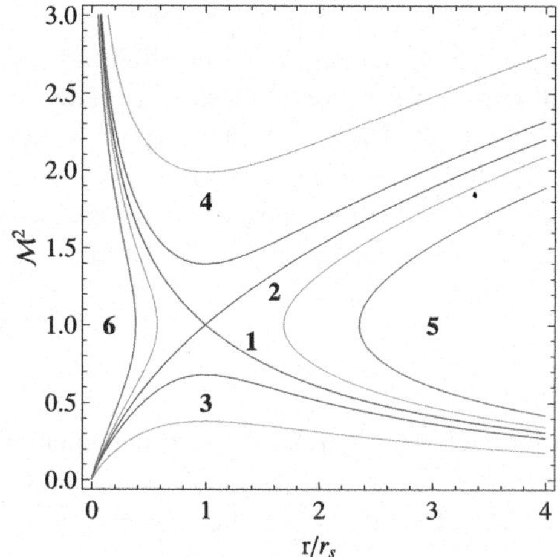

Fig. 4.1 Mach number squared $\mathcal{M}^2 = (v/a_\mathrm{s})^2$ as a function of the adimensional radial coordinate r/r_s, for $\gamma = 7/5$

We shall call r_s the *sonic radius*. The left-hand side of Eq. (4.21) must also vanish at $r = r_\mathrm{s}$. This implies that either

$$v^2(r_\mathrm{s}) = a_\mathrm{s}^2(r_\mathrm{s}) \tag{4.23}$$

or

$$\frac{dv^2}{dr}(r_\mathrm{s}) = 0. \tag{4.24}$$

The characteristics of the solution of Eq. (4.21) depend on the behavior at the sonic radius and the boundary conditions imposed at $r \to \infty$ and $r \to 0$. There are six types of solutions:

1. $v^2(r_\mathrm{s}) = a_\mathrm{s}^2(r_\mathrm{s})$ and $v^2 \to 0$ for $r \to \infty$
2. $v^2(r_\mathrm{s}) = a_\mathrm{s}^2(r_\mathrm{s})$ and $v^2 \to 0$ for $r \to 0$
3. $\frac{dv^2}{dr}(r_\mathrm{s}) = 0$ and $v^2(r) < a_\mathrm{s}^2(r)$
4. $\frac{dv^2}{dr}(r_\mathrm{s}) = 0$ and $v^2(r) > a_\mathrm{s}^2(r)$
5. $\frac{dv^2}{dr}(r_\mathrm{s}) \to \infty$ at $v^2 = a_\mathrm{s}^2(r_\mathrm{s})$ and $r > r_\mathrm{s}$
6. $\frac{dv^2}{dr}(r_\mathrm{s}) \to \infty$ at $v^2 = a_\mathrm{s}^2(r_\mathrm{s})$ and $r < r_\mathrm{s}$

Some solutions for different values of $\dot{M}$ are shown in Fig. 4.1. At fixed γ, there is only one solution of types 1 and 2. As we shall see, the value of the accretion rate in these cases is not a free parameter, but it is uniquely determined by the boundary conditions and the mass of the accretor; we call this critical value $\dot{M}_\mathrm{cr}$. For $\dot{M} < \dot{M}_\mathrm{cr}$ there is an infinite number of solutions of type 3 and 4, and infinite solutions of type 5 and 6 for $\dot{M} > \dot{M}_\mathrm{cr}$.

Solutions of type 5 and 6 are double-valued and therefore not physically meaningful. They can represent, however, part of the solution for some range of r if a shock front develops in the flow. Since we look for a solution with $v_\infty = 0$, we also exclude types 2 and 4. Only solutions of type 1 and 3 satisfy the boundary condition that the flow is at rest at infinity. Solutions of type 3 are subsonic, that is $v^2 < a_s^2$ for all radius. They describe a slow accretion flow that settles to equilibrium, but also a slow wind if $v < 0$. That of type 1 is the only transonic accretion solution—the velocity becomes supersonic for $r < r_s$. We can find the expression for the accretion rate $\dot{M}_{cr}$ that corresponds to this solution as follows.

For $\gamma > 1$, the integration of the Euler equation (4.17) yields

$$\frac{v^2}{2} + \frac{a_s^2}{\gamma - 1} - \frac{GM}{r} = \frac{a_\infty^2}{\gamma - 1}, \tag{4.25}$$

where we used that $v_\infty = 0$ and the equation of state. Equation (4.25) is called the Bernoulli equation or Bernoulli integral, and states the conservation of energy for the fluid. Evaluating it at the sonic radius we find that

$$a_s^2(r_s) = a_\infty^2 \left(\frac{2}{5 - 3\gamma} \right). \tag{4.26}$$

To obtain the density at the sonic radius we use Eq. (4.19). Then

$$\rho(r_s) = \rho_\infty \left[\frac{a_s(r_s)}{a_\infty} \right]^{\frac{2}{\gamma - 1}} = \rho_\infty \left(\frac{2}{5 - 3\gamma} \right)^{\frac{1}{\gamma - 1}}. \tag{4.27}$$

Notice that for $\gamma = 5/3$ the density and the speed of sound tend to infinity at the sonic radius, that shifts to the origin. The critical accretion rate now follows from Eq. (4.20),

$$\dot{M}_{cr} = \pi G^2 M^2 \rho_\infty a_\infty^{-3} \left(\frac{2}{5 - 3\gamma} \right)^{\frac{5 - 3\gamma}{2(\gamma - 1)}}. \tag{4.28}$$

The accretion rate is then fixed by M and only two boundary conditions, ρ_∞ and a_∞ for instance; its dependence on γ is rather weak.

Combining Eqs. (4.19) and (4.20) gives the velocity of the gas in terms of the speed of sound,

$$v(r) = \frac{\dot{M}}{4\pi \rho_\infty r^2} \left[\frac{a_\infty}{a_s(r)} \right]^{\frac{2}{\gamma - 1}}. \tag{4.29}$$

Inserting this in the Bernoulli equation yields an algebraic equation for $a_s(r)$. Once the speed of sound is known, the density and the velocity profiles follow from Eqs. (4.19) and (4.29), respectively.

The case of an isothermal flow with $\gamma = 1$ must be analyzed separately. The Bernoulli equation now involves a logarithmic term

$$\frac{v^2}{2} + a_\infty^2 \ln \rho - \frac{GM}{r} = a_\infty^2 \ln \rho_\infty. \tag{4.30}$$

It can be shown (see Bondi 1952) that the critical accretion rate in this case is simply the limit of Eq. (4.28) when $\gamma \to 1$,

$$\dot{M}_{\mathrm{cr}}^{(\gamma=1)} = e^{3/2} \pi G^2 M^2 \rho_\infty a_\infty^{-3}. \tag{4.31}$$

Without solving the problem completely some general conclusions may be drawn. Let us define the *accretion radius* as

$$r_{\mathrm{accr}} = \frac{2GM}{a_\infty^2}. \tag{4.32}$$

For $r \gg r_{\mathrm{accr}}$ the velocity, the speed of sound, and the density behave as

$$v \approx \frac{a_\infty}{16} \left[\frac{2}{5-3\gamma} \right]^{\frac{1}{2}\left(\frac{5-3\gamma}{\gamma-1}\right)} \left(\frac{r_{\mathrm{accr}}}{r} \right)^2 \left(1 - \frac{1}{2} \frac{r_{\mathrm{accr}}}{r} \right) \approx 0, \tag{4.33}$$

$$a_{\mathrm{s}} \approx a_\infty \left[1 + \left(\frac{\gamma-1}{4} \right) \frac{r_{\mathrm{accr}}}{r} \right] \approx a_\infty, \tag{4.34}$$

$$\rho \approx \rho_\infty \left(1 + \frac{1}{2} \frac{r_{\mathrm{accr}}}{r} \right) \approx \rho_\infty. \tag{4.35}$$

In the region $r \gtrsim r_{\mathrm{accr}}$, then, the influence of the gravitational potential is weak and the variables approximately keep their values at infinity. We can interpret the accretion radius as that where the thermal and the gravitational energy of the gas are comparable,

$$\left[\frac{\rho a_{\mathrm{s}}^2 (r_{\mathrm{accr}})}{2} \right] \left[\frac{\rho GM}{r_{\mathrm{accr}}} \right]^{-1} \sim 1. \tag{4.36}$$

For $r \ll r_{\mathrm{s}}$ the flow is supersonic. From the Bernoulli equation we now get

$$v \approx \left(\frac{2GM}{r} \right)^{1/2} = v_{\mathrm{ff}}, \tag{4.37}$$

where v_{ff} is the free-fall velocity. This is because the underlying layers do not affect the entrained matter. The density profile in this region follows from the continuity equation

$$\rho \approx \rho(r_{\mathrm{s}}) \left(\frac{r_{\mathrm{s}}}{r} \right)^{3/2}, \tag{4.38}$$

whereas the temperature can be obtained from the equation of state of an ideal gas

$$T \approx T(r_{\mathrm{s}}) \left(\frac{r_{\mathrm{s}}}{r} \right)^{\frac{3}{2(\gamma-1)}}. \tag{4.39}$$

Notice, however, that when the temperature is high enough the flow will start to radiate and cool. If the effect of radiation is to be taken into account, then we must add the second law of thermodynamics to Eqs. (4.16) and (4.17). The variation of the internal energy per unit mass of the gas is

$$de = dQ - PdV, \tag{4.40}$$

where $V = 1/\rho$ is the specific volume and dQ is the heat exchanged per unit mass. For an ideal monoatomic gas or a fully ionized plasma we have

$$\frac{3k}{2\mu m_p}\frac{dT}{dt} = \frac{k}{\mu m_p}\frac{T}{\rho}\frac{d\rho}{dt} - \alpha_{\rm ff} T^{1/2}\rho + \frac{dQ}{dt}. \tag{4.41}$$

Here k is the Boltzmann constant, m_p is the mass of the proton, and μ is the mean molecular weight of the gas. The second term on the right-hand side is due to Bremsstrahlung (free-free) radiation ($\alpha_{\rm ff} \approx 5 \times 10^{20}$ erg cm^3 g^{-2} s^{-1} K$^{-1/2}$ for Hydrogen), and the third term takes into account other possible radiative losses. Using $dr = vdt$ and recalling Eqs. (4.37) and (4.38), we can obtain the equation for the temperature distribution in a steady-state spherically symmetric accretion flow

$$\frac{dT}{dr} = -\frac{T}{r} - \alpha_{\rm ff}\,\rho(r_{\rm s})\left(\frac{r_{\rm s}}{2GM}\right)^{1/2}\frac{T^{1/2}}{r} + \frac{2\mu m_p}{3k}\frac{dQ}{dr}. \tag{4.42}$$

If there are no additional radiation losses besides free-free radiation, Eq. (4.42) can be solved to obtain

$$T = \left[K\ln\left(\frac{r}{R_{\rm G}}\right) + T_\infty^{1/2} \right]^2, \tag{4.43}$$

where we have assumed that at $R = R_{\rm G}$ the temperature is T_∞. Equation (4.43) shows that under such conditions the temperature *decreases* as the flow approaches the black hole. A flow that behaves in this way is called a *cooling flow*.

The radial free fall time is

$$t_{\rm ff} \approx \frac{r}{v_{\rm ff}} \propto r^{3/2}, \tag{4.44}$$

whereas the cooling time for Bremsstrahlung losses ($dQ/dT \propto T^{1/2}\rho$) is[3]

$$t_{\rm cool,Br} \approx \frac{3kT/2\mu m_p}{\alpha_{\rm ff}T^{1/2}\rho} \propto \frac{\sqrt{T}}{\rho} \approx r. \tag{4.45}$$

Comparing both timescales we see that the relative role of cooling decreases as the black hole is approached.

------

[3]Notice that $T \propto r^{-1}$ and $\rho \propto r^{3/2}$.

4.2.2 The Eddington Limit

Close to the black hole there could be sources of radiation, for example if a magnetic field and dissipation of angular momentum are involved. The outgoing radiation will pass through the accretion flow and may influence its dynamics. Consider a gas formed by ionized Hydrogen, and let σ be the cross section of interaction of the emitted radiation with matter. If the luminosity of the source is L, the number of photons that cross a sphere of radius r per unit area per unit time is $L/4\pi r^2 E_\gamma$, where E_γ is the mean energy of the photons. This radiation field carries an outgoing flux of momentum $L/4\pi r^2 c$. The number of collisions between photons and particles at r is proportional to σ. Each particle experiences then a force

$$F_{\mathrm{rad}} = \frac{\sigma L}{4\pi r^2 c},\tag{4.46}$$

equal to the rate at which it absorbs momentum from the photons.

We can, to proceed further, assume that radiation interacts mainly with the electrons via Compton scattering, so that $\sigma = \sigma_{\mathrm{T}} \approx 0.66 \times 10^{-24}$ cm^2. Protons are dragged outwards by the electrons since the two species are coupled through the Coulombian attraction. The attractive gravitational force on each proton (we neglect the contribution of electrons since they are much less massive) at r is

$$F_{\mathrm{grav}} = \frac{GMm_p}{r^2}.\tag{4.47}$$

Then, if the luminosity equals

$$L_{\mathrm{Edd}} \equiv \frac{4\pi GMm_p c}{\sigma}\tag{4.48}$$

the gravitational force and the radiation force are balanced and spherical accretion is stopped. This critical luminosity is called the *Eddington luminosity* of the accreting source. In the case of Thomson scattering ($\sigma = \sigma_{\mathrm{T}}$) it takes the value

$$L_{\mathrm{Edd}} \approx 1.3 \times 10^{38} \left(\frac{M}{M_\odot}\right) \text{ erg s}^{-1}.\tag{4.49}$$

Associated with the critical luminosity we can define the *Eddington accretion rate* as

$$\dot{M}_{\mathrm{Edd}} = \frac{L_{\mathrm{Edd}}}{c^2} \approx 0.2 \times 10^{-8} \left(\frac{M}{M_\odot}\right) M_\odot \text{ yr}^{-1}.\tag{4.50}$$

The *Eddington temperature* T_{Edd} is the characteristic temperature of a blackbody of radius equal to the Schwarzschild radius that radiates at $L = L_{\mathrm{Edd}}$,

$$T_{\mathrm{Edd}} = \left(\frac{L_{\mathrm{Edd}}}{4\pi \sigma_{\mathrm{SB}} R_{\mathrm{Schw}}^2}\right) \approx 6.6 \times 10^7 \left(\frac{M}{M_\odot}\right)^{-1/4} \text{ K}.\tag{4.51}$$

Another way to inhibit spherical accretion is the production of winds or particle ejections in the inner regions of the accretion flow. If L_{ej} is the power carried away by the ejected particles and v_{ej} is their velocity, the pressure exerted on the infalling matter will be

$$P_{\mathrm{ej}} = \frac{L_{\mathrm{ej}}}{4\pi R^2 v_{\mathrm{ej}}}. \tag{4.52}$$

If the central source ejects particles before the onset of the spherical accretion, we can estimate the critical luminosity in ejected particles equating pressures at the accretion radius. This gives

$$\frac{L_{\mathrm{ej}}}{4\pi r_{\mathrm{accr}}^2 v_{\mathrm{ej}}} \approx \rho_\infty a_\infty^2 \approx \frac{\dot{M} a_\infty}{\pi r_{\mathrm{accr}}^2}. \tag{4.53}$$

From here we get

$$L_{\mathrm{ej}}^{\mathrm{crit}} \approx 4\dot{M} a_\infty v_{\mathrm{ej}}. \tag{4.54}$$

In general, only a fraction η of the accretion power is released as radiation, i.e.

$$L = \eta \dot{M} c^2, \tag{4.55}$$

then

$$L_{\mathrm{ej}}^{\mathrm{crit}} \approx 4\frac{L}{\eta}\left(\frac{a_\infty v_{\mathrm{ej}}}{c^2}\right). \tag{4.56}$$

We see that a weak wind can halt spherical accretion.

4.3 Cylindrical Accretion

4.3.1 Bondi-Hoyle-Lyttleton Model

The problem of axisymmetric accretion (usually called Bondi-Hoyle or Bondi-Hoyle-Lyttleton accretion) is that of determining the accretion rate onto a moving gravitating center. Unlike the case of spherical accretion, the problem is quite complex and there are no analytical solutions to the hydrodynamic equations. [4]

The first estimates of the accretion rate were obtained by Hoyle and Lyttleton (1939) and Bondi and Hoyle (1944). To work out the problem it is convenient to adopt a reference frame fixed to the accretor, a body of mass M that for the moment we assume to be point-like. The uniform velocity and density of the flow at infinity are v_∞ and ρ_∞, respectively. The symmetry axis of the system is determined by the direction of v_∞ (or equivalently that of the direction of motion of the body). We shall work in polar coordinates (r, θ) with $r = 0$ at the position of the accretor, see Fig. 4.2.

---

[4]See, however, Foglizzo and Ruffert (1997, 1999) for some general analytical results.

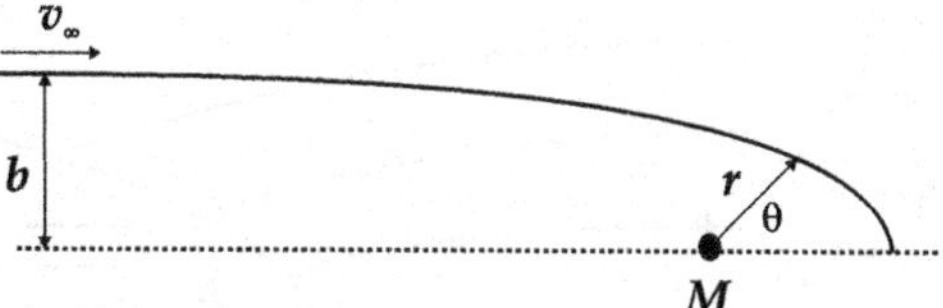

Fig. 4.2 Sketch of the geometry of a Bondi-Hoyle-Lyttleton axisymmetric accretion flow. The path of an element of fluid (in the ballistic approximation) is shown and the relevant parameters of the problem are indicated

In the approximation of Hoyle and Lyttleton (1939) the pressure of the gas is neglected,[5] so the trajectories of the elements of matter can be calculated using the classical equations of motion for a particle in a Newtonian gravitational field,

$$\frac{d^2 r}{dt^2} - r\left(\frac{d\theta}{dt}\right)^2 = -\frac{GM}{r^2}, \tag{4.57}$$

$$r^2\frac{d\theta}{dt} = bv_\infty. \tag{4.58}$$

The second equation states the conservation of angular momentum; b is the impact parameter. The solution for the orbit $r(\theta)$ is (e.g. Bisnovatyi-Kogan et al. 1979)

$$r(\theta) = \left[\frac{GM}{b^2 v_\infty^2}(1 + \cos\theta) - \frac{1}{b}\sin\theta\right]^{-1}, \tag{4.59}$$

whereas the two components of the velocity are given by

$$v_r = -\left(v_\infty^2 + \frac{2GM}{r} - \frac{b^2 v_\infty^2}{r^2}\right)^{1/2} \tag{4.60}$$

and

$$v_\theta = \frac{bv_\infty}{r}. \tag{4.61}$$

It follows that matter reaches the $\theta = 0$ axis at a radius

$$r = \frac{b^2 v_\infty^2}{2GM} \tag{4.62}$$

with a velocity

$$\mathbf{v} = -v_\infty \hat{r} + \frac{2GM}{bv_\infty}\hat{\theta}. \tag{4.63}$$

---

[5]These amounts to assuming that the mean free path of the particles is much larger than the characteristic length scale of the problem, given by the accretion radius and the size of the accretor.

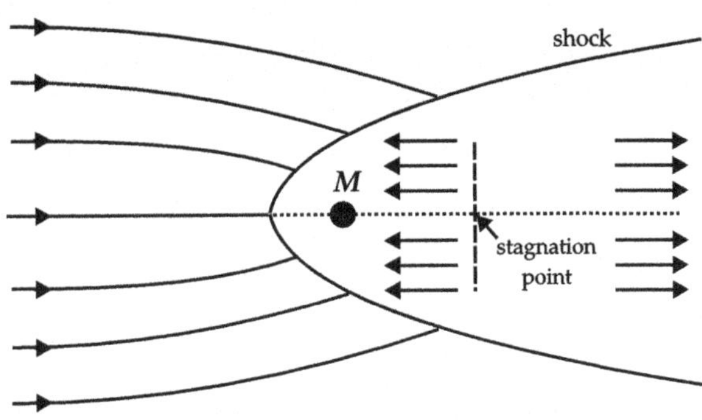

Fig. 4.3 Sketch of the geometry of an axisymmetric accretion flow in the Bondi-Hoyle model. The *arrows* indicate the direction of the flow

Now, as matter coming from both sides approaches the symmetry axis, the density in this region increases and the effects of gas pressure cannot be ignored. Hoyle and Lyttleton (1939) made the simplifying assumption that the transverse component of the velocity vanishes on the $\theta = 0$ axis due to collisions, whereas the radial velocity is unchanged. If this is true, the energy per unit mass of the particles on the axis is

$$E = \frac{1}{2}v_\infty^2 - \frac{2G^2M^2}{b^2 v_\infty^2}. \tag{4.64}$$

For $E < 0$ the material cannot escape the gravitational pull of M. This means that all matter on streamlines with impact parameter

$$b < R_{\mathrm{HL}} \equiv \frac{2GM}{v_\infty^2} \tag{4.65}$$

is accreted. The critical impact parameter R_{HL} is called the Hoyle-Lyttleton radius. The accretion rate then equals the mass flux at infinity through a circle of area πR_{HL}^2 centered on the symmetry axis. This gives the Hoyle-Lyttleton accretion rate

$$\dot{M}_{\mathrm{HL}} = \pi R_{\mathrm{HL}}^2 v_\infty \rho_\infty = 4\pi G^2 M^2 \rho_\infty v_\infty^{-3}. \tag{4.66}$$

In the Hoyle-Lyttleton approximation accretion proceeds along an "accretion line" (the $\theta = 0$ axis) of infinite density. Bondi and Hoyle (1944) developed an improved treatment of the problem in which the accretion region is not a line but an "accretion column", a wake formed upstream as seen from the gravitating body, see Fig. 4.3. Outside the accretion column the gas is in ballistic motion, but the pressure cannot be neglected inside the wake. The characteristic size of the accretion column is determined by the balance between the pressure exerted by the fluid on its boundary from the inside and the transverse momentum flux of the matter entering the column. The thickness of the transition region depends on ρ_∞, and for large densities it becomes a surface of discontinuity.

Bondi and Hoyle (1944) assumed that the mass per unit length per unit time entering the accretion column is the same as in the Hoyle-Lyttleton approach. If we denote this quantity by A, then from Eq. (4.62)

$$2\pi b \rho_\infty v_\infty db = 2\pi \rho_\infty \frac{GM}{v_\infty} dr \equiv A dr. \tag{4.67}$$

Furthermore, it is assumed that the radial velocity of the matter when it enters the column is still v_∞. Two more suppositions are made: first, that the pressure gradient along the symmetry axis can be neglected compared to the gravitational force and, second, that the velocity v of the flow in the wake is uniform on any cross section and parallel to the symmetry axis.

Under these approximations, Bondi and Hoyle (1944) showed that the equations for the conservation of mass and the component of the momentum parallel to the symmetry axis are

$$\frac{d}{dr}(mv) = A \tag{4.68}$$

and

$$\frac{d}{dr}\left(mv^2\right) = A v_\infty - \frac{mMG}{r^2}, \tag{4.69}$$

where mdr is the mass per unit length on the symmetry axis. Appropriate boundary conditions are $v = v_\infty$ for $r \to \infty$, $v = 0$ at $r = r_0$, and that v remains bounded for $r \neq 0$.

Integration of Eq. (4.68) yields

$$mv = A(r - r_0). \tag{4.70}$$

The point r_0 is an stagnation point, $v(r_0) = 0$. Since $m > 0$, the flow is directed towards M for $r < r_0$ and away from M for $r > r_0$. Thus only the material entering the wake at $r < r_0$ will be accreted. The accretion rate is then approximately given by

$$\dot{M}_{\mathrm{BH}} \approx A r_0 = 2\pi \alpha G^2 M^2 \rho_\infty v_\infty^{-3}, \tag{4.71}$$

where we have defined $\alpha = r_0 v_\infty^2 / GM$. Equations (4.68) and (4.69) plus the imposed boundary conditions are satisfied for any value of α. If, however, the extra condition that the velocity $v(r)$ is monotonic is demanded, then it follows from Eq. (4.69) that $\alpha > 1$.[6]

Based on the similarity of Eqs. (4.28) and (4.71), Bondi (1952) proposed an interpolation expression for the accretion rate

$$\dot{M} \approx \frac{2\pi G^2 M^2 \rho_\infty}{(a_\infty^2 + v_\infty^2)^{3/2}}. \tag{4.72}$$

For $a_\infty^2 \gg v_\infty^2$ it reduces to the accretion rate in the spherically symmetric case (for $\gamma = 3/2$), whereas for $v_\infty^2 \gg a_\infty^2$ it matches the result of Bondi-Hoyle (for $\alpha = 1$).

---

[6]See Edgar (2004) for a proof of this statement.

Equation (4.72) is an estimate for $\dot{M}$ in an intermediate case. Other interpolation formulas have been proposed based on the results of numerical simulations (e.g. Shima et al. 1985).

4.3.2 Results of Numerical Simulations

A complete hydrodynamic treatment of axisymmetric accretion must be carried out numerically. Numerical simulations are useful to address issues such as the dependence of the flow pattern on the relevant parameters (the Mach number, the size of the accretor, the polytropic index γ), and its stability under perturbations.

Despite the many simplifications implied in its deduction, the Hoyle-Lyttleton formula for the accretion rate, Eq. (4.66), is confirmed to order of magnitude by the results of simulations of axisymmetric accretion flows under different physical conditions.

The dynamics of the flow depend mainly on the value of the Mach number at infinity $M_\infty = v_\infty/a_\infty$. Subsonic flows ($M_\infty < 1$) reach the steady state and after crossing the sonic surface fall freely onto the accretor, much as in the case of Bondi accretion. No shocks fronts develop. The behavior of supersonic flows at infinity ($M_\infty > 1$) is far more complex. The central result is that a shock wave develops; its shape and position depend on the value of M_∞ and γ.

Early 3D simulations (Ruffert and Arnet 1994; Ruffert 1994, 1995, 1996) showed that for $4/3 \le \gamma \le 5/3$ the shock is in front of the accretor and detached; it as a bow shock. For $\gamma \sim 1$, the shock moves towards the rear side of the accretor and gets attached to it; it becomes a tail shock.

The size of the accretor is the third key parameter in the characterization of the flow. If a large accretor (with a size comparable to R_{HL}) is considered, Ruffert (1994, 1995) showed that for $\gamma = 4/3$–$5/3$ the shock is not a bow shock but a tail shock.[7]

Numerical calculations are limited in general by the size of the accretor: the smaller the accretor the larger the velocity of the flow is in its surroundings, so that smaller time steps are required and the simulations become very time-consuming. The most recent 3D simulations of axisymmetric supersonic flows (Blondin and Raymer 2012) considered accretor sizes of $0.05R_{\mathrm{HL}}$ and $0.01R_{\mathrm{HL}}$. Color maps for the Mach number and the mass flux onto the surface of accretor are shown in Fig. 4.4.

The stability of a Bondi-Hoyle-Lyttleton accretion flow is an issue still under discussion. Several types of instabilities have been observed in numerical simulations, but the physical mechanisms leading to some of them are not well understood. Furthermore, it is not clear if they are real or a numerical artifact. A critical review of the different instabilities that have been proposed to operate in a Bondi-Hoyle-Lyttleton flow is presented in Foglizzo et al. (2005).

---

[7]In 2D simulations the shock is always attached to the accretor, although calculations by Foglizzo et al. (2005) suggest that it should get detached for $\gamma \sim 3$.

Fig. 4.4 Mach number downstream the shock front in two orthogonal planes containing the symmetry axis, for sizes of the accretor $0.05 R_{HL}$ (*top*) and $0.01 R_{HL}$ (*bottom*). The flow is incident from the right. The colors on the surface of the accretor represent the value of the mass flux on it. Notice that most of the accretion occurs downstream the accretor. In the case of the smallest accretor the position of the bow shock oscillates quasi-periodically, although without disrupting the axial symmetry of the flow. From Blondin and Raymer (2012). Reproduced by permission of the AAS

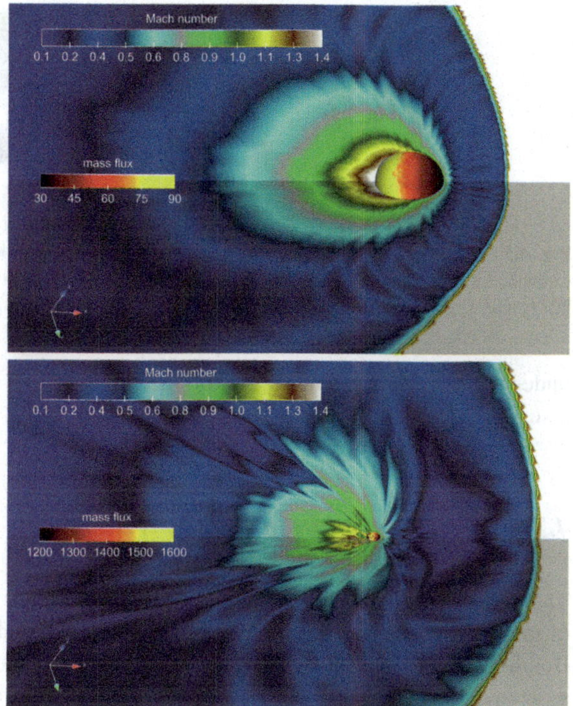

Perhaps the most spectacular (and much studied) type of instability is the so-called "flip-flop" instability: the accretion wake flips from side to side of the accretor with the formation of a transient accretion disk (see Fig. 4.5). There is a peak in the accretion rate every time the disk disappears. This instability has been observed in axisymmetric and non axisymmetric planar simulations in 2D (e.g. Matsuda et al. 1987; Fryxell and Taam 1988; Shima et al. 1998; Blondin and Pope 2009). A weaker version appeared in 3D simulations for small accretors performed by Ruffert and Arnet (1994), but not in those of Blondin and Raymer (2012). As pointed out by Edgar (2004), it is possible that the flip-flop instability is an artifact of 2D planar numerical simulations (in which the accretor is a cylinder and not a sphere) and does not occur in 3D.

If the accretor is a black hole the simulations must be carried out in the fully relativistic regime, at least close to the event horizon. This accretion regime has been studied by Petrich et al. (1989) and Font and Ibáñez (1998a, 1998b) in the Schwarzschild space-time, and by Font et al. (1999) and Penner (2011) for a Kerr black hole. The main difference with simulations in a Newtonian gravitational field is that the flow always settles to the steady state and no instabilities are found, even for non-axisymmetric asymptotic boundary conditions. It must be noticed, however, that for the large accretor sizes these authors considered ($\sim$0.5–2R_{HL}) the flow is expected to be stable even in the non-relativistic case. Recently, Zanotti et al. (2011) performed simulations in the relativistic regime including radiation effects where,

Fig. 4.5 Flip-flop instability in a planar flow. The accretion shock flips from one side to another of the accretor in the first two frames. Eventually a transient accretion disk is formed as seen in the last frame. From Blondin and Pope (2009). Reproduced by permission of the AAS

under some particular initial conditions, flow patters reminiscent of the flip-flop instability develop.

4.3.3 Wind Accretion in Binary Systems

The original motivation for the study of axisymmetric accretion was the accretion of matter onto the Sun due to its motion in the interstellar medium. Since then, the Bondi-Hoyle-Lyttletlon model has been applied to study many other astrophysical systems. These include accretion onto protostars (e.g. Bonnell et al. 2001; Bonnell and Bate 2002; Padoan et al. 2005) and accretion by galaxies inside clusters (e.g. Stevens et al. 1999; Sakelliou 2000; Schulreich and Breitschwerdt 2011). Accretion is also expected to be approximately axisymmetric in binaries embedded in a common envelope (see Taam and Sandquist 2000 for a review) and wind-fed binaries (e.g. Shapiro and Lightman 1976; Theuns et al. 1996; Soker 2004; de Val-Borro et al. 2009).

Wind accretion takes place, for example, in binary systems formed by an O or B-type star and a compact object (neutron star or black hole). In such systems the compact object accretes matter from the wind of the non-collapsed star. The typical velocity of the wind of an early-type star is a few times $10^3 \, \mathrm{km \, s^{-1}}$, much larger than the speed of sound in the interstellar medium $a_\mathrm{s} \sim 100 \, \mathrm{km \, s^{-1}}$. Under these conditions, the effects of pressure can be neglected and the accretion radius approximated by the Hoyle-Lyttleton formula (Shapiro and Lightman 1976),

$$R_\mathrm{accr} \approx \frac{2GM}{v_\mathrm{rel}^2},$$ (4.73)

where M is the mass of the compact object and v_rel its velocity relative to the wind

$$v_\mathrm{rel}^2 \approx v_\mathrm{c}^2 + v_\mathrm{w}^2.$$ (4.74)

Here v_c is the velocity of the compact object relative to the donor star and v_w the velocity of the wind. Because of the orbital motion, the wake develops in the direction of the relative velocity, see Fig. 4.6.

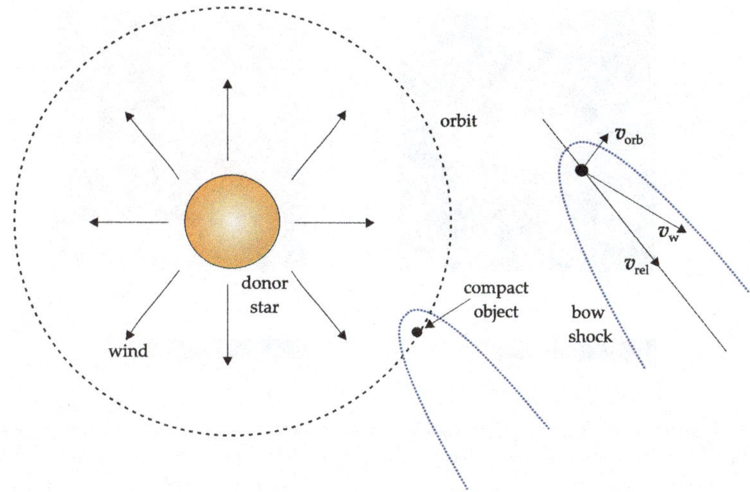

Fig. 4.6 Wind accretion in a binary system in the Bondi-Hoyle-Lyttleton regime. Adapted from Frank et al. (2002)

Matter within a cylinder of radius $\sim R_{\mathrm{accr}}$ centered in the symmetry axis is captured by the compact object, so we can estimate the accretion rate as

$$\dot{M} \approx \pi R_{\mathrm{accr}}^2 \rho_{\mathrm{w}} v_{\mathrm{rel}} = 4\pi G^2 M^2 \rho_{\mathrm{w}} v_{\mathrm{rel}}^{-3}, \tag{4.75}$$

where ρ_{w} is the mass density of the wind at the position of the compact object. It is interesting to compare this accretion rate with the mass loss rate $\dot{M}_*$ of the donor star,

$$\frac{\dot{M}}{\dot{M}_*} \approx \frac{\pi R_{\mathrm{accr}}^2 \rho_{\mathrm{w}} v_{\mathrm{rel}}}{4\pi R_{\mathrm{orb}}^2 \rho_{\mathrm{w}} v_{\mathrm{w}}} = \frac{1}{4}\left(\frac{R_{\mathrm{accr}}}{R_{\mathrm{orb}}}\right)^2\left(\frac{v_{\mathrm{rel}}}{v_{\mathrm{w}}}\right). \tag{4.76}$$

Here R_{orb} is the orbital separation of the binary. We can estimate this ratio for the black hole binary Cygnus X-1. This system is composed by a black hole of mass $M \approx 14.5 M_\odot$, and an O-type star of mass $M_* \approx 19.2 M_\odot$ and radius $R_* \approx 16.4 R_\odot$ (Orosz et al. 2011). The orbital period is $P_{\mathrm{orb}} \approx 5.6$ d (Brocksopp et al. 1999). Using Kepler's law to calculate the orbital separation (assuming a circular orbit) and taking the wind velocity equal to the escape velocity of the star, we get $\dot{M} \approx 0.01 \dot{M}_*$ for Cygnus X-1. In general, $\dot{M}/\dot{M}_* \approx 10^{-4}$–$10^{-3}$ for typical parameters of X-ray binaries, so that a small fraction of the matter in the wind is accreted.

Detailed 3D Smoothed Particle Hydrodynamics (SPH) simulations of wind accretion in binaries have been carried out by Okazaki et al. (2008). They applied the model to estimate the accretion rate in the gamma-ray binary LS 5039 and assess the possibility that the matter accreted from the wind might feed a jet launched from the black hole. Figure 4.7 shows the spatial distribution of mass density in some of the simulations. A strong bow-show forms behind the black hole; its shape is distorted, specially at periastron, because of the large velocity of the black hole relative

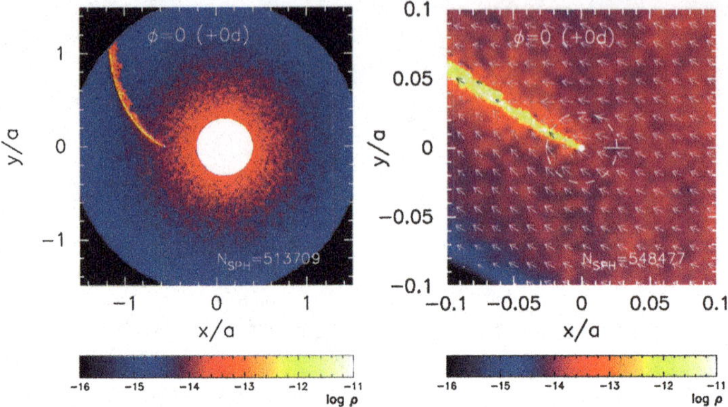

Fig. 4.7 Logarithmic mass density distribution in a simulation of wind accretion in the gamma-ray binary LS 5039 at periastron (orbital phase $\phi = 0$). In the *left panel*, the *large white circle* represents the donor star and the *small white dot* the black hole. The *right panel* shows a close-up of the surroundings of the black hole. The *dashed circle* is the accretion radius, calculated as in Eq. (4.73) with $v_w = 1200$ km s^{-1} and $M = 3.7 M_\odot$. The *white line* indicates the direction towards the star. The *arrows* represent the velocity field of the flow. Lengths are measured in units of the semi-major axis of the orbit a. The number of particles in each simulations is N_{SPH}. From Okazaki et al. (2008)

to the star. The velocity field reveals that the flow is essentially as in the Bondi-Hoyle-Lyttletlon model: matter coming closer to the black hole than the accretion radius is focused and accreted. The mass accretion rate predicted by the simulations ($\sim 2 \times 10^{15}$–10^{16} g s^{-1}) is in very good agreement (only slightly lower) with the Hoyle-Lyttleton value, Eq. (4.75). The difference has to do with the increase of v_{rel} because of the acceleration of the wind by the gravitational pull of the compact object.

Wind accretion in binaries, however, cannot be purely of the Bondi-Hoyle-Lyttleton type: due to the orbital motion the accretion flow has a net angular momentum different from zero. This might lead to the formation of an accretion disk, as we discuss next.

4.4 Disk Accretion

In most realistic astrophysical situations the matter captured by a gravitational field has a total non-zero angular momentum. The accretion of matter with angular momentum onto a black hole may lead to the formation of an accretion disk. The main difficulty in the formulation of a consistent theory of accretion disks lies in the lack of knowledge on the nature of turbulence in the disk and, therefore, in the estimate of the dynamic viscosity.

We shall consider accretion disks in steady state where the accretion rate is an external parameter and the characterization of the turbulence is provided by a single parameter: the so-called α parameter, introduced by Shakura (1972) and Shakura and Sunyaev (1973).

4.4.1 Basic Equations

We start by revising the conditions under which an accretion disk may form. Let J be the angular momentum per unit mass of an element of plasma when it gets trapped in the gravitational field of an accreting body of mass M. Assuming that the plasma loses energy faster than angular momentum, matter will drift to the orbit with the lowest energy compatible with the value of J. This is a circular Keplerian orbit of radius

$$R_{\mathrm{circ}} = \frac{J^2}{GM},\qquad(4.77)$$

called the *circularization radius*. An accretion disk can form if R_{circ} is larger than the effective size of the accretor, for example the radius of the innermost stable circular orbit around a black hole or the radius of the magnetosphere in a neutron star.

As it falls onto the accretor the plasma heats at the expense of its rotational and gravitational energy, but it also transfers angular momentum outwards due to internal torques. The characteristics of the disk strongly depend on the efficiency of the dissipation of energy and angular momentum. Collectively, the dissipation mechanisms are loosely termed "viscosity". If the typical timescale of energy dissipation is much shorter than the timescale of angular momentum redistribution, matter will slowly spiral towards the compact object in a series of approximately circular orbits, forming an accretion disk.

The velocity of an element of plasma in the disk has a tangential component v_ϕ and a small radial component v_R. The angular velocity of the disk is $\Omega(R) = v_\phi/R$, not necessarily equal to the Keplerian value

$$\Omega_{\mathrm{K}}(R) = \left(\frac{GM}{R^3}\right)^{1/2}.\qquad(4.78)$$

The complete structure of the disk is found solving the hydrodynamic equations for mass, energy, and momentum conservation. We shall adopt the following simplifying assumptions:

1. the disk is axisymmetric, i.e. $\partial/\partial\phi = 0$,
2. the disk is thin, i.e. its characteristic size scale in the z-axis is $H \ll R$ (see Fig. 4.8),
3. the matter in the disk is in hydrostatic equilibrium in the z-direction, and
4. the self-gravitation of the disk is negligible.

The thin disk approximation greatly reduces the complexity of the problem. Under this approximation the dependence on z of all variables except the mass density

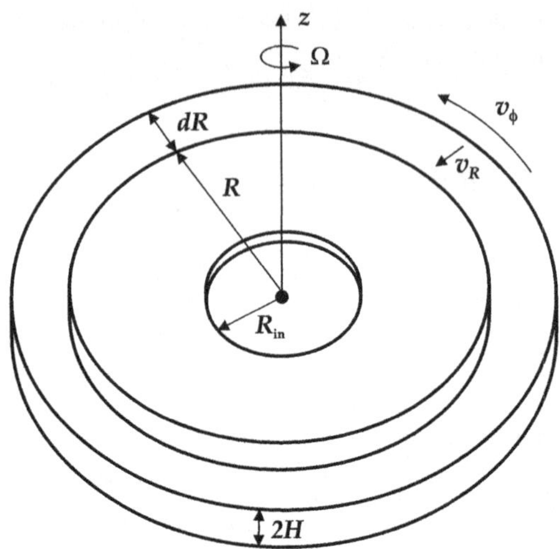

Fig. 4.8 Sketch of a thin accretion disk. The *black dot* indicates the position of the black hole

$\rho(R, z)$ may be ignored. The hydrodynamic equations are then integrated in the z-direction to write them in terms of the surface density

$$\Sigma(R) = 2 \int_0^H \rho(R, z)\,dz. \tag{4.79}$$

This procedure applied to the equation for the conservation of mass gives

$$\frac{\partial \Sigma}{\partial t} + \frac{1}{R}\frac{\partial}{\partial R}(R\Sigma v_R) = 0. \tag{4.80}$$

The equation for momentum conservation in the z-direction is very simple due to the assumption of hydrostatic equilibrium and the thin disk approximation. The only external force on the plasma is the gravitational attraction of the mass M, then

$$\frac{1}{\rho}\frac{\partial P}{\partial z} = -\frac{\partial \Phi}{\partial z} \approx -\frac{GM}{R^3}z. \tag{4.81}$$

Here we have used that the gravitational potential is $\Phi = -GM/\sqrt{R^2 + z^2}$. The half-thickness H of the disk may be estimated from Eq. (4.81). Writing $\Delta z \approx H$ and $P = \rho a_{\mathrm{s}}^2$, we get

$$H \approx \frac{a_{\mathrm{s}}}{\Omega_{\mathrm{K}}}. \tag{4.82}$$

The condition $H \ll R$ thus implies that the Keplerian tangential velocity must be supersonic

$$a_{\mathrm{s}} \ll \sqrt{\frac{GM}{R}}. \tag{4.83}$$

The z-integrated equation for the conservation of momentum in the radial direction reads

$$\frac{\partial v_R}{\partial t} + v_R \frac{\partial v_R}{\partial R} = -\frac{1}{\Sigma}\frac{\partial P}{\partial R} + \frac{v_\phi^2}{R} - \frac{\partial \Phi}{\partial R} + \frac{1}{\Sigma}\left[\frac{\partial}{\partial R}(RT_{RR}) - T_{\phi\phi}\right], \qquad (4.84)$$

where T_{RR} and $T_{\phi\phi}$ are two components of the stress tensor averaged over z. An usual approximation is to neglect the pressure gradient in the radial direction. Furthermore, since we expect that $v_R \ll v_\phi$, the strongest viscous forces are exerted between two adjacent annulus of the disk. This component of the force per unit area is represented by the $T_{R\phi}$ component of the stress tensor, so we shall assume that $T_{RR} = T_{\phi\phi} = 0$. Equation (4.84) then simplifies to

$$\frac{\partial v_R}{\partial t} + v_R \frac{\partial v_R}{\partial R} = \frac{v_\phi^2}{R} - \frac{\partial \Phi}{\partial R}. \qquad (4.85)$$

Finally, we have to write down the z-averaged momentum equation in the ϕ direction. This is

$$\frac{\partial v_\phi}{\partial t} + \frac{v_R}{R}\frac{\partial}{\partial R}(Rv_\phi) = \frac{1}{R^2\Sigma}\frac{\partial}{\partial R}\left(R^2 T_{R\phi}\right). \qquad (4.86)$$

The left-hand side represents the variation of the angular momentum per unit mass and the right-hand side the internal torques generated by viscous forces.

The form of the stress tensor depends on the mechanism of angular momentum dissipation. If the torques are generated only by shear viscosity

$$T_{R\phi} = \nu \Sigma R \frac{\partial \Omega}{\partial R}, \qquad (4.87)$$

where ν is the coefficient of kinematic viscosity.[8] This is not expected to be a realistic approximation, since accretion disks are prone to develop instabilities and become turbulent. In particular, the magneto-rotational instability studied by Balbus and Hawley (1991) might be a very efficient mechanism of angular momentum transport.

We can try to guess a general expression for the stress tensor based on dimensional arguments. The z-integrated component of the stress tensor has dimensions of pressure times length. The simplest expression for $T_{R\phi}$ is then

$$T_{R\phi} \approx -\alpha \Sigma a_{\rm s}^2, \qquad (4.88)$$

where α is a constant and $\Sigma a_{\rm s}$ is the z-integrated isothermal pressure. Equation (4.88) is the famous "α-prescription" introduced by Shakura (1972) and

---

[8]Notice that for a disk in rigid rotation $\partial\Omega/\partial R = 0$ and the internal torques vanish.

Shakura and Sunyaev (1973). It is possible to relate α to an effective viscosity. Under the thin disk approximation and assuming that the angular velocity is Keplerian, equating Eqs. (4.87) and (4.88) yields

$$\nu \approx \alpha a_s H. \tag{4.89}$$

Notice that since we are dealing with z-averaged equations, what we denote by ν is in fact an averaged kinematic viscosity

$$\nu \equiv \langle \nu \rangle = \frac{2}{\Sigma} \int_0^H \rho(R,z)\nu(R,z)dz. \tag{4.90}$$

Equation (4.89) is useful to put a loose constraint on the value of α. In a turbulent flow the kinematic viscosity is approximately given by (e.g. Landau and Lifshitz 1987)

$$\nu \approx v_{\text{turb}} l_{\text{turb}}, \tag{4.91}$$

where v_{turb} is the velocity and l_{turb} the largest size of the turbulent cells. In an accretion disk l_{turb} cannot exceed the height scale H, and the speed v_{turb} is expected to be subsonic—otherwise turbulence would be likely dissipated through shocks. This implies that $\alpha \lesssim 1$.

An expression for the stress tensor (the α-prescription or other) must be supplied to solve the set of hydrodynamic equations for an accretion disk. Some general results, however, can be obtained without doing this explicitly. For example, we can solve Eqs. (4.80) and (4.86) assuming that the accretion flow has reached the steady state. From the continuity equation we immediately obtain that

$$\dot{M} = -2\pi R \Sigma v_R, \tag{4.92}$$

where, as usual, $\dot{M}$ is the mass accretion rate, and we have taken $v_R < 0$ inwards. Integration of Eq. (4.86) gives

$$\dot{M}\Omega R^2 = -2\pi R^2 T_{R\phi} + C. \tag{4.93}$$

The constant C is fixed by the boundary conditions imposed on the stress tensor. If there exists a radius R_{in} where $T_{R\phi}(R_{\text{in}}) = 0$, then $C = \dot{M}\Omega(R_{\text{in}})R_{\text{in}}^2$ and

$$T_{R\phi} = -\frac{1}{2\pi}\dot{M}\Omega_K\left(1 - \sqrt{\frac{R_{\text{in}}}{R}}\right), \tag{4.94}$$

where we have assumed that the angular velocity is Keplerian. If the accretor is a black hole we can take R_{in} as the radius of the of innermost circular stable orbit, $R_{\text{in}} = R_{\text{isco}}$. Another case where the zero-torque boundary condition can be applied is when the accretor is a rotating star of radius R_*, such that its equatorial angular velocity Ω_* satisfies $\Omega_*(R_*) < \Omega(R_*)$. The angular velocity of the disk (that increases inwards) must reach a maximum at a certain R_{in} and then decrease to match $\Omega_*(R_*)$. At $R = R_{\text{in}}$ the component $T_{R\phi}$ of the stress tensor vanishes (since $\partial\Omega/\partial R(R_{\rm in}) = 0$) and so we again recover the result in Eq. (4.94). The region $R_* < R < R_{\rm in}$ is called a boundary layer. If the disk is thin the width of the boundary layer is $\Delta R \ll R_*$, so in practice we can use that $R_{\rm in} = R_* + \Delta R \approx R_*$, neglecting corrections of order $\Delta R/R_*$.

4.4.2 *The Radiative Spectrum of a Thin Disk*

The most important result we can obtain without a detailed model for the viscosity is the value of the total luminosity of the disk. Writing the stress tensor as in Eq. (4.87), the expression for the rate of energy dissipation per unit volume due to viscous forces that results from Eq. (4.11) is

$$q^+ = \rho v \left(R \frac{d\Omega}{dR} \right)^2 . \tag{4.95}$$

Assuming that $\Omega = \Omega_{\rm K}$ and using Eq. (4.90) we can find the rate of energy dissipation per unit area on each face of the disk,

$$Q^+ = \int_0^H q^+ dz = \frac{9}{8}\frac{GM}{R^3} v\Sigma . \tag{4.96}$$

Now, if the stress tensor is parameterized as in Eq. (4.87) and using Eq. (4.94) we get

$$Q^+(R) = \frac{3GM\dot{M}}{8\pi R^3}\left(1 - \sqrt{\frac{R_{\rm in}}{R}} \right). \tag{4.97}$$

The rate of energy dissipation due to viscous forces is then independent of the viscosity. If all this energy is radiated away, integrating Q^+ over the two faces of the disk from the inner radius $R_{\rm in}$ to the outer radius $R_{\rm out}$, gives the total luminosity

$$L_{\rm d} = 2 \times 2\pi \int_{R_{\rm in}}^{R_{\rm out}} Q^+(R)R\,dR = \frac{3GM\dot{M}}{2R_{\rm in}}\left[\frac{1}{3} - \frac{R_{\rm in}}{R_{\rm out}}\left(1 - \frac{2}{3}\sqrt{\frac{R_{\rm in}}{R_{\rm out}}} \right) \right]. \tag{4.98}$$

In the limit $R_{\rm out} \gg R_{\rm in}$,

$$L_{\rm d} \approx \frac{GM\dot{M}}{2R_{\rm in}} . \tag{4.99}$$

This represents exactly a half of the gravitational energy lost by the accretion flow when it reaches the inner radius of the disk. The other half is still retained by the matter as kinetic energy; it is eventually liberated when the flow impacts on the surface of the star, or engulfed if the accretor is a black hole.

Adopting $R_{\rm in} = 6R_{\rm grav}$ and dividing by $\dot{M}c^2$ we get that the efficiency of energy release of disk accretion onto a Schwarzschild black hole is $\sim 8\,\%$. For a co-rotating disk around an extreme Kerr black hole, where $R_{\rm in} = R_{\rm grav}$, the efficiency reaches $\sim 42\,\%$.

It is interesting to examine the local energy balance in the disk. The energy per unit time dissipated in an annulus of width ΔR is

$$\Delta L_{\rm d} = 4\pi\, Q^+ R \Delta R = \frac{3GM\dot{M}}{2R^2}\left(1 - \sqrt{\frac{R_{\rm in}}{R}}\right)\Delta R, \tag{4.100}$$

whereas the gravitational energy lost by the matter per unit time in the same annulus is $(GM\dot{M}/R^2)\Delta R$. A half of this energy turns into kinetic energy, so only a fraction

$$\Delta L_{\rm grav} = \frac{GM\dot{M}}{2R^2}\Delta R \tag{4.101}$$

is available to be dissipated. Then we have

$$\Delta L_{\rm d} - \Delta L_{\rm grav} = \frac{GM\dot{M}}{R^2}\left(1 - \frac{3}{2}\sqrt{\frac{R_{\rm in}}{R}}\right)\Delta R. \tag{4.102}$$

For $R < (9/4)R_{\rm in}$ the energy flux through the surface of the annulus is less than a half of the gravitational energy lost by the accretion flow. The excess is transported outwards as work exerted by viscous forces and dissipated in the region $R > (9/4)R_{\rm in}$, where the surface energy flux is larger than a half the rate of change of gravitational energy. When the whole disk is considered, the energy balance in both regions compensate and we get exactly the result in Eq. (4.99).

If the disk is optically thick in the z-direction, every element of area on its surface radiates as a blackbody at the local *effective surface temperature* $T(R)$. The temperature profile is found recalling that for a blackbody the energy radiated per unit area is $\sigma_{\rm SB}T^4$, where $\sigma_{\rm SB}$ is the Stefan-Boltzmann constant. Then

$$T(R) = \left[\frac{Q^+(R)}{\sigma_{\rm SB}}\right]^{1/4} = \left(\frac{3GM\dot{M}}{8\pi\,\sigma_{\rm SB}R^3}\right)^{1/4}\left(1 - \sqrt{\frac{R_{\rm in}}{R}}\right)^{1/4}. \tag{4.103}$$

For $R \gg R_{\rm in}$ the temperature has the characteristic dependence

$$T(R) \approx T_{\rm d}\left(\frac{R}{R_{\rm in}}\right)^{-3/4}, \tag{4.104}$$

where

$$T_{\rm d} = \left(\frac{3GM\dot{M}}{8\pi\,\sigma_{\rm SB}R_{\rm in}^3}\right)^{1/4}. \tag{4.105}$$

Sticking to the blackbody assumption, the emissivity per unit frequency I_ν of each element of area on the disk is described by the Planck function

$$I_\nu(\nu, R) = B_\nu(\nu, R) \equiv \frac{2h\nu^3}{c^2[\exp(h\nu/kT) - 1]}. \tag{4.106}$$

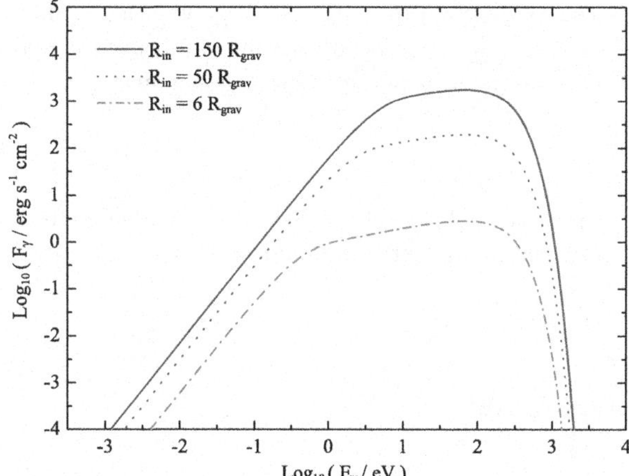

Fig. 4.9 Spectral energy distribution of a geometrically thin, optically thick accretion disk as a function of the inner radius, for $M = 10 M_\odot$, $T(R_{\mathrm{in}}) = 10^6$ K, $\theta_{\mathrm{d}} = 30°$, and $d = 2$ kpc. The inner radius of the disk is given in units of the gravitational radius of the accretor, $R_{\mathrm{grav}} = GM/c^2$

Here h and k are the Planck constant and the Boltzmann constant, respectively. The total flux at frequency ν detected by an observer at a distance d whose line of sight forms and angle θ_{d} with the normal to the disk is then

$$F_\nu(\nu) = \frac{\cos\theta_{\mathrm{d}}}{d^2} \int_{R_{\mathrm{in}}}^{R_{\mathrm{out}}} 2\pi R I_\nu \, dR. \tag{4.107}$$

The shape of the spectral energy distribution predicted by Eq. (4.107) is shown in Fig. 4.9. It is a superposition of blackbody spectra of temperature $T(R)$. The flux grows as $F_\nu \propto \nu^2$ for photon energies $h\nu \ll kT(R_{\mathrm{out}})$, and decreases exponentially for $h\nu \gg kT(R_{\mathrm{in}})$. For intermediate energies the spectrum has the characteristic dependence $F_\nu \propto \nu^{1/3}$. As $T(R_{\mathrm{out}})$ approaches $T(R_{\mathrm{in}})$ this part of the spectrum narrows, and it becomes similar to that of a simple blackbody.

For a blackbody radiator the mean energy of the emitted photons is $E_\gamma \approx 2.7 kT$. The position of the peak of the spectral energy distribution of the disk is then fixed by the value of the maximum temperature T_{max} on the surface. From Eq. (4.103) it is easily checked that $T_{\mathrm{max}} \approx 0.5 T_{\mathrm{d}}$ at $R_{\mathrm{max}} = (46/39) R_{\mathrm{in}}$. For typical parameters of a stellar-mass black hole binary ($M = 10 M_\odot$, $\dot{M} = 10^{-8} M_\odot$ yr^{-1}) and taking $R_{\mathrm{in}} = 6 R_{\mathrm{grav}}$, we get $T_{\mathrm{max}} \approx 3 \times 10^6$ K and $E_\gamma \approx 1.6$ keV. The model then predicts that geometrically thin, optically thick disks in black hole binaries radiate X-rays. This result is in very good agreement with observations, and constitutes the fundamental base of the success of the model.

4.4.3 The Local Structure of Thin Disks in the α-Prescription

Let us now revise the series of equations and approximations of the accretion disk model we have outlined in the previous section.

The assumption that the disk is geometrically thin, i.e. $H \ll R$, allowed us to work with height-averaged versions of most of the hydrodynamic equations. These are written in terms of the surface mass density

$$\Sigma \approx 2H\rho. \tag{4.108}$$

The half-thickness of the disk can be estimated from the condition of hydrostatic equilibrium in the vertical direction. This yields

$$H \approx \frac{a_{\mathrm{s}}}{\Omega_{\mathrm{K}}}, \tag{4.109}$$

where we used

$$P = \rho a_{\mathrm{s}}^2 \tag{4.110}$$

for the pressure.

In steady state, the integration of the averaged continuity equation is straightforward. If $\dot{M}$ is the mass accretion rate across an annulus of the disk, then

$$\dot{M} = -2\pi R \Sigma v_{\mathrm{R}}. \tag{4.111}$$

From the conservation of momentum in the azimuthal direction and assuming that $\Omega = \Omega_{\mathrm{K}}$, we obtained an expression for the $T_{R\phi}$ (the only non-zero) component of the height-averaged stress tensor, Eq. (4.94). Remember that this expression is strictly valid under the zero-torque condition at the disk inner radius. The exact mechanism responsible for the dissipation of energy and angular is unknown, but to proceed further we may attribute it to the action of an effective kinematic viscosity v. Then we can write $T_{R\phi} = v\Sigma R(\partial\Omega_{\mathrm{K}}/\partial R)$, and so

$$v\Sigma = \frac{\dot{M}}{3\pi}\left(1 - \sqrt{\frac{R_{\mathrm{in}}}{R}}\right). \tag{4.112}$$

We have five equations, (4.108) to (4.112), for seven unknowns ρ, Σ, H, a_{s}, P, v_{R}, and v, in terms of the parameters M, $\dot{M}$, R, and R_{in}. They must be supplemented with an equation of state, an equation for the energy balance, and a prescription for the viscosity.

Following Shakura (1972) and Shakura and Sunyaev (1973) we adopt the α-prescription for the viscosity,

$$v = \alpha a_{\mathrm{s}} H. \tag{4.113}$$

Recall that this is equivalent to choosing Eq. (4.88) for the stress tensor under the think disk approximation.

We have often assumed that in a thin disk the angular velocity is Keplerian and the radial velocity very small compared to the tangential velocity. Before proceeding, let us now check it, at least in the α-prescription. From Eqs. (4.111) and (4.112) we have

$$v_R = -\frac{3}{2}\frac{\nu}{R}\left(1 - \sqrt{\frac{R_{\rm in}}{R}}\right). \tag{4.114}$$

Replacing α from Eq. (4.89) and since $H \ll R$, we get $v/R \approx \alpha a_s H/R \ll \alpha a_s$. For $\alpha \lesssim 1$ the radial velocity is then highly subsonic and so very small compared to the tangential velocity. Back to Eq. (4.85) for the radial momentum conservation, we can now neglect the term $v_R \partial v_R/\partial R$. Thus

$$v_\phi \approx \left(\frac{GM}{R}\right)^{1/2}, \tag{4.115}$$

and we confirm that the tangential velocity is indeed approximately Keplerian.

The total pressure in the plasma is the sum of the gas pressure and the radiation pressure

$$P = P_{\rm gas} + P_{\rm rad} = \frac{\rho k T}{\mu m_p} + \frac{4\sigma_{\rm SB}}{3c}T, \tag{4.116}$$

where m_p is the mass of the proton and μ the mean molecular weight. We have neglected in Eq. (4.116) any contribution of the magnetic field to the pressure.

Finally, we have to discuss the radiative transport in the disk. Since the disk is geometrically thin the radiation is transported from the interior to the surface mainly in the vertical direction. We shall also assume that the disk is optically thick, so that the radiative transport proceeds through diffusion. This means that the optical depth in the z-direction must satisfy

$$\tau = \int_0^H \kappa\rho dz \gg 1, \tag{4.117}$$

where κ is the opacity.

The appropriate equation for the energy flux in the z-direction in an optically thick disk is

$$Q^- = -\frac{16\sigma_{\rm SB}T^3}{3\kappa\rho}\frac{\partial T}{\partial z}. \tag{4.118}$$

We expect that the temperature is maximum at $z = 0$ and decreases towards the surface, so that $T_c \equiv T(z = 0) \gg T(z = H)$. Then we can approximate the derivative as $\partial T/\partial z \approx -T_c/H$ to obtain

$$Q^- \approx \frac{16\sigma_{\rm SB}T_c^4}{3\kappa\rho H}. \tag{4.119}$$

In steady state all the energy dissipated by viscous forces per unit area must be released as radiation, so that $Q^- = Q^+$. Therefore

$$\frac{16\sigma_{\rm SB}T_c^4}{3\kappa\rho H} \approx \frac{3GM\dot{M}}{8\pi R^3}\left(1 - \sqrt{\frac{R_{\rm in}}{R}}\right). \tag{4.120}$$

This is the equation for the central temperature of the disk. To complete the system we must provide an expression for the opacity

$$\kappa = \kappa(\rho, T).\tag{4.121}$$

The set of algebraic equations (4.108) to (4.112) plus (4.113), (4.116), (4.120), and (4.121) may be solved to find Σ, H, ρ, a_s, P, v_R, v, T_c, and κ as a function of M, $\dot{M}$, α, R, and R_{in}.

The solution was found by Shakura and Sunyaev (1973) (see also Novikov and Thorne 1973 for the relativistic version); its characteristics depend on which is the main contribution to the pressure and which process dominates the opacity to the escape of radiation. According to this, the disk can be divided into three regions:

1. an outer region (large R) in which gas pressure dominates over radiation pressure and the opacity is due to free-free absorption,
2. a middle region (smaller R) in which gas pressure dominates over radiation pressure but opacity is due to Thomson scattering off electrons, and
3. an inner region (small R) in which radiation pressure dominates over gas pressure and opacity is mainly due to scattering.

The transition from the outer to the middle region occurs where the opacity by free-free absorption and electron scattering become comparable, $\kappa_{ff} \sim \kappa_{es}$. The transition to the middle to the inner region occurs where $P_{gas} \sim P_{rad}$. Notice that the middle and inner regions may not exist depending on the value of $\dot{M}$.

As long as absorption dominates the opacity the spectrum of the disk can be approximated as a blackbody. In those regions where opacity is mainly due to electron scattering the spectrum is modified with respect to that given in Eqs. (4.106) and (4.107). As shown by Shakura and Sunyaev (1973) and Novikov and Thorne (1973) the emissivity is of the form

$$I_\nu \propto \frac{x^{3/2} e^{-x/2}}{(e^{-x} - 1)^{1/2}},\tag{4.122}$$

where $x = h\nu/kT$. This is called a "modified" blackbody spectrum. The result is that the region of the spectrum $F_\nu \propto \nu^{1/3}$ (see Fig. 4.9) becomes approximately flat, $F_\nu \propto \nu^0$.

4.4.4 Accretion Disks in Strong Gravitational Fields

The general theory of accretion disks in a strong gravitational field was developed by Page and Thorne (1974). They found the vertically and time-averaged structure of a thin disk of negligible self-gravity in a space-time with a stationary, axially symmetric, asymptotically flat metric.

Page and Thorne (1974) wrote the interval ds^2 in a general manner as

$$ds^2 = -e^{2\nu}dt^2 + e^{2\psi}(d\phi - \omega dt)^2 + e^{2\mu}dr^2 + dz^2,\tag{4.123}$$

where t is the temporal coordinate and (r, ϕ, z) have the usual meaning. The functions v, ψ, ω, and μ depend only on r since the disk is, by assumption, geometrically thin. Equation (4.123) is then valid only near the plane of the disk (that lies in the equatorial plane of the black hole); otherwise corrections must be introduced to the coefficients of the metric tensor to account for their dependence on z.

A further assumption of the model is that matter moves in circular geodesic orbits in the equatorial plane of the black hole. Then, the 4-velocity $\mathbf{u}$ of a particle is approximately equal to the 4-velocity of a geodesic orbit on the equatorial plane, $\mathbf{u}(r) \approx \mathbf{w}(r)$. This assumption implies that the gravitational attraction of the black hole dominates over the pressure gradient; otherwise the paths would not be geodesics. The specific energy at infinity, angular momentum, and angular velocity of a particle in a circular geodesic orbit are

$$\tilde{E} = -w_t(r), \qquad \tilde{L} = w_\phi(r), \qquad \Omega = w^\phi/w^t, \qquad (4.124)$$

respectively.

From the relativistic equations for the conservation of mass, energy, and momentum, Page and Thorne (1974) derived the expressions for the time-averaged mass accretion rate $\dot{M}$ and energy flux $Q^+(r)$ per unit proper time per unit area of the disk. In terms of the coefficients of the metric and the surface mass density $\Sigma(r)$ these read

$$\dot{M} = -2\pi e^{v+\psi+\mu} \Sigma u^r, \qquad (4.125)$$

$$Q^+(r) = -\frac{\dot{M}}{4\pi} e^{-(v+\psi+\mu)} \frac{\Omega_{,r}}{(\tilde{E} - \Omega \tilde{L})^2} \int_{r_{\mathrm{isco}}}^r (\tilde{E} - \Omega \tilde{L}) \tilde{L}_{,r} dr. \qquad (4.126)$$

Here r_{isco} is the radius of the innermost stable circular geodesic orbit, where the zero-torque condition was applied to obtain Eq. (4.126). As in the model of Shakura and Sunyaev (1973), the radiation is assumed to escape in the vertical direction only.

Once a specific form for the metric tensor is chosen, $\dot{M}$ and Q^+ can be explicitly calculated. The expressions for a Kerr black hole, for example, are given in Novikov and Thorne (1973) and Page and Thorne (1974).

4.4.5 Self-gravitating Accretion Disks

The study of self-gravitating disks is traditionally associated to the problem of formation of galactic spiral arms (e.g. Goldreich and Lynden-Bell 1965) and planets (e.g. Boss 1998). Self-gravity is also expected to be relevant in the outer regions of accretion disks in AGN (e.g. Shlosman and Begelman 1989; Goodman 2003) and some types of young stellar objects (e.g. Lodato and Bertin 2003b). We discuss next some relevant results of the theory and simulations on self-gravitating disks; for further reading we refer the reader to the book by Binney and Tremaine (1987) and the comprehensive review by Karas et al. (2004).

Choosing a criterion to quantify the importance of the self-gravity of an accretion disk is not absolutely straightforward. An obvious first choice is to compare the mass of the compact object M and the mass of the disk

$$M_{\rm d}(R) = 4\pi \int_0^R \Sigma\left(R'\right) R' dR'. \tag{4.127}$$

We expect that the effects of self-gravity become relevant when $M_{\rm d}$ exceeds some arbitrary fraction of M. A more reliable criterion is to compare the values of the gravitational potentials generated by the central object and the disk, that we shall denote by $\Phi_{\rm BH}$ and $\Phi_{\rm d}$, respectively. Self-gravity in some spatial direction i is important in those regions where $|\nabla_i \Phi_{\rm d}| \gg |\nabla_i \Phi_{\rm BH}|$. Notice that the gravitational acceleration on an element of fluid is $\mathbf{g} = -\nabla\Phi$.

A third way of assessing the relevance of self-gravity is by analyzing the stability of the disk against perturbations. Useful in this respect is Toomre's parameter (Toomre 1964)

$$Q \equiv \frac{a_{\rm s}\kappa}{\pi G \Sigma}, \tag{4.128}$$

where

$$\kappa^2 = \frac{2\Omega}{R} \frac{d}{dR}\left(R^2 \Omega\right) \tag{4.129}$$

is the *epicyclic frequency*. This is the frequency of oscillation of an element of fluid subject to a radial displacement from its equilibrium orbit. For Keplerian rotation in particular, $\kappa = \Omega_{\rm K}$. An infinite thin disk with negligible viscosity becomes locally unstable under axisymmetric perturbations for $Q \lesssim 1$.[9] The disk becomes globally unstable against non-axisymmetric perturbations for larger values of Toomre's parameter, approximately in the range $1 \lesssim Q \lesssim 2$.

We have seen that the half-thickness of a thin disk is $H \sim a_{\rm s}/\Omega_{\rm K}$. If we further approximate $M_{\rm d} \sim \pi R^2 \Sigma$, then

$$Q \sim \left(\frac{M_{\rm BH}}{M_{\rm d}}\right)\left(\frac{H}{R}\right). \tag{4.130}$$

This simple analysis reveals that the stability not only depends on the mass ratio (massive disks and/or systems with low-mass central objects are more unstable) but on the geometry of the disk through H/R—the outer boundaries of a thin disk are more prone to become unstable.

If the conditions for the onset of instability are satisfied, self-gravity clearly cannot be neglected. It has been argued that self-gravitating disks "auto-regulate" to a marginally stable steady state with $Q = $ constant (e.g. Bertin and Lodato 1999; Gammie 2001). The auto-regulation mechanism works essentially in this way: if

---

[9]The same criterion applies to the global stability of the disk under axisymmetric perturbations with $kR \gg 1$, where k is the wavenumber of the perturbation.

the flow cools too effectively so that $H/R \ll 1$, the disk grows unstable—to regain stability, the flow becomes turbulent and heats. Turbulence driven by gravitational instabilities may be dominant in cold and dense regions of the disk where the gas is not highly ionized, so that the development of the magnetorotational instability is suppressed.

The perturbations in the density and the velocity field of the flow induced by gravitational instabilities contribute to the transport of angular momentum outwards, thus enhancing accretion. Furthermore, the force of self-gravity also generates torques that redistribute angular momentum. Both effects may be incorporated preserving the form of Eq. (4.86) through an appropriate expression for the component of the stress tensor,

$$T_{R\phi} = \Sigma \delta v_R \delta v_\phi + \int \frac{g_R g_\phi}{4\pi G} dz. \tag{4.131}$$

The first term (usually called "Reynolds stress") represents the contribution of gravitational instabilities; $\delta \mathbf{v}$ is the fluctuation of the velocity field. The second term accounts for the torques exerted by the disk's self-gravitational force; it was deduced by Lynden-Bell and Kalnajs (1972). Once an expression for $T_{R\phi}$ is given, it is possible to associate it with an effective viscosity applying the α-prescription. This is the approach followed by Lin and Pringle (1987) and many others. Notice, nevertheless, that alternative parameterizations for the viscosity have been proposed, see for instance Duschl et al. (2000).

A fundamental property of self-gravitating disks is that their angular velocity profile differs from Keplerian. Consider the radial component of the momentum equation, Eq. (4.85). In steady state and neglecting the radial velocity, we get $\Omega^2 R \sim \nabla_R \Phi_{\mathrm{BH}} + \nabla_R \Phi_{\mathrm{d}}$ so, in general, we expect that $\Omega \neq \Omega_{\mathrm{K}}$. This provides a way of testing the influence of self-gravity in accretion disks. Indeed, sub-Keplerian rotation velocities were measured in the nucleus of the active galaxies NGC 1068 (Greenhill and Gwinn 1997) and IC 1481 (Mamyoda et al. 2009). The observations can be successfully explained when the gravity of the disk is considered, see for example the models by Lodato and Bertin (2003a) and Huré et al. (2011).

The gravitational potential of the disk satisfies Poisson's equation

$$\nabla^2 \Phi_{\mathrm{d}}(\mathbf{r}) = 4\pi G \rho(\mathbf{r}), \tag{4.132}$$

where $\rho(\mathbf{r})$ is the mass density at a point $\mathbf{r}$. Its general solution in terms of Green's function is

$$\Phi_{\mathrm{d}}(\mathbf{r}) = -G \int_{V_{\mathrm{d}}} \frac{\rho(\mathbf{r}')}{|\mathbf{r} - \mathbf{r}'|} d\mathbf{r}', \tag{4.133}$$

where the integration is carried out over the volume V_{d} occupied by the disk. The acceleration $\mathbf{g}_{\mathrm{d}} = -\nabla \Phi_{\mathrm{d}}$ due to the gravitational pull of the disk can be formally calculated differentiating Eq. (4.133),

$$\mathbf{g}_{\mathrm{d}}(\mathbf{r}) = -G \int_{V_{\mathrm{d}}} \frac{\rho(\mathbf{r}')(\mathbf{r} - \mathbf{r}')}{|\mathbf{r} - \mathbf{r}'|^3} d\mathbf{r}'. \tag{4.134}$$

Adding Eqs. (4.132) or (4.133) to those for the conservation of mass, energy, and momentum, yields a coupled system that must be solved to calculate self-consistently the structure of the disk. This is, however, no easy task.

Exact analytical solutions of Poisson's equation are known for a limited number of prescribed mass density profiles. Solutions obtained in this way are usually called "potential-density pairs" (e.g. Evans and de Zeeuw 1992; Huré et al. 2007, Schulz 2009, 2012; see also references in Trova et al. 2012). Paczyński (1978a, 1978b) solved a simplified version of the equations for the vertical structure of the disk. In Paczyński's model, matter is assumed to be distributed on an infinite slab of constant half-thickness H about the plane $z = 0$, axially symmetric and homogeneous in the radial direction. In this configuration the $z-$component of the gravitational acceleration exerted by the disk is

$$g_z^{\mathrm{d}} = 2\pi G \Sigma_z^{\mathrm{d}}, \tag{4.135}$$

where

$$\Sigma_z^{\mathrm{d}} = 2 \int_0^z \rho(R, z')dz'. \tag{4.136}$$

The condition of hydrostatic equilibrium in the z-direction then reads

$$\frac{dP}{dz} = -\rho \left[\frac{GM_{\mathrm{BH}}z}{(R^2 + z^2)^{3/2}} + 2\pi G \Sigma_z^{\mathrm{d}} \right]. \tag{4.137}$$

Two more equations complete the system, one for the surface density and a polytropic equation of state,

$$\frac{d\Sigma_z^{\mathrm{d}}}{dz} = 2\rho, \tag{4.138}$$

$$P = K\rho^{1+1/n}, \tag{4.139}$$

with constant K and n. Solutions for different values of n are given in Paczyński (1978a).

Paczyński's model is a good approximation as long as the disk can be considered homogeneous in the radial direction, i.e. as long as the radial derivatives in Poisson's equation can be neglected compared to the derivatives in the vertical direction

$$\frac{1}{R}\frac{\partial}{\partial R}\left(R\frac{\partial \Phi_{\mathrm{d}}}{\partial R} \right) \ll \frac{\partial^2 \Phi_{\mathrm{d}}}{\partial z^2}. \tag{4.140}$$

The drawback of the model is that, since it says nothing about the radial structure of the flow, the condition in Eq. (4.140) cannot actually be checked.

Several methods and their numerical implementations have been developed to solve Eq. (4.133); these include the series expansion of $1/|\mathbf{r} - \mathbf{r}'|$ in different basis (e.g. Müller and Steinmetz 1995; Cohl and Tohline 1999; Chan et al. 2006; Trova et al. 2012), Fast Fourier Transform (e.g. Binney and Tremaine 1987), and others (e.g. Huré 2004; Huré and Pierens 2005). This approach poses some difficulties, mainly the singularity of the integrand at $\mathbf{r}' \to \mathbf{r}$ (although there exist some tricks to avoid it, see e.g. Huré and Pierens 2005; Huré and Dieckmann 2012). Aside this issue, solving Eq. (4.133) is numerically very accurate although computationally expensive. The calculation of the potential of the disk from Poisson's differential equation (e.g. Bodo and Curir 1992; Störzer 1993) is computationally less time-consuming.

Nowadays the problem of calculating the structure of self-gravitating disk is directly tackled using numerical simulations. The results indicate that the structure of the disk emerges from an interplay between the effects of cooling and self-gravity.

Performing two-dimensional simulations, Gammie (2001) found that if the cooling timescale of the flow is $t_{\mathrm{cool}} = \beta \Omega^{-1}$ with $\beta \lesssim 3$, gravitational instabilities lead to fragmentation and collapse of parts of the disk. In the opposite limit, if the typical cooling time of the gas is $t_{\mathrm{cool}} \gtrsim 3\Omega^{-1}$, the disks settles to a state characterized by fluctuations in the surface mass density but stable against fragmentation. Recent 3D SPH simulations by Rice et al. (2012) appear to corroborate this result, although for slightly larger values of the critical β. Figure 4.10 shows the surface density profiles for different values of β in simulations with 5×10^5 particles. Fragmentation is observed for $\beta \lesssim 6 - 7$; notice in all cases the global spiral structure of the disk. Depending on the context, bound objects formed by this mechanism may be the seeds of planets, brown dwarfs, and stars.

4.5 Advection-Dominated Accretion Flows

The geometrically thin, optically thick accretion disk model allows several generalizations. In particular, the assumption that all the heat generated by viscosity is radiated away does not hold for all accretion rates. As we shall see, under some conditions the radial velocity of the accretion flow becomes large and the heat cannot be transformed into radiation and emitted fast enough. A significant fraction of the heat is stored as kinetic energy in the flow and advected onto the accretor. At the same time the disk "inflates", so that the thin disk assumption breaks down.

The existence of another accretion regime different from a thin disk is required to account for the observational spectrum of X-ray binaries. In some of these systems the emission extends into the hard X-rays ($\gtrsim 100$ MeV), well beyond the frequencies predicted by any physically reasonable thin disk model. Besides, the spectral shape of the hard X-ray component is not that of a blackbody but a power-law, what hints to a non-thermal radiative mechanism. The idea of a geometrically thick, hot "corona" with electron temperature $T_e \sim 10^9$–10^{10} K was first suggested by Bisnovatyi-Kogan and Blinnikov (1976, 1977) to explain the variability and the X-ray spectrum of the black hole X-ray binary Cygnus X-1. These authors argued that the Compton up-scattering of photons from the disk off the electrons in the corona would produce a power-law spectrum like the observed.

A second piece of evidence suggesting the existence of radiatively inefficient accretion regimes comes from the observation of low-luminosity galactic nuclei.

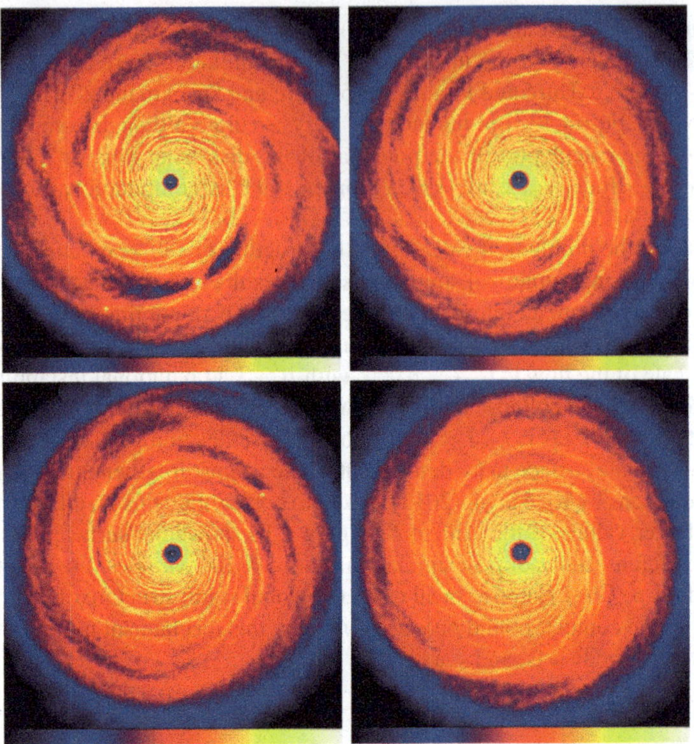

Fig. 4.10 Surface density profile of a self-gravitating disk with a characteristic cooling time $t_{\rm cool} = \beta \Omega^{-1}$. *Top left*: $\beta = 4$. *Top right*: $\beta = 5$. *Bottom left*: $\beta = 6$. *Bottom left*: $\beta = 7$. From Rice et al. (2012). Reproduced by permission of Oxford University Press on behalf of the Royal Astronomical Society

These include the compact source at the Galactic Center, Sgr A*, that hosts a black hole of $\sim 4 \times 10^6 M_\odot$. The value of the accretion rate in Sgr A* is estimated to be about 10^{-4}–$10^{-3} M_\odot$ yr^{-1}. The luminosity predicted by the thin disk model for such M and $\dot{M}$ is larger than 10^{40} erg s^{-1}, whereas the observed luminosity of Sgr A* is only about $\sim 10^{37}$ erg s^{-1}. Furthermore, the shape of the spectrum (approximately flat from radio to X-rays) is not consistent with that of an optically thick, geometrically thin disk.

Besides that of Shakura and Sunyaev (1973), we can name other solutions to the problem of accretion onto a mass M including the effect of viscosity and rotation. One is the introduced by Shapiro et al. (1976). They developed a model where the only source of pressure is the gas but allowed the temperature of ions and electrons to be different, thus

$$P = \frac{\rho k}{m_p}(T_e + T_i).\tag{4.141}$$

It is assumed that the mechanism of viscous dissipation preferably heats the ions, that ions and electrons are only weakly coupled (they do not thermalize on relevant timescales), and that electrons cool much more efficiently. Under these conditions the temperature of ions naturally turns out to be much higher than that of electrons, $T_i \approx 10^{11}$ K $\gg T_e \approx 10^{8-9}$ K. Notice that due to the large value of T_i the flow is not strictly geometrically thin but has a height scale $H/R \sim 0.2$. The remaining two suppositions of the model are that the plasma is optically thin to absorption and its main radiative cooling process is inverse Compton scattering off electrons. The emerging spectrum is a power-law in the energy range of the X-rays and soft gamma rays. Shapiro et al. (1976) suggested that a two-temperature accretion flow like this might exist in the inner region of a standard thin disk. The solution, however, is thermally unstable (e.g. Pringle 1976) and therefore unlikely to describe any real astrophysical flow.

In the models of Shakura and Sunyaev (1973) and Shapiro et al. (1976) all the energy dissipated by viscosity is radiated. It is possible to construct other solutions with a different energy balance, in particular allowing for a fraction of the dissipated energy to remain in the plasma and be advected. These types of flows are called *advection-dominated accretion flows* (ADAFs).

There are two types of advection-dominated accretion flows. *Optically thick* ADAFs develop at very high accretion rates, typically larger than the Eddington value. In this limit the radiation gets trapped in the accretion flow and advected because the optical depth is very large. The properties of these ADAFs have been studied, for example, by Begelman (1978), Begelman and Meier (1982), and Abramowicz et al. (1988). *Optically thin* ADAFs occur in the opposite limit of sufficiently low accretion rates. In this regime the cooling timescale of the flow is longer than the accretion timescale, resulting again in a significant fraction of the energy being advected. The theory of optically thin ADAFs was pioneered by Ichimaru (1977), and later developed in detail by Narayan and Yi (1994, 1995a, 1995b), Abramowicz et al. (1995), Chen (1995), and Chen et al. (1995). It has received much attention because of its success in explaining the radiative properties of black hole binaries in certain spectral states and those of low-luminosity active galactic nuclei. We shall focus our discussion on ADAFs in the optically thin regime, with emphasis on two-temperature models.

4.5.1 *The Equations of an ADAF*

The equations that describe an ADAF are the usual hydrodynamic equations for a viscous accretion flow. It is convenient now to work in spherical coordinates (r, θ, ϕ). We shall assume that the system has azimuthal symmetry, so that $\partial/\partial\phi = 0$, and has reached the steady state.

The continuity equation then reads

$$\frac{1}{r^2}\frac{\partial}{\partial r}\left(r^2 \rho v_r\right) + \frac{1}{r\sin\theta}\frac{\partial}{\partial\theta}(\sin\theta \rho v_\theta) = 0, \tag{4.142}$$

where, as usual, ρ is the mass density, and v_r and v_θ are the radial and polar components of the velocity of the flow, respectively.

As we shall show below there is a possible solution with $v_\theta = 0$. In that case, the three components of the momentum conservation equation become

$$\rho\left(v_r\frac{\partial v_r}{\partial r} - \frac{v_\phi^2}{r}\right) = -\frac{GM\rho}{r^2} - \frac{\partial P}{\partial r} + \frac{\partial}{\partial r}\left[2\nu\rho\frac{\partial v_r}{\partial r} - \frac{2}{3}\nu\rho\left(\frac{2v_r}{r} + \frac{\partial v_r}{\partial r}\right)\right]$$
$$+ \frac{1}{r}\frac{\partial}{\partial\theta}\left(\frac{\nu\rho}{r}\frac{\partial v_r}{\partial\theta}\right) + \frac{\nu\rho}{r}\left[4r\frac{\partial}{\partial r}\left(\frac{v_r}{r}\right) + \frac{\cot\theta}{r}\frac{\partial v_r}{\partial\theta}\right],$$

$$\tag{4.143}$$

$$\rho\left(-\frac{\cot\theta}{r}v_\phi^2\right) = -\frac{1}{r}\frac{\partial P}{\partial\theta} + \frac{\partial}{\partial r}\left(\frac{\nu\rho}{r}\frac{\partial v_r}{\partial\theta}\right) +$$
$$+ \frac{1}{r}\frac{\partial}{\partial\theta}\left[\frac{2\nu\rho}{r}v_r - \frac{2\nu\rho}{3}\left(\frac{2v_r}{r} + \frac{\partial v_r}{\partial r}\right)\right] + \frac{3\nu\rho}{r^2}\frac{\partial v_r}{\partial\theta},$$

$$\tag{4.144}$$

$$\rho\left(v_r\frac{\partial v_\phi}{\partial r} + \frac{v_\phi v_r}{r}\right) = \frac{\partial}{\partial r}\left[\nu\rho r\frac{\partial}{\partial r}\left(\frac{v_\phi}{r}\right)\right] + \frac{1}{r}\frac{\partial}{\partial\theta}\left[\frac{\nu\rho\sin\theta}{r}\frac{\partial}{\partial\theta}\left(\frac{v_\phi}{\sin\theta}\right)\right]$$
$$+ \frac{\nu\rho}{r}\left[3r\frac{\partial}{\partial r}\left(\frac{v_\phi}{r}\right) + \frac{2\cot\theta\sin\theta}{r}\frac{\partial}{\partial\theta}\left(\frac{v_\phi}{\sin\theta}\right)\right]. \tag{4.145}$$

In writing Eqs. (4.143) to (4.145) we have neglected the self-gravity of the fluid, and we have used the standard form of the stress tensor for a viscous flow given in Eq. (4.5).

These four equations must be complemented with one for the energy balance. Since the form of this equation is the fundamental difference between an ADAF and a thin disk, we shall develop it in certain detail. Let T be the temperature and s the entropy per unit mass per unit volume of the gas; then

$$T\rho\frac{ds}{dt} = T\rho\left[\frac{\partial s}{\partial t} + (\mathbf{v}\cdot\boldsymbol{\nabla})s\right] = q^+ - q^-. \tag{4.146}$$

Here q^+ is the rate of heating due to viscosity per unit volume per unit mass and q^- the rate of energy loss due to radiation per unit volume per unit mass. In a thin disk $q^+ = q^-$, but this is generalized in an ADAF to account for advection. It is convenient to write

$$q^- = (1 - f)q^+, \tag{4.147}$$

so that

$$T\rho\frac{ds}{dt} = fq^+. \tag{4.148}$$

The parameter f measures the degree up to which the flow is advection-dominated since, by definition,

$$f = \frac{q^+ - q^-}{q^+} \equiv \frac{q_{\mathrm{adv}}}{q^+}. \tag{4.149}$$

The regime $f \ll 1$ corresponds to $q^+ \approx q^- \gg q_{\mathrm{adv}}$; this describes standard thin disks and the two-temperature disks of Shapiro et al. (1976). The case $|f| \gg 1$ occurs when $-q_{\mathrm{adv}} \sim q^- \gg q^+$. These are flows where the viscous dissipation is negligible and all the entropy of the gas is converted into radiation; an example is Bondi accretion. Finally, the case $f \approx 1$ where $q_{\mathrm{adv}} \approx q^+ \gg q^-$ describes an ADAF—cooling is inefficient and most of the heat is advected with the matter.

It is useful to write Eq. (4.146) in terms of the energy ε per unit volume of the gas. According to the first law of thermodynamics

$$d\varepsilon = T ds - P dV, \tag{4.150}$$

where $V = 1/\rho$ is the specific volume. Then

$$\rho \frac{d\varepsilon}{dt} - \frac{P}{\rho} \frac{d\rho}{dt} = f q^+. \tag{4.151}$$

In steady state and using Eq. (4.11) for q^+, the energy equation finally reads

$$\rho \left(v_r \frac{\partial \varepsilon}{\partial r} - \frac{P}{\rho^2} v_r \frac{\partial \rho}{\partial r} \right) = -\frac{2 f v \rho}{3} \left[\frac{1}{r^2} \frac{\partial}{\partial r} \left(r^2 v_r \right) \right]^2 + 2 f v \rho \left\{ \left(\frac{\partial v_r}{\partial r} \right)^2 + 2 \left(\frac{v_r}{r} \right)^2 \right.$$

$$+ \frac{1}{2} \left(\frac{1}{r} \frac{\partial v_r}{\partial \theta} \right)^2 + \frac{1}{2} \left[r \frac{\partial}{\partial r} \left(\frac{v_r}{r} \right) \right]^2$$

$$\left. + \frac{1}{2} \left[\frac{\sin \theta}{r} \frac{\partial}{\partial \theta} \left(\frac{v_\phi}{\sin \theta} \right) \right]^2 \right\}. \tag{4.152}$$

The set of Eqs. (4.142), (4.145), and (4.152) admits a self-similar solution as shown by Narayan and Yi (1995a). The velocity, the speed of sound, and the mass density are given by

$$v_r = v_{\mathrm{ff}}(r) v(\theta), \tag{4.153}$$

$$v_\theta = 0, \tag{4.154}$$

$$v_\phi = r \Omega_{\mathrm{K}}(r) \Omega(\theta), \tag{4.155}$$

$$a_s = \sqrt{P/\rho} = r \Omega_{\mathrm{K}}(r) a_s(\theta), \tag{4.156}$$

$$\rho = r^{-3/2} \rho(\theta), \tag{4.157}$$

where v_{ff} is the free-fall velocity

$$v_{\mathrm{ff}} = \left(\frac{GM}{r} \right)^{1/2}. \tag{4.158}$$

The dimensionless functions $v(\theta)$, $\Omega(\theta)$, $a_s(\theta)$, and $\rho(\theta)$ fix the angular dependence of the variables. Notice that the expression for the mass density follows from v_r and the constancy of the accretion rate

$$\dot{M} = -2\pi \int_\pi^0 \rho r^2 v_r \sin(\theta)d\theta. \tag{4.159}$$

Using that $v_r \propto r^{-1/2}$ immediately yields $\rho \propto r^{-3/2}$. Then $r^2 v_r \rho$ is independent of r and from Eq. (4.142) one concludes that $v_\theta = 0$ is, indeed, a possible solution.

Inserting Eqs. (4.153) to (4.157) in (4.145) and (4.152) (the continuity equation is automatically satisfied) we get a system of coupled algebraic equations in the variable θ for the functions $v(\theta)$, $\Omega(\theta)$, $a_s(\theta)$, and $\rho(\theta)$. To solve it, we need to provide an expression for the viscosity. Narayan and Yi (1995a) adopted the α-prescription

$$\nu = \frac{\alpha a_s^2}{\Omega_K} \tag{4.160}$$

with α a constant.

With this expression for the viscosity, the set of equations to be solved is

$$-\frac{1}{2}v^2 - \sin^2\theta\,\Omega^2 = -1 + a_s^2\left(\frac{5}{2} - \alpha v + \alpha\cot\theta\frac{dv}{d\theta}\right) + \frac{1}{\rho}\frac{d}{d\theta}\left(\alpha\rho a_s^2\frac{dv}{d\theta}\right),$$

$$-\cos\theta\sin\theta\,\Omega^2 = -\frac{1}{\rho}\frac{d}{d\theta}\left(\rho a_s^2\right) + \frac{\alpha a_s^2}{2}\frac{dv}{d\theta} + \frac{1}{\rho}\frac{d}{d\theta}\left(\alpha a_s^2\rho v\right),$$

$$\frac{1}{2}\sin\theta\, v\Omega = -\frac{3}{4}\sin\theta\, a_s^2\Omega + \frac{1}{\rho}\frac{d}{d\theta}\left(\alpha\sin\theta\rho a_s^2\frac{d\Omega}{d\theta}\right) + 2\alpha\cos\theta a_s^2\frac{d\Omega}{d\theta},$$

$$-\frac{3}{2}\frac{\varepsilon' v}{\alpha} = 3v^2 + \frac{9}{4}\sin^2\theta\,\Omega^2 + \sin^2\theta\left(\frac{d\Omega}{d\theta}\right)^2 + \left(\frac{dv}{d\theta}\right)^2. \tag{4.161}$$

Here we introduced the parameter

$$\varepsilon' \equiv \frac{\varepsilon}{f} = \frac{1}{f}\left(\frac{5/3 - \gamma}{\gamma - 1}\right), \tag{4.162}$$

where γ is the ratio of the specific heats of the gas. To completely determine the solution boundary conditions must be chosen. One is the value of $\dot{M}$; the rest are imposed at $\theta = 0$ and $\theta = \pi/2$. At $\theta = \pi/2$ we demand reflection symmetry with respect to the equatorial plane, then

$$\frac{dv}{d\theta} = \frac{d\Omega}{d\theta} = \frac{da_s}{d\theta} = \frac{d\rho}{d\theta} = 0. \tag{4.163}$$

At the axis $\theta = 0$ we demand that the solutions are smooth and non-singular, so

$$\frac{dv}{d\theta} = \frac{d\Omega}{d\theta} = \frac{da_s}{d\theta} = \frac{d\rho}{d\theta} = 0, \quad v(\theta = 0) = 0. \tag{4.164}$$

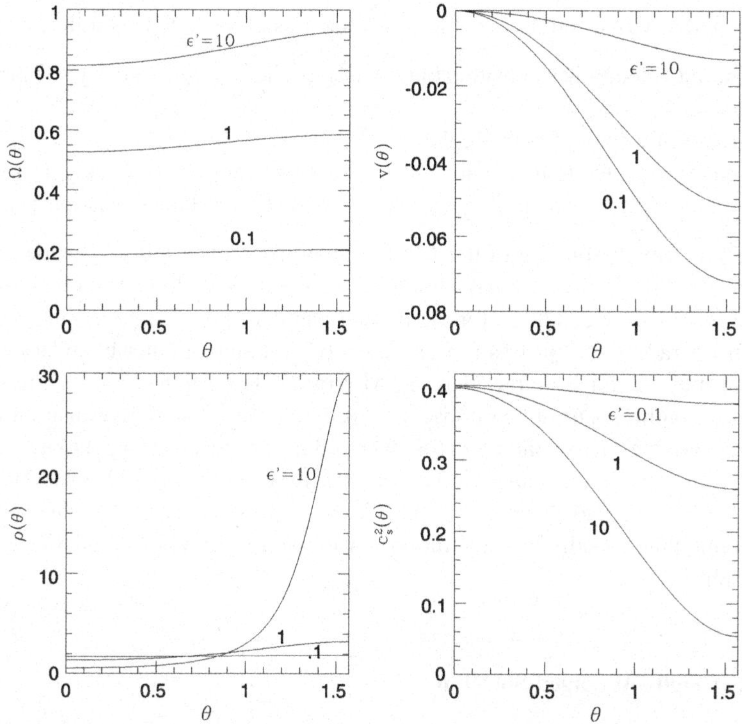

Fig. 4.11 Angular velocity Ω, radial velocity v, mass density ρ, and speed of sound c_s at fixed radius as a function of the polar angle θ, for $\alpha = 0.1$ and several values of the parameter ε'. From Narayan et al. (1998b)

The condition $v(\theta = 0) = 0$ follows directly from the third of Eqs. (4.161).[10] Not all these boundary conditions are independent, but only a subset of them must be chosen.

Figure 4.11 shows the solutions of Eqs. (4.161) as a function of the polar angle for different values of $\varepsilon' = 1, 0.1, 10$. At fixed γ, decreasing ε' implies increasing f, so that a smaller value of ε' correspond to a more advection-dominated flow.

For $\varepsilon' = 0.1, 1$ the angular velocity is almost independent of θ. It follows from Eqs. (4.153) and (4.155) that the tangential velocity is sub-Keplerian whereas the radial velocity is not at all negligible—for $\varepsilon' \ll 1$ we have $v_r \sim 0.1 v_{\mathrm{ff}}$ at the equator. Furthermore, $\rho(\theta)$ is almost constant. These three properties of the flow totally distinguish it from a thin disk. For $\varepsilon' = 10$, however, $\Omega(\theta) \sim 1$ so that $v_\phi \sim r\Omega_{\mathrm{K}}$, and the mass density sharply peaks at the equator. These are exactly two expected characteristics of a thin disk. In any case the radial velocity is maximum (in absolute value) at $\theta = \pi/2$, so most of the accretion proceeds along the equatorial plane.

---

[10]There is another possible solution with $v(\theta = 0) = -\varepsilon'/2\alpha$. This is a generalization of the spherically symmetric Bondi accretion to viscous flows.

In general, we can summarize the characteristics of an ADAF as follows:

1. the radial velocity is a considerable fraction of the free-fall velocity so accretion is fast,
2. the rotation velocity is sub-Keplerian,
3. the gas is expected to be hot since it has no time to cool before being accreted,
4. the typical height scale is $H \sim a_s/\Omega_K \sim r$—the flow is then quasi-spherical.

The self-similar solution of the ADAF equations is expected to provide a good description of the flow far from the boundaries. A complete global solution must satisfy some boundary condition for the flow of angular momentum onto the accretor at the inner radius R_{in} (just as in thin disk models), and an outer boundary condition that may be, for example, that the ADAF matches a thin disk at some radius R_{tr}. Extra conditions must be imposed, as well, to ensure a smooth transition of the physical variables across the radius R_s where the flow becomes supersonic. Global ADAF solutions were calculated, for example, by Chen et al. (1997) and Narayan et al. (1997a); see also Yuan et al. (2008) for a simplified treatment of the problem. They found that the self-similar solution is accurate far from the boundaries and the sonic radius.

4.5.1.1 Height-Averaged Solution

One way to simplify the full set of hydrodynamic equations of the previous section is to average them over z, just as it is done in thin disk models. You can see Narayan and Yi (1994) for the complete expressions of the height-averaged equations; here we only quote the results for reference. As a function of the cylindrical radius R they read

$$v_R = -\frac{1}{3\alpha}(5 + 2\varepsilon')g(\alpha, \varepsilon')v_{ff} \approx -\frac{3\alpha}{(5 + 2\varepsilon')}v_{ff},$$

$$\Omega = \left[\frac{2\varepsilon'(5 + 2\varepsilon')}{9\alpha^2}g(\alpha, \varepsilon')\right]^{1/2}\Omega_K \approx \left(\frac{2\varepsilon'}{5 + 2\varepsilon'}\right)^{1/2}\Omega_K, \qquad (4.165)$$

$$a_s^2 = \frac{2(5 + 2\varepsilon')}{9\alpha^2}g(\alpha, \varepsilon')v_{ff}^2 \approx \frac{2\varepsilon'}{5 + 2\varepsilon'}v_{ff}^2,$$

where

$$g(\alpha, \varepsilon') \equiv \left[1 + \frac{18\alpha^2}{(5 + 2\varepsilon')^2}\right]^{1/2} - 1. \qquad (4.166)$$

The approximations are valid in the limit $\alpha^2 \ll 1$. The height-averaged mass density follows from the radial velocity and the accretion rate

$$\dot{M} = -4\pi\rho R H v_R. \qquad (4.167)$$

Narayan and Yi (1995a) showed that the solutions of the exact equations of Sect. 4.5.1, when averaged over θ, differ no more than $\sim$20 % from those of the height-averaged equations. These suggests that the height-integrated equations must be interpreted not as averages over z, but as averages over θ at fixed r. Notice, however, that the accuracy of the approximation has been proofed specifically for the self-similar solution and it is not a general result.

4.5.2 Two-Temperature ADAFs

Two-temperature ADAF-like flows were described for the first time by Ichimaru (1977), and later extensively studied by Narayan and Yi (1994, 1995a, 1995b), Abramowicz et al. (1995), Chen (1995), and Chen et al. (1995) among others.

Two-temperature *optically thin* ADAF models are based on a series of assumptions on the thermodynamics of the gas, the coupling between ions and electrons, and the mechanisms of cooling. We can summarize them as follows.

First, the total pressure $P = \rho a_{\rm s}^2$ is considered as the sum of the pressure of a two-temperature gas and the magnetic pressure

$$P = P_{\rm gas} + P_{\rm mag} = \frac{\rho k}{m_p} \left(\frac{T_e}{\mu_e} + \frac{T_i}{\mu_i} \right) + \frac{B^2}{8\pi}. \tag{4.168}$$

Here $\mu_{i,e}$ are the mean molecular weights of ions an electrons. Notice that radiation pressure has been neglected. This is justified since in the optically thin ADAF model we seek to construct the flow is radiatively inefficiently. This is not true in ADAFs at high accretion rates. The magnetic and the gas pressure are assumed to be related to the total pressure as

$$P_{\rm mag} = (1 - \beta)P, \qquad P_{\rm gas} = \beta P, \tag{4.169}$$

with β a constant. Typical ADAF models fix $\beta \sim 0.5$.

The second fundamental hypothesis is that the heat generated by viscosity is preferably transferred to ions. Only a small fraction $\delta \ll 1$ heats the electrons; a value of the order of $\delta \sim m_e/m_p \sim 10^{-3}$ is usually adopted. This naturally results in $T_i \gg T_e$. Notice that depending on the mechanism of energy dissipation this hypothesis might not be true at all. Since, however, electrons cool much more efficiently than ions we still can expect that $T_i > T_e$ even if $\delta \sim 1$.

It is further supposed that the only process of coupling between ions and electrons is Coulomb scattering. The rate of energy transfer per unit volume from ions to electrons $q_{ie}(T_i, T_e)$ is given, for example, in Stepney and Guilbert (1983).

Assuming that electrons cool completely, so that $q_e^- = q_{ie}$, we can now write the equation for the energy balance in an ADAF. In steady state all the energy deposited in the ions must be transferred to electrons or advected so, for $\delta = 0$,

$$q^+ = q_{\rm adv} + q_{ie} = f q^+ + q_{ie}. \tag{4.170}$$

At the same time, all the energy transferred from ions to electrons is radiated, and since electron cooling is the only source of radiation we have that

$$q_{ie} = q_e^- = q^- . \tag{4.171}$$

The main difference between a two-temperature ADAF an other accretion regimes lies in the last two equations.

For fixed M, $\dot{M}$, α, β, and r, Eqs. (4.169) to (4.171) may be solved to find T_i, T_e, and f. The rest of the parameters (ρ, a_s, v, etc.) involved in the expressions for q^-, q_e^-, q_{ie}, etc., are obtained from the equations in Sect. 4.5.1, or from their averaged solutions, Eqs. (4.165) and (4.167).

The temperature of ions behaves roughly as $T_i \approx 10^{12}\beta\tilde{r}^{-1}$ K, where $\tilde{r}$ is the radius in units of the Schwarzschild radius of the accretor. The electron temperature also increases inwards for large radius, but then saturates to $T_e \approx 10^9$–10^{10} K in the region $\tilde{r} \lesssim 10^2$–10^3.

An interesting result is that, at fixed r and $\dot{M}$, one or more solutions exist depending on the value of f (Narayan and Yi 1995b). For $\dot{M} \leq \dot{M}_{\rm crit}$ there is one solution with $f \approx 1$; this describes an advection-dominated flow. There is a second solution with $f \ll 1$ that exists only for $\dot{M} \geq \dot{M}'_{\rm crit}$; this is a cooling-dominated flow, i.e. a thin disk. For $\dot{M}'_{\rm crit} \leq \dot{M} \leq \dot{M}_{\rm crit}$ both solutions are possible. Furthermore, there exists a third branch for intermediate values of $f < 1$. This solution is unstable and corresponds to the two-temperature disk model of Shapiro et al. (1976). For accretion rates $\dot{M} > \dot{M}_{\rm crit}$ only the thin disk solution remains, and only the ADAF solutions exists for $\dot{M} < \dot{M}'_{\rm crit}$.

The maximum accretion rate $\dot{M}_{\rm crit}$ for an ADAF may be estimated as the value of $\dot{M}$ above which the energy dissipated by viscous forces can be balanced only by radiative losses without the need of advection. A detailed calculation of $\dot{M}_{\rm crit}$ was performed by Narayan and Yi (1995b). Defining $\dot{m} = \dot{M}/\dot{M}_{\rm Edd}$,[11] they found that $\dot{m}_{\rm crit} \sim 0.3\alpha^2$ for $\tilde{r} \lesssim 10^3$, whereas $\dot{m}_{\rm crit} \sim 0.3\alpha^2(\tilde{r}/10^3)^{-1/2}$ for $\tilde{r} \gtrsim 10^3$.

4.5.3 *The Radiative Spectrum of an ADAF and Its Applications*

The radiative spectrum of an ADAF results from the action of different cooling processes as seen in Fig. 4.12. For electrons the most relevant mechanisms are synchrotron radiation, Bremsstrahlung, and inverse Compton scattering. The total cooling rate of electrons is given by the added contribution of all these processes

$$q_e^- = q_{\rm Br}^- + q_{\rm synchr}^- + q_{\rm IC}^- . \tag{4.172}$$

---

[11]Narayan and Yi (1995b) adopted a value of the Eddington accretion rate of $\dot{M}_{\rm Edd} = L_{\rm Edd}/0.1c^2 \approx 1.39 \times 10^{18} M/M_\odot$ g s^{-1}.

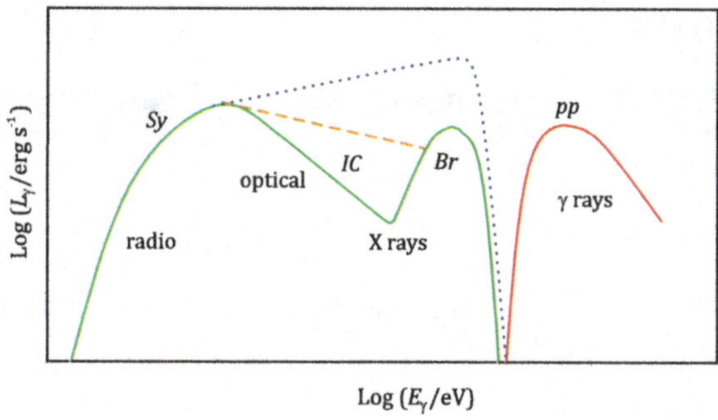

Fig. 4.12 A generic radiative spectrum from an optically thin advection-dominated accretion flow. *Sy*, *IC*, *Br*, and *pp* mark the regions of the spectrum dominated by synchrotron, inverse Compton, Bremsstrahlung, and decay of neutral pions created in proton-proton collisions, respectively. The solid line corresponds to a model with low $\dot{M}$, the dashed line to an intermediate $\dot{M}$, and the dotted line to a model with a high accretion rate near the critical value above which the ADAF solution does not exist. Adapted from Narayan et al. (1998b)

The target photons for inverse Compton scattering are the Bremsstrahlung and synchrotron radiation fields, plus some possible external source of radiation (the radiation of the donor star in a binary, for example). Then

$$q_{\mathrm{IC}}^- = q_{\mathrm{IC,Br}}^- + q_{\mathrm{IC,\,synchr}}^- + q_{\mathrm{IC,ext}}^-. \tag{4.173}$$

As first pointed out by Mahadevan et al. (1977), ions may also contribute to the radiative spectrum of the accretion flow mainly through inelastic collisions of two protons. This process creates neutral pions that decay into two gamma-rays, $\pi^0 \to \gamma + \gamma$. If the protons are thermal the gamma-ray spectrum is basically a peak at $E_\gamma \sim m_{\pi^0} c^2/2 \sim 70$ MeV. If, on the other hand, protons are non-thermal, the shape of the gamma-ray spectrum depends on the details of the proton distribution. A power-law energy distribution of protons, for example, yields a power-law gamma-ray spectrum. We shall discuss non-thermal radiative processes in more detail in Sect. 5.9.

Accretion flows around supermassive black holes, specially in low-luminosity galactic nuclei, have been successfully modeled as ADAFs; see Yuang (2007) for a review. Specific applications include a variety of systems such as Sgr A* in the Galactic Center (e.g. Narayan et al. 1995, 1998a, 1998b; Yuan et al. 2002, 2003, 2004), nuclei of giant elliptical galaxies such as M87 (e.g. Fabian and Rees 1995; Reynolds et al. 1996; di Matteo and Fabian 1997), Fanaroff-Riley type I galaxies (e.g. Reynolds et al. 1996; Wu et al. 2007), and low-ionization nuclear emission-line galaxies (e.g. Lasota et al. 1996; Quataert et al. 1999).

Thin disk + ADAF models have been applied to explain the radiative spectrum of soft X-ray transients in quiescence. These are binary systems formed by an accreting black hole and a low-mass donor star, that after spending long periods in a very low luminosity state (quiescence) suddenly enter in outburst. Classical works in this field are those of Narayan et al. (1996, 1997b) for the sources A06200-00 and V404 Cyg.

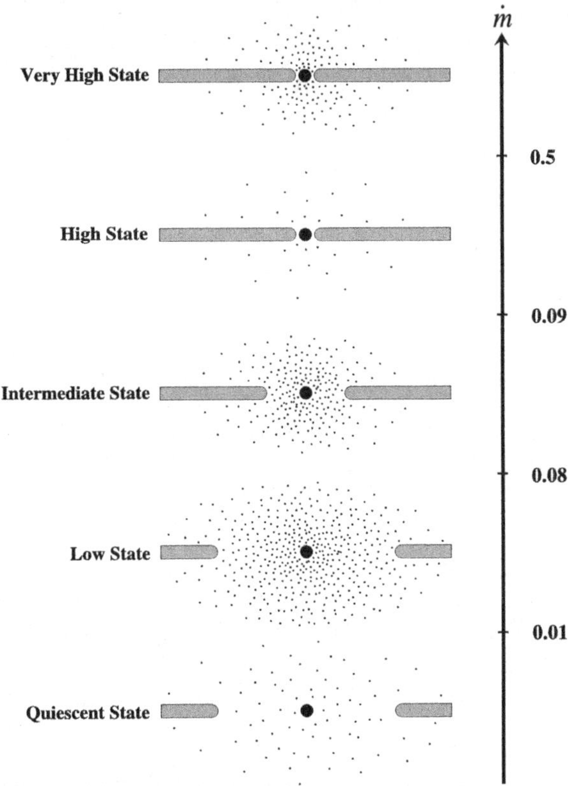

Fig. 4.13 Geometry of the accretion flow around an accreting stellar-mass black hole in different spectral states as a function of the accretion rate in Eddington units, $\dot{m} = \dot{M}/\dot{M}_{\mathrm{Edd}}$. From Esin et al. (1997). Reproduced by permission of the AAS

ADAF models have been applied as well to reproduce the spectrum of black hole X-ray binaries. These sources go through a series of states characterized by different temporal and spectral properties of the X-ray spectrum. The two main spectral states are the so-called *high-soft* and *low-hard* states. Very roughly, the spectrum in the high-soft state is dominated by a thermal component that peaks in the soft X-rays, and in the low-hard state by a non-thermal power-law spectrum that extends up to the hard X-rays/soft gamma rays.

The transition between states is attributed to the change in the accretion regime from an ADAF to a thin disk due to the variation of the accretion rate (e.g. Narayan 1996; Esin et al. 1997, 1998, 2001). The cycle is sketched in Fig. 4.13. For low accretion rates the system is in quiescence/low-hard state. The surroundings of the compact object are filled by a quasi-spherical "corona" of hot gas with the properties of an ADAF. There exists a thin accretion disk quite detached from the compact object; the transition radius between the corona and the disk depends on $\dot{M}$. During this phase the emission is dominated by the radiation from the corona, that is a power-law spectrum mainly due to Compton scattering. As the accretion rate increases the source goes through intermediate states of mixed characteristics. Finally, $\dot{M}$ exceeds the critical value above which the ADAF solution cannot exist. The corona disappears and the inner radius of the disk approaches the compact ob-

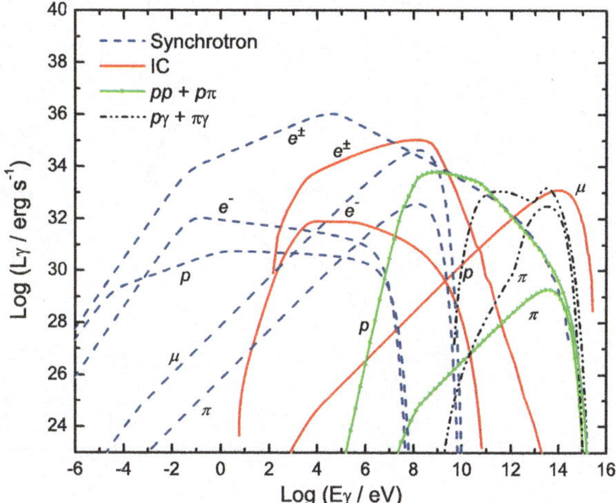

Fig. 4.14 Non-thermal contributions to the radiative spectrum in a corona model with a relativistic proton-to-electron power ratio of 100. From Romero et al. (2010), reproduced with permission ©ESO

ject. The source enters the high-soft state, when the spectrum is dominated by the blackbody emission from the disk.

Related to self-consistent ADAF + thin disk models are the "disk+corona" models (see Poutanen 1998 for a review). In these models the corona is simply added as a separate component of the accretion flow besides the accretion disk. Some geometry for the corona is chosen—it may be a quasi-spherical cloud that partially overlaps with the disk or some other configuration. The characteristics of the corona (the electron temperature T_e and the optical depth τ) and its radiative spectrum must be found solving the coupled kinetic equations for particles and radiation. Several effects add complexity to the model. For example, if T_e is large, the high-energy tail of the electron distribution is expected to deviate from a Maxwellian; this population of non-thermal particles can produce significant radiation above $m_e c^2 \sim 500$ keV. Furthermore, if a population of non-thermal very energetic protons develops, the corona might be a site of significant gamma-ray emission. Recently, Romero et al. (2010) and Vieyro and Romero (2012) have considered the effect of non-thermal particle populations (electrons and protons) in the corona around a stellar-mass black hole. In such a case relativistic protons cool by synchrotron, photo-pair, photo-meson, and inelastic proton-proton interactions. The full non-thermal spectrum can be quite complicated. A generic spectral energy distribution obtained with such model is shown in Fig. 4.14. Figure 4.15 shows an application to the famous source Cygnus X-1 in low-hard state. Notice that the soft tail observed in the spectrum above 1 MeV is explained as non-thermal radiation from the corona. The results of some models predict that the emission from the corona might be also detectable at very high-energies. The absorption gap between $E_\gamma \sim 100$ MeV and $E_\gamma \sim 100$ GeV arises by creation of electron-positron pairs via two-photon annihilation.

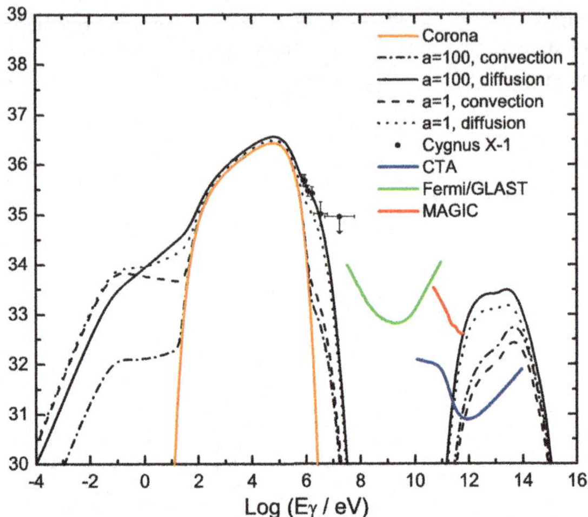

Fig. 4.15 Non-thermal contributions to the radiative spectrum in a corona model for Cygnus X-1. Cases are considered with different relativistic proton-to-electron power in the corona, given by the parameter "a". The curved labeled "corona" is the characteristic Comptonization power-law spectrum of the corona. The sensitivity curves of the gamma-ray satellite *Fermi* and the Cherenkov telescopes MAGIC and CTA (to operate in coming years) are shown. Figure from Romero et al. (2010), reproduced with permission ©ESO; observational data from McConnell et al. (2000)

4.5.4 *Other Radiatively Inefficient Accretion Regimes*

Several radiatively inefficient accretion flows (RIAFs) different from ADAFs have been introduced over the years. One of such models is the *advection-dominated inflow outflow solution* (ADIOS) developed by Blandford and Begelman (1999, 2004); see also Begelman (2012) for a recent reformulation. The key characteristic of an ADIOS is that it allows the removal of matter in the form of a wind at all radii, that carries away energy and angular momentum. The idea is based on the observation that in an ADAF the Bernoulli constant of the gas is positive,

$$\mathrm{Be} = \frac{1}{2}v^2 - \frac{GM}{r} + \frac{5}{2}a_{\mathrm{s}}^2 > 0. \qquad (4.174)$$

This means that if the flow somehow reverted its direction of motion it could escape to infinity with positive energy. Narayan and Yi (1995a) already suggested that this feature of ADAFs make them prone to launch outflows.

The basic hydrodynamic equations of an ADIOS are the same than those of an ADAF, but the accretion rate is not a constant. Blandford and Begelman (1999) parameterized it as

$$\dot{M} \propto r^p \qquad (4.175)$$

with $0 \leq p < 1$. This gives a mass density profile on the equator that scales as $\rho \propto r^{-3/2+p}$. The ADAF solution is recovered when $p = 0$. In order to yield Be > 0, Blandford and Begelman (1999) showed that $p \approx 0.5$–1.

A second feature of ADAFs, already pointed out by Narayan and Yi (1994), is that they are convectively unstable for low values of the parameter α. Convection may substantially modify the structure of the flow since it provides a mechanism to transport angular momentum inwards, as opposite to the outwards transport associated to viscosity. *Convection-dominated accretion flows* (CDAFs) have been extensively studied analytically (e.g. Narayan et al. 2000; Quataert and Gruzinov 2000; Abramowicz et al. 2002; Igumenshchev 2002; Narayan et al. 2002) and numerically (e.g. Stone et al. 1999, Igumenshchev and Abramowicz 1999, 2000, Igumenshchev et al. 2000). For values of $\alpha \lesssim 0.1$, the characteristics of a CDAF significantly differ from those of an ADAF. First, the mass density scales as $\rho \propto r^{-1/2}$, whereas in a pure ADAF $\rho \propto r^{-3/2}$. And second, in CDAFs solutions the accretion rate is much smaller than in an ADAF. In fact, a CDAF resembles a static envelope more than an accretion flow.

We have up to now only spoken superficially of the role of the magnetic field in the accretion flow. The magnetic field is, however, an essential component in any theory of accretion. We have already mentioned that the magnetorotational instability is likely the mechanism responsible for the angular momentum distribution in an accretion disk. We shall see in the next chapter that the magnetic field also plays a fundamental role in the launching and collimation of outflows. There are a large number of works in the literature that study RIAFs including magnetic fields; you can see Bisnovatyi-Kogan and Lovelace (2001) for a review on this topic.

Meier (2005; see also Fragile and Meier 2009) introduced the idea of *magnetically-dominated accretion flows* (MDAFs), flows where the magnetic forces dominate over thermal and radiation forces. According to these works, MDAFs could develop in the inner region ($r \lesssim 100 R_{\mathrm{grav}}$) of an ADAF. The suggested mechanism for the formation of a MDAF is the radiative cooling of both electrons and ions in a plasma with $T_e \approx T_i$. If the plasma cools the thermal pressure and the height scale of the flow decrease, with the consequent increase of the magnetic-to-thermal pressure ratio. This stabilizes the flow against the development of magnetic turbulence, so that an ordered magnetosphere can develop.

In the model of Meier (2005) the magnetosphere inside the transition radius is formed by closed field lines that connect to the black hole. In this region the flow is non-turbulent and almost radial along the field lines. In the ADAF/MDAF transition region, however, there are magnetic field lines that connect the transition region with infinity. This configuration is, in principle, appropriate for the launching of jets.

Finally, let us briefly comment on the *dissipationless disk* model developed by Bogovalov and Kelner (2005, 2010). The model is based on the idea that energy and angular momentum are removed from the accretion flow by a wind. In this sense it is related to an ADIOS, but it is otherwise very different. In a dissipationless disk the viscosity of the flow is ignored but the magnetic field is included, and the dynamics of the plasma are described by the equations of ideal magnetohydrodynamics. The mechanism that launches the wind is related to the geometry of the magnetic field lines close to surface of the disk; we shall look deeper into this in the following chapter.

The most important result of the dissipationless disk model is that the accretion rate vanishes at the position of the accretor: this implies that all the accreted matter is ejected. The gravitational energy of the inflow is then transferred with a very high efficiency to the outflow. This type of model could explain the observations of systems with very low luminosity disks and powerful jets, such as the radio galaxy M87.

4.5.5 Neutrino-Cooled Accretion Flows

The currently accepted models for both long and short gamma-ray bursts (GRBs; see Chap. 6 and Mészáros 2002, Piran 2004, and Ghisellini 2011 for reviews) involve the formation of a transient, hyperdense, hyperaccreting accretion disk/torus. In the collapsar model, for example, long GRBs are driven by accretion onto a black hole formed in the core of a collapsing massive star. The accretion rates involved are huge, of the order of $\dot{M} \sim 10^{-2}$–$1 M_\odot$ s^{-1}, and the progenitor star is swallowed in a timescale of ~ 10 s. All electromagnetic radiation produced in the accretion disk formed during the collapse is trapped and advected. The disk cools, instead, more efficiently by emission of neutrinos.

The properties of accretion flows in the central engines of GRBs including neutrino emission have been studied among many others by Popham et al. (1999), Ruffert and Janka (1999), Narayan et al. (2001), di Matteo et al. (2002), Kohri and Mineshige (2002), Aloy et al. (2005), Kohri et al. (2005), Lee et al. (2005), Reynoso et al. (2006), Chen and Beloborodov (2007), Metzger et al. (2008), and Zhang and Dai (2009); a review on hyperaccreting disks is presented in Beloborodov (2008). In these works the accretion flow is essentially modeled as a Shakura-Sunyaev disk with some modifications to account for the extreme conditions in the source. Besides cooling by radiation of neutrinos, these refinements include advection, photodissociation of nuclei, corrections to the pressure because of the dense radiation field and the degeneracy of matter, and effects of the strong gravitational field.

Hyperaccreting disk can be broadly divided into a number of regions depending on their properties regarding the emission and propagation of neutrinos. Chen and Beloborodov (2007) estimated the boundaries between these zones as a function of $\dot{M}$ and R; the plots are shown in Fig. 4.16. The production of neutrinos is switched on abruptly at a certain radius R_ν. This happens where the density and the temperature are high enough for the mean energy of electrons to be of the order of $\sim (m_n - m_p)c^2$ (m_n is the mass of the neutron), thus triggering neutrino production by capture—see Eq. (4.181) below. Approaching the inner edge of the disk, at radius R_{τ_ν} the plasma becomes opaque to the propagation of neutrinos and they thermalize with matter. At fixed $\dot{M}$, the disk becomes opaque for anti-neutrinos at a different radius $R_{\tau_{\bar{\nu}}} < R_{\tau_\nu}$. Despite the optical depth being large, the disk still cools efficiently by emission of neutrinos, since the time it takes neutrinos to diffuse outwards is shorter than the advection time. Eventually, in the inner regions of the flow, neutrinos are mostly advected. Notice that for large $\dot{M}$ and R there is a region where the disk becomes prone to gravitational instabilities.

Fig. 4.16 Regions of a neutrino-cooled accretion disk around a rotating black hole of mass $M_{\mathrm{BH}} = 3M_\odot$ and spin $a_* = 0.95$, for values of the viscosity parameter $\alpha = 0.1$ (*top*) and $\alpha = 0.01$ (*bottom*). Cooling by neutrino emission is relevant between the curves labeled "ν-cooled" and "trapped". Nuclei (mostly alpha particles) are photodissociated in the region to the left of the curve "free n, p/α". For large accretion rates, the disk becomes opaque to neutrino propagation at small radius, and gravitationally unstable at large radius. From Chen and Beloborodov (2007). Reproduced by permission of the AAS

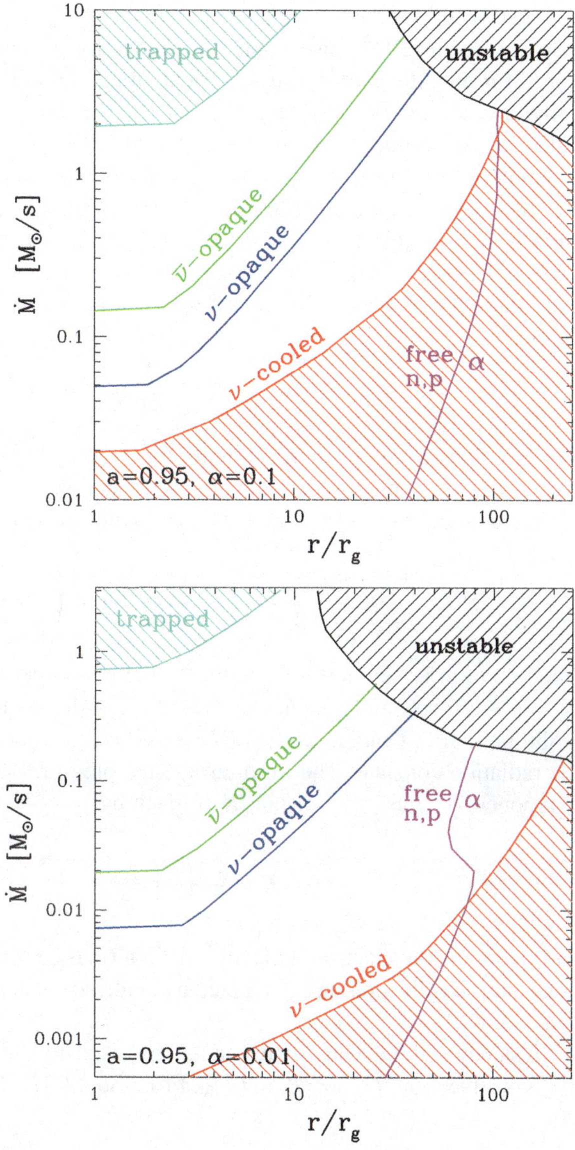

We briefly review the basics of neutrino-cooled disk models as presented in di Matteo et al. (2002) and Reynoso et al. (2006). These authors considered an accretion disk in steady state[12] rotating with Keplerian angular velocity Ω_K, and

---

[12]The viscous timescales in the inner regions of the disk, where the neutrinos are emitted, are much shorter than the typical timescales of variation of the accretion rate in the outer disk (di Matteo et al. 2002). Furthermore, Kohri and Mineshige (2002) have shown that for sufficiently large temperature

adopted the α-prescription to characterize the dissipation of energy by viscosity. In Reynoso et al. (2006) the classical formulas for a Shakura-Sunyaev disk are corrected for the effects of the strong gravitational field around a Kerr black hole. The correction factors depend on the radial coordinate, the mass, and the spin of the hole; they are deduced in Riffert and Herold (1995).

Cooling by neutrino emission is relevant in the inner and hottest ($kT \gtrsim 1$ MeV) part of a hyperdense disk. The accretion rate may be roughly approximated as constant in that region. From Eq. (4.92), $\dot{M} = -2\pi R \Sigma v_R$, where $\Sigma \approx 2\rho H$ is the surface density, ρ is the mass density, and H is the half-thickness of the disk. The expression for H modified to include general relativistic effects is

$$H \approx \frac{a_\mathrm{s}}{\Omega_\mathrm{K}} \sqrt{\frac{B}{C}}. \tag{4.176}$$

The functions B and C approach unity for large R, so Eq. (4.82) is recovered in the region where strong gravitational effects are negligible.

The total pressure $P = \rho a_\mathrm{s}^2$ has several contributions; it may be written as

$$P = \rho \frac{kT}{m_\mathrm{N}} \left(\frac{1 + 3X_\mathrm{nuc}}{4} \right) + \frac{11}{2} a T^4 + K \left(\frac{\rho}{\mu_e} \right)^{4/3} + \frac{U_\nu}{3}. \tag{4.177}$$

The first term is the pressure exerted by the non-degenerate component of the gas; X_nuc is the fraction of free nucleons and m_N is the nucleon mass. The second term is the pressure of radiation plus relativistic electron-positron pairs; $a = 4\sigma_\mathrm{SB}/c$ is the radiation constant. The third term is the pressure of the degenerate, relativistic component the gas; the constant K is given by

$$K = \frac{2\pi hc}{3} \left(\frac{3}{8\pi m_\mathrm{N}} \right)^{4/3}, \tag{4.178}$$

and $\mu_e = 2$ is the electron molecular weight (mass per electron). Finally, the fourth contribution is the pressure generated by neutrinos, where U_ν is the neutrino density (see below).

In steady state, the energy dissipated by viscosity equals the total energy loss including advection, $Q^+ = Q^- + Q_\mathrm{adv}$. From Eq. (4.97), duly corrected by relativistic effects,

$$Q^+ \approx \frac{3G\dot{M}M_\mathrm{BH}}{8\pi R^3} \left(\frac{D}{B} \right), \tag{4.179}$$

where $D \to 1$ for large R. The cooling rate Q^- must account for all the relevant energy loss processes. In an hyperdense accretion disk, the most important mecha-

---

and density ($T \sim 10^{11}$ K, $\rho \sim 10^{13}$ g cm^{-3}) the timescale of positron/electron capture by nucleons (see text) and the timescale for neutrino diffusion in the vertical direction are both shorter than the accretion timescale. Under these conditions, the steady-state approximation is justified.

nisms are—besides electromagnetic radiation—neutrino emission and photodissociation of nuclei (mainly alpha particles into free nucleons). Then,

$$Q^- \approx Q^-_{\mathrm{rad}} + Q^-_{\mathrm{photo}} + Q^-_\nu. \tag{4.180}$$

We have already discussed several mechanisms of radiative cooling in Sect. 4.5.3.
Appropriate expressions for Q_{adv} and Q^-_{photo} as a function of the temperature, density, and composition of the plasma are collected in Reynoso et al. (2006) and refer-
ences therein. Notice that energy is advected in the form of radiation and neutrinos
as well as nucleons.[13]

   A number of channels of neutrino emission may be relevant in the environment
of a hyperaccreting disk. The most efficient is the capture of electrons/positrons by
nucleons,

$$e^- + p \rightarrow n + \nu_e, \qquad e^+ + n \rightarrow p + \bar{\nu}_e. \tag{4.181}$$

Other mechanisms include electron/positron annihilation

$$e^- + e^+ \rightarrow \nu_i + \bar{\nu}_i \tag{4.182}$$

and nucleon-nucleon Bremsstrahlung

$$N + N \rightarrow N + N + \nu_i + \bar{\nu}_i, \tag{4.183}$$

where $N = p, n$, and $i = e, \mu, \tau$ denote the three neutrino flavors. Useful expressions for the corresponding neutrino emissivities can be found in di Matteo et al.
(2002) and Kohri and Mineshige (2002).

   At high accretion rates ($\dot{M} \gtrsim 1 M_\odot$ s^{-1}) the material of the disk is not completely transparent to the propagation of neutrinos. The inverse of the processes in
Eqs. (4.182), (4.182), and (4.183), together with the scattering off nucleons, repre-
sent sources of opacity. If τ_s and τ_{abs} are the optical depths due to neutrino scattering
and absorption, an interpolating formula for Q^-_ν that is approximately valid both in
the optically thin and the optically thick limit is (e.g. Popham and Narayan 1995)

$$Q^-_\nu \approx \frac{7}{8}\sigma_{\mathrm{SB}} T^4 \sum_i \left[\frac{3}{4}\left(\frac{\tau_i}{2} + \frac{1}{\sqrt{3}} + \frac{1}{3\tau_{i,\mathrm{abs}}} \right) \right]^{-1}. \tag{4.184}$$

Here $\tau_i = \tau_{i,\mathrm{abs}} + \tau_{i,\mathrm{s}}$ is the total optical depth for each neutrino species i; see
Reynoso et al. (2006) a compilation of the formulas. From the opacities we can
also estimate the density of neutrinos U_ν (Popham and Narayan 1995),

$$U_\nu \approx \sum_i \frac{\tau_i/2 + 1/\sqrt{3}}{\tau_i/2 + 1/\sqrt{3} + 1/3\tau_{i,\mathrm{abs}}}. \tag{4.185}$$

---

[13]Degenerate matter does not add to the entropy of the flow.

Fig. 4.17 *Top*: disk temperature, density, and half-thickness as a function of radius for two values of the accretion rate, $\dot{M} = 0.1 M_\odot$ s^{-1} (*solid line*) and $\dot{M} = 1 M_\odot$ s^{-1} (*dashed line*). *Bottom*: relative importance of neutrino cooling and advection as a function of radius for two values of the accretion rate, $\dot{M} = 0.1 M_\odot$ s^{-1} (*top panel*) and $\dot{M} = 1 M_\odot$ s^{-1} (*bottom panel*). In all models $M = 3 M_\odot$, $a_* = 0.9$, and $\alpha = 0.1$. From Reynoso et al. (2006), reproduced with permission ©ESO

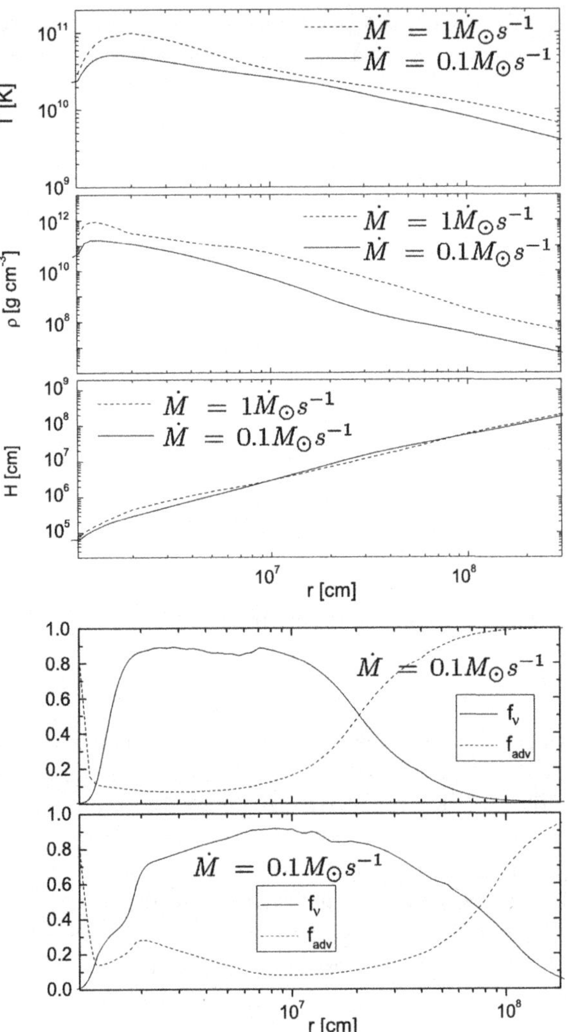

Figure 4.17 shows the temperature, density, and disk half-thickness profiles as a function of R for $\dot{M} = 0.1$ and $1 M_\odot$ s^{-1}, obtained by Reynoso et al. (2006). Also shown are the ratios $f_\nu \equiv Q_\nu/Q^+$ and $f_{\rm adv} \equiv Q_{\rm adv}/Q^+$, that quantify the fraction of the dissipated energy that is released as neutrinos and advected, respectively. Neutrino cooling dominates over a large range of radii. Advection is more important in the innermost (this is a consequence of the relativistic corrections) and in the outer regions of the disk. Similar results were found by di Matteo et al. (2002).

The typical energy released during a GRB (assuming relativistic beaming) is 10^{51} erg (e.g. Piran 2003). Such a large amount of energy has two possible sources: the rotational energy of the black hole and the binding energy of the accretion flow. The rotational energy of a black hole may be extracted via the Blandford-Znajek mechanism; we shall discuss it in the next chapter. The gravitational energy of the infalling matter, on the other hand, is expected to be released in the form of neutrinos that subsequently annihilate into pairs, $\nu + \bar{\nu} \to e^+ + e^-$.[14] This process might load the relativistic jets as first suggested by Eichler et al. (1989).

There are some factors that limit the efficiency of neutrino annihilation. In the first place, neutrinos and antineutrinos must be able to escape the disk before being advected. Furthermore, the cross section for this process is very small ($\sim 10^{-44}$ cm^2) and only a fraction 10^{-2}–10^{-3} of the energy in neutrinos is converted to pairs. Recent calculations by Zalamea and Beloborodov (2011) predict a typical annihilation luminosity of $L_{\nu\bar{\nu}} \approx 10^{52}$ erg s^{-1}. The exact result depends on the values of α, $\dot{M}$, M, and strongly on the spin parameter a_*. For $M = 3M_\odot$, $\dot{M} = 0.3M_\odot$ s^{-1}, and $a_* = 0.95$, for example, these authors find that neutrino annihilation can easily supply the observed luminosity of GRBs.

To conclude, we notice that the copious emission of neutrinos may also drive a mildly relativistic, baryon-loaded wind from the disk (e.g. Levinson 2006). This wind could help to collimate the ultra relativistic jet thought to be launched from the central core.

4.5.6 *Note on the Calculation of a Comptonization Spectrum*

If a population of electrons and a radiation field spatially coexist, unless they are in thermal equilibrium, there will be a net exchange of energy because of elastic collisions between the two species. When the photon transfers energy to the electron we speak of Compton scattering, whereas in the opposite case the process is called *inverse* Compton scattering.

The modification of the photon distribution due to repeated scatterings off thermal electrons is called *Comptonization*. This process is of interest since the characteristic power-law spectrum of black hole coronae is usually interpreted as arising from Comptonization of photons from the accretion disk by electrons in the corona.

We can estimate quantitatively the relevance of Comptonization introducing a parameter y defined as

$$y = \left(\frac{\Delta \varepsilon}{\varepsilon} \right) N_{\mathrm{s}}, \qquad (4.186)$$

where ε is the energy of the photon, $\Delta \varepsilon$ is its average energy change in a single collision, and N_{s} is the mean number of scatterings.

An approximate expression for N_{s} can be calculated as follows. Let L be the typical size of the source and n_e the number density of electrons. The mean free path of a photon will be $\lambda_{\mathrm{s}} = 1/n_e \sigma_{\mathrm{IC}}$, where σ_{IC} the cross section for Compton scattering. The optical depth of the source is then $\tau_{\mathrm{s}} \approx L/\lambda_{\mathrm{s}}$.

---

[14]In the strong magnetic field ($\sim 10^{15}$ G) expected to exist in the central engines of GRBs, neutrinos can also create pairs through the reaction $\nu \to \nu + e^+ + e^-$, see e.g. Gvozdev and Ognev (2001).

If the source is optically thin $N_s \approx \tau_s \ll 1$, but if the source is optically thick the photons will diffuse in a random walk. The distance covered by a photon after N scatterings is $\langle x^2 \rangle^{1/2} \sim \sqrt{N}\,\lambda_s$. Taking $L = \langle x^2 \rangle^{1/2}$ and $N = N_s$, we obtain that in the optically thick case $N_s \approx \tau_s^2$. Considering the two limits, we can approximate the number of scatterings as

$$N_s \approx \max\left\{\tau_s, \tau_s^2\right\}. \tag{4.187}$$

There is no general expression for the fractional energy change $\Delta\varepsilon/\varepsilon$ of the photon since this depends on the form of the energy distribution of the electrons. In the special case of a thermal, non-relativistic distribution of electrons of temperature $kT_e < m_e c^2$,

$$\frac{\Delta\varepsilon}{\varepsilon} = \frac{4kT_e - \varepsilon}{m_e c^2} \tag{4.188}$$

per scattering. When $4kT_e \gg \varepsilon$, on average the electron transfers energy to the photon. When $4kT_e \ll \varepsilon$, the electron's thermal energy is much smaller than the initial photon energy, and the photon loses energy on average.

Suppose that the initial energy of the photon is $\varepsilon_i \ll 4kT_e < m_e c^2$. After one scattering the energy of the photon will be ε_2 such that

$$A \equiv \frac{\varepsilon_2}{\varepsilon_i} = \frac{\varepsilon_i + \Delta\varepsilon}{\varepsilon_i} \approx 1 + \frac{4kT_e}{m_e c^2}. \tag{4.189}$$

After N_s collisions the final energy ε_f of the photon is then amplified in a factor

$$\frac{\varepsilon_f}{\varepsilon_i} \approx \left(1 + \frac{4kT_e}{m_e c^2}\right)^{N_s} \approx \exp\left(N_s \frac{4kT_e}{m_e c^2}\right) \approx \exp(y). \tag{4.190}$$

For $y \ll 1$, then, the primary photon spectrum remains largely unaffected by scattering. When $y \gg 1$ photons reach equilibrium with the electrons. The modified photon spectrum depends basically on T_e and has the shape of a Wien distribution $\propto e^{-\varepsilon/kT_e}$. In this limit the process is called *saturated* Comptonization. The intermediate regime $y \sim 1$ is known as *unsaturated* Comptonization. In this regime the Comptonized spectrum is the result of successive scatterings, and depends both on the electron temperature and the optical depth.

In an optically thin medium of optical depth $\tau_s \ll 1$, the shape of the Comptonized spectrum can be estimated as follows. The probability of a photon undergoing k scatterings before escaping the source is $\sim \tau_s^k$. If $I(\varepsilon_i)$ is the initial intensity of photons at energy ε_i, then after Comptonization the intensity at energy ε_f will be approximately

$$I(\varepsilon_f) \approx I(\varepsilon_i)\tau_s^k \approx N(\varepsilon_i)\left(\frac{\varepsilon_f}{\varepsilon_i}\right)^{-\alpha}, \tag{4.191}$$

where $\alpha = -\ln\tau_s/\ln A$. The Comptonized spectrum takes the form of a power-law.

In the general case, the evolution of the (isotropic) photon distribution $n(\nu)$ is described by the Boltzmann equation for the process $e^-(\mathbf{p}) + \gamma(\nu) \rightleftharpoons e^-(\mathbf{p}') + \gamma(\nu')$, where $\mathbf{p}$ and $\mathbf{p}'$ are the initial and final momenta of the electron, and v and v' the initial and final frequencies of the photon, respectively. If $f(\mathbf{p})$ is the Maxwellian distribution of electrons, the Boltzmann equation is

$$\frac{\partial n}{\partial t} = c \int d\mathbf{p} \int d\Omega \frac{d\sigma}{d\Omega} \{ f(\mathbf{p}')n(v')[1+n(v)] - f(\mathbf{p})n(v)[1+n(v')] \}. \quad (4.192)$$

The first term represents the contribution to $n(v)$ due to scatterings of photons with frequency v', whereas the second term accounts for the decrease in $n(v)$ due to photons scattered into frequency v'. The factors in square brackets appear because of the use of the Bose-Einstein statistics: the probability of scattering from frequency v' to v is increased by the factor $[1+n(v)]$, because photons tend toward mutual occupation of the same quantum state. Finally, $d\sigma/d\Omega$ is the differential Compton cross section.

Equation (4.192) can be solved analytically only in a few cases or under some approximations. In particular, if the electrons are non-relativistic the energy transfer $h\Delta v$ to the photons per collision is small. The Boltzmann equation may then be expanded in powers of Δv. The expansion up to second order yields the so-called Kompaneets equation (Kompaneets 1957),

$$\frac{\partial n}{\partial t_{\mathrm{c}}} = \left(\frac{kT_e}{m_e c^2} \right) \frac{1}{x^2} \frac{\partial}{\partial x} \left[x^4 \left(\frac{\partial n}{\partial x} + n + n^2 \right) \right]. \quad (4.193)$$

Here $x = hv/kT_e$, and $t_{\mathrm{c}} = ct/\lambda_{\mathrm{s}}$ is the mean time interval between successive scatterings. Recall that Eq. (4.193) correctly describes the evolution of the photon distribution under the assumptions that the electrons are non-relativistic and have a thermal distribution.

The photon distribution in equilibrium must be the Bose-Einstein distribution. You can check that setting the time derivative to zero, the solution of the Kompaneets equation is indeed of the form $n = [\exp(x - x_0) - 1]^{-1}$.

We are interested in finding a solution for the case in which $y \sim 1$ is reached for a certain frequency $\tilde{v}$ such that $h\tilde{v} \sim kT_e$. This situation describes a regime of unsaturated Comptonization, since the electrons are not energetic enough to saturate the photon spectrum to a Wein distribution. With some simplifying assumptions, under these conditions an approximate solution of the Kompaneets equation may be found. If the input photon spectrum is $I_0(v)$ is different from zero only for $v < v_0 \ll kT_e/h$, the approximate solution of the Kompaneets equation in steady state is (see e.g. Rybicki and Lightman 1979)

$$I(v) \approx I_0 \left(\frac{v}{v_0} \right)^{3+m}, \quad hv \ll kT_e,$$

$$\quad (4.194)$$

$$I(v) \approx v^3 \exp\left(-\frac{hv}{kT_e} \right), \quad hv \gg kT_e.$$

The spectral index is given by

$$m = -\frac{3}{2} \pm \sqrt{\frac{9}{4} + \frac{4}{y}}, \tag{4.195}$$

with a plus sign for $y \gg 1$ and a minus sign for $y \ll 1$. The photon spectrum is, again, a power-law. This mechanism is though to produce the observed power-law X-ray spectrum from stellar-mass black hole coronae.

4.6 Accretion in Binary Systems: Roche Lobe Overflow

We have already discussed some details of accretion in wind-fed binaries. There is another accretion mechanism through which matter can be transferred from a star to a compact object, namely the overflow of the Roche lobe.

Let us consider a binary system formed by black hole of mass M and a star of mass M_* in a circular orbit. It is easier to treat the problem from a reference frame rotating with the angular velocity of the binary, since in this frame the two masses are fixed. The origin of the coordinate frame is at the center of mass of the system. In the plane of the orbit, the effective potential on a test particle resulting from the gravitational and centrifugal forces[15] is

$$\Phi = -\frac{GM_*}{R_1} - \frac{GM}{R_2} - \frac{\Omega^2(x^2 + y^2)}{2}, \tag{4.196}$$

where $\mathbf{r} = (x, y)$ is the position of the particle, $R_1 = |\mathbf{r} - \mathbf{r}_1|$ and $R_2 = |\mathbf{r} - \mathbf{r}_2|$ indicate the distances of the particle to M_* and M, respectively, and Ω is the angular velocity associated to the radius of the orbit a,

$$a = \left[\frac{G(M + M_*)}{\Omega^2}\right]^{1/3}. \tag{4.197}$$

The total energy of the particle is

$$\Phi + \frac{1}{2}v^2 = E_0. \tag{4.198}$$

For low energies, the particle, if emitted by the star, will fall back to it. At the turning point its velocity will be $v = 0$. In such a case, $\Phi = E_0$. This condition defines equipotential surfaces (Hill's surfaces) that limit the motion of particles of energy E_0. Some equipotentials curves in the plane of the orbit are plotted in Fig. 4.18. The function Φ has two deep valleys centered at the positions of M and M_*; in these regions the equipotentials are approximately circular. Particles with energy

---

[15]The Coriolis force is not included in Eq. (4.196).

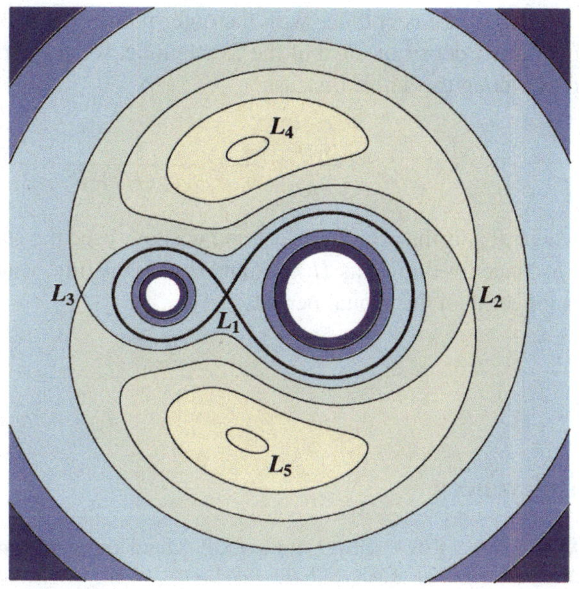

Fig. 4.18 Equipotential curves for a mass ratio $q = M/M_* = 0.3$. The *thick curve* defines the Roche lobes of the two components of the binary, that come in contact at the Lagrangian point L_1. The points L_1, L_2, and L_3 are saddle points of Φ, whereas L_4 and L_5 are local maxima

corresponding to such equipotentials are trapped in the gravitational well of one of the stars. At a certain energy $E_R = \Phi_R$, the Hill's surfaces of both masses come in contact; the part of the separatrix curve that surrounds each star is called the *Roche lobe*. The point of contact of the two Roche lobes is the *inner Lagrangian point L_1* (a saddle point of Φ) where the sum of all forces equals zero,

$$\nabla \Phi = 0. \tag{4.199}$$

At some stage of its evolution the donor star may increase its radius and fill its Roche lobe. It then starts to inject matter onto the black hole through the inner Lagrangian point. There, a particle can go from one lobe to the other without losing energy. The specific angular momentum l_Ω of the injected gas is related to the angular momentum of the orbital motion

$$l_\Omega \approx \Omega a^2. \tag{4.200}$$

The thickness of the gas jet crossing the Lagrange point is $\sim 0.1 R_*$, where R_* is the radius of the donor star. The flow rate is determined by the state of evolution of the star and the ratio $q = M/M_*$. For $q \ll 1$, the flow occurs on a time scale

$$\tau = \frac{GM_*}{R_* L_*}, \tag{4.201}$$

where L_* is the total luminosity of the star. Then, the matter flow rate is

$$\dot{M} = \frac{M_*}{\tau}. \tag{4.202}$$

The jet of gas collides with the outer part of the disk. Since the size of the disk is of the order of the size of the Roche lobe, we can write the time the gas takes to move along the whole disk as

$$t_R \approx \frac{R_{\text{out}}}{v_R} \approx \frac{a}{v_r} \approx \frac{a}{\alpha v_\phi (H/R)^2} \approx \frac{P_{\text{orb}}}{2\pi\alpha}\left(\frac{R}{H}\right)^2, \qquad (4.203)$$

where P_{orb} is the orbital period, and we have used the standard thin disk solution for v_R. Since for thin disks $H \ll R$ and $\alpha < 1$, the time scale of radial motion of matter is longer than the orbital period

$$\frac{t_R}{P_{\text{orb}}} \gg 1. \qquad (4.204)$$

References

M. Abramowicz, B. Czerny, J.-P. Lasota, E. Szuszkiewicz, Astrophys. J. **332**, 646 (1988)

M. Abramowicz, X. Chen, S. Kato, J.-P. Lasota, O. Regev, Astrophys. J. Lett. **438**, L37 (1995)

M.A. Abramowicz, I.V. Igumenshchev, E. Quataert, R. Narayan, Astrophys. J. **565**, 1101 (2002)

M.A. Aloy, H.-T. Janka, E. Müller, Astron. Astrophys. **436**, 273 (2005)

S.A. Balbus, J.F. Hawley, Astrophys. J. **376**, 214 (1991)

M.C. Begelman, Mon. Not. R. Astron. Soc. **184**, 53 (1978)

M.C. Begelman, Mon. Not. R. Astron. Soc. **420**, 2912 (2012)

M.C. Begelman, D.L. Meier, Astrophys. J. **253**, 873 (1982)

A.M. Beloborodov, AIP Conf. Proc. **1054**, 51 (2008)

G. Bertin, G. Lodato, Astron. Astrophys. **350**, 694 (1999)

J. Binney, S. Tremaine, *Galactic Dynamics* (Princeton University Press, Princeton, 1987)

G.S. Bisnovatyi-Kogan, S.I. Blinnikov, Sov. Astron. Lett. **2**, 191 (1976)

G.S. Bisnovatyi-Kogan, S.I. Blinnikov, Astron. Astrophys. **59**, 111 (1977)

G.S. Bisnovatyi-Kogan, R.V.E. Lovelace, New Astron. Rev. **45**, 663 (2001)

G.S. Bisnovatyi-Kogan, Y.M. Kazhdan, A.A. Klypin, A.E. Lutskii, N.I. Shakura, Sov. Astron. **23**, 201 (1979)

R.D. Blandford, M.C. Begelman, Mon. Not. R. Astron. Soc. **303**, L1 (1999)

R.D. Blandford, M.C. Begelman, Mon. Not. R. Astron. Soc. **349**, 68 (2004)

J.M. Blondin, T.C. Pope, Astrophys. J. **700**, 95 (2009)

J.M. Blondin, E. Raymer, Astrophys. J. **752**, 30 (2012)

G. Bodo, A. Curir, Astron. Astrophys. **253**, 318 (1992)

S.V. Bogovalov, S.R. Kelner, Astron. Rep. **49**, 57 (2005)

S.V. Bogovalov, S.R. Kelner, Int. J. Mod. Phys. D **19**, 339 (2010)

H. Bondi, Mon. Not. R. Astron. Soc. **112**, 95 (1952)

H. Bondi, F. Hoyle, Mon. Not. R. Astron. Soc. **104**, 273 (1944)

I.A. Bonnell, M.R. Bate, Mon. Not. R. Astron. Soc. **336**, 659 (2002)

I.A. Bonnell, C.J. Clarke, M.R. Bate, J.E. Pringle, Mon. Not. R. Astron. Soc. **324**, 573 (2001)

A.P. Boss, Astrophys. J. **503**, 923 (1998)

C. Brocksopp, R.P. Fender, V. Larionov, V.M. Lyuty, A.E. Tarasov, G.G. Pooley, W.S. Paciesas, P. Roche, Mon. Not. R. Astron. Soc. **309**, 1063 (1999)

C. Chan, D. Psaltis, F. Özel, Astrophys. J. **645**, 506 (2006)

W.-X. Chen, A.M. Beloborodov, Astrophys. J. **657**, 383 (2007)

X. Chen, Mon. Not. R. Astron. Soc. **275**, 641 (1995)

X. Chen, M. Abramowicz, J.-P. Lasota, R. Narayan, I. Yi, Astrophys. J. **443**, L61 (1995)

X. Chen, M. Abramowicz, J.-P. Lasota, Astrophys. J. **476**, 61 (1997)

H.S. Cohl, J.E. Tohline, Astrophys. J. **527**, 86 (1999)

M. de Val-Borro, M. Karovska, D. Sasselov, Astrophys. J. **700**, 1148 (2009)

T. di Matteo, A.C. Fabian, Mon. Not. R. Astron. Soc. **286**, 393 (1997)

T. di Matteo, R. Perna, R. Narayan, Astrophys. J. **579**, 706 (2002)

W. Duschl, P.A. Strittmatter, P.L. Biermann, Astron. Astrophys. **357**, 1123 (2000)

R. Edgar, New Astron. Rev. **48**, 843 (2004)

D. Eichler, M. Livio, T. Piran, D.N. Schramm, Nature **340**, 126 (1989)

A.A. Esin, J.E. McClintock, R. Narayan, Astrophys. J. **489**, 865 (1997)

A.A. Esin, R. Narayan, W. Cui, J.E. Grove, S.-N. Zhang, Astrophys. J. **505**, 854 (1998)

A.A. Esin, J.E. McClintock, J.J. Drake, M.R. Garcia, C.A. Haswell, R.I. Hynes, M.P. Muno, Astrophys. J. **555**, 483 (2001)

N.W. Evans, P.T. de Zeeuw, Mon. Not. R. Astron. Soc. **257**, 152 (1992)

A.C. Fabian, M.J. Rees, Mon. Not. R. Astron. Soc. **277**, L5 (1995)

T. Foglizzo, M. Ruffert, Astron. Astrophys. **320**, 342 (1997)

T. Foglizzo, M. Ruffert, Astron. Astrophys. **347**, 901 (1999)

T. Foglizzo, P. Galletti, M. Ruffert, Astron. Astrophys. **435**, 397 (2005)

J.A. Font, J.M. Ibáñez, Astrophys. J. **494**, 297 (1998a)

J.A. Font, J.M. Ibáñez, Mon. Not. R. Astron. Soc. **298**, 835 (1998b)

J.A. Font, J.M. Ibáñez, P. Papadopoulos, Mon. Not. R. Astron. Soc. **305**, 920 (1999)

P.C. Fragile, D.L. Meier, Astrophys. J. **693**, 771 (2009)

J. Frank, A. King, D. Raine, *Accretion Power in Astrophysics* (Cambridge University Press, Cambridge, 2002)

V. Frolov, I. Novikov, *Black Hole Physics: Basic Concepts and New Developments* (Springer, Berlin, 1998)

B.A. Fryxell, R.E. Taam, Astrophys. J. **335**, 862 (1988)

C.F. Gammie, Astrophys. J. **553**, 174 (2001)

G. Ghisellini, in *Jets at All Scales*, ed. by G.E. Romero, R.A. Sunyaev, T. Belloni. Proceedings of the IAU Symposium, vol. 275 (2011), p. 335

P. Goldreich, D. Lynden-Bell, Mon. Not. R. Astron. Soc. **130**, 125 (1965)

J. Goodman, Mon. Not. R. Astron. Soc. **339**, 937 (2003)

L.J. Greenhill, C.R. Gwinn, Astrophys. Space Sci. **248**, 261 (1997)

A.A. Gvozdev, I.S. Ognev, JETP Lett. **74**, 298 (2001)

F. Hoyle, R.A. Lyttleton, Proc. Camb. Philos. Soc. **35**, 405 (1939)

J.-M. Huré, Astron. Astrophys. **434**, 1 (2004)

J.-M. Huré, A. Dieckmann, Astron. Astrophys. **541**, A130 (2012)

J.-M. Huré, A. Pierens, Astrophys. J. **624**, 289 (2005)

J.-M. Huré, D. Pelat, A. Pierens, Astron. Astrophys. **475**, 401 (2007)

J.-M. Huré, F. Hersant, C. Surville, N. Nakai, T. Jacq, Astron. Astrophys. **530**, A145 (2011)

S. Ichimaru, Astrophys. J. **214**, 840 (1977)

I.V. Igumenshchev, Astrophys. J. **577**, L31 (2002)

I.V. Igumenshchev, M.A. Abramowicz, Mon. Not. R. Astron. Soc. **303**, 309 (1999)

I.V. Igumenshchev, M.A. Abramowicz, Astrophys. J. Suppl. Ser. **130**, 463 (2000)

I.V. Igumenshchev, M.A. Abramowicz, R. Narayan, Astrophys. J. **537**, L27 (2000)

V. Karas, J.-M. Huré, O. Semerák, Class. Quantum Gravity **21**, R1 (2004)

K. Kohri, S. Mineshige, Astrophys. J. **577**, 311 (2002)

K. Kohri, R. Narayan, T. Piran, Astrophys. J. **629**, 341 (2005)

A.S. Kompaneets, Sov. Phys. JETP **4**, 730 (1957)

L.D. Landau, E.M. Lifshitz, *Fluid Mechanics*, 2nd English edn. (Pergamon, New York, 1987)

J.P. Lasota, M.A. Abramowicz, X. Chen, J. Krolik, R. Narayan, I. Yi, Astrophys. J. **461**, 142 (1996)

W.H. Lee, E. Ramirez-Ruiz, D. Page, Astrophys. J. **632**, 421 (2005)

A. Levinson, Astrophys. J. **648**, 510 (2006)

D.N.C. Lin, J.E. Pringle, Mon. Not. R. Astron. Soc. **225**, 607 (1987)

G. Lodato, G. Bertin, Astron. Astrophys. **398**, 517 (2003a)

G. Lodato, G. Bertin, Astron. Astrophys. **408**, 1015 (2003b)

D. Lynden-Bell, A.J. Kalnajs, Mon. Not. R. Astron. Soc. **157**, 1 (1972)

R. Mahadevan, R. Narayan, J. Krolik, Astrophys. J. **486**, 268 (1977)

K. Mamyoda, N. Nakai, A. Yamauchi, P. Diamond, J.-M. Hureé, Publ. Astron. Soc. Jpn. **61**, 1143 (2009)

T. Matsuda, M. Inoue, K. Sawada, Mon. Not. R. Astron. Soc. **226**, 785 (1987)

M.L. McConnell, J.M. Ryan, W. Collmar, V. Schönfelder, H. Steinle, A.W. Strong, H. Bloemen, W. Hermsen, L. Kuiper, K. Bennett, B.F. Phlips, C. Ling, Astrophys. J. **543**, 928 (2000)

D.L. Meier, Astrophys. Space Sci. **300**, 55 (2005)

P. Mészáros, Annu. Rev. Astron. Astrophys. **40**, 137 (2002)

B.D. Metzger, A.L. Piro, E. Quataert, Mon. Not. R. Astron. Soc. **390**, 781 (2008)

E. Müller, M. Steinmetz, Comput. Phys. Commun. **89**, 45 (1995)

R. Narayan, Astrophys. J. **462**, 136 (1996)

R. Narayan, I. Yi, Astrophys. J. **490**, 605 (1994)

R. Narayan, I. Yi, Astrophys. J. **444**, 231 (1995a)

R. Narayan, I. Yi, Astrophys. J. **445**, 710 (1995b)

R. Narayan, I. Yi, R. Mahadevan, Nature **374**, 623 (1995)

R. Narayan, J.E. McClintock, I. Yi, Astrophys. J. **457**, 821 (1996)

R. Narayan, S. Kato, F. Honma, Astrophys. J. **476**, 49 (1997a)

R. Narayan, D. Barret, J.E. McClintock, Astrophys. J. **482**, 448 (1997b)

R. Narayan, R. Mahadevan, J.E. Grindlay, R.G. Popham, C. Gammie, Astrophys. J. **492**, 554 (1998a)

R. Narayan, R. Mahadevan, E. Quataert, in *Theory of Black Hole Accretion Disks*, ed. by M.A. Abramowicz, G. Bjornsson, J.E. Pringle (Cambridge University Press, Cambridge, 1998b), p. 148

R. Narayan, I.V. Igumenshchev, M.A. Abramowicz, Astrophys. J. **529**, 798 (2000)

R. Narayan, T. Piran, P. Kumar, Astrophys. J. **557**, 949 (2001)

R. Narayan, E. Quataert, I.V. Igumenshchev, M.A. Abramowicz, Astrophys. J. **577**, 295 (2002)

I.D. Novikov, K.S. Thorne, in *Black Holes*, ed. by C. DeWitt, B. DeWitt (Gordon and Breach, Paris, 1973), p. 343

A.T. Okazaki, G.E. Romero, S.P. Owocki, PoS(Integral08) 074 (2008)

J.A. Orosz, J.E. McClintock, J.P. Aufdenberg, R.A. Remillard, M.J. Reid, R. Narayan, G. Lijun, Astrophys. J. **742**, 084 (2011)

B. Paczyński, Acta Astron. **28**, 91 (1978a)

B. Paczyński, Acta Astron. **28**, 241 (1978b)

P. Padoan, A. Kritsuk, M.L. Norman, A. Nordlund, Astrophys. J. **622**, L61 (2005)

D.N. Page, K.S. Thorne, Astrophys. J. **191**, 499 (1974)

A.J. Penner, Mon. Not. R. Astron. Soc. **414**, 1467 (2011)

L.I. Petrich, S.L. Shapiro, R.F. Stark, S.A. Teukolsky, Astrophys. J. **336**, 313 (1989)

T. Piran, ASP Conf. Ser. **308**, 355 (2003)

T. Piran, Rev. Mod. Phys. **76**, 1143 (2004)

R. Popham, R. Narayan, Astrophys. J. **442**, 337 (1995)

R. Popham, S.E. Woosley, C. Fryer, Astrophys. J. **518**, 356 (1999)

J. Poutanen, in *Theory of Black Hole Accretion Disks*, ed. by M.A. Abramowicz, G. Bjornsson, J.E. Pringle (Cambridge University Press, Cambridge, 1998), p. 100

J.E. Pringle, Mon. Not. R. Astron. Soc. **177**, 65 (1976)

E. Quataert, A. Gruzinov, Astrophys. J. **539**, 809 (2000)

E. Quataert, T. di Matteo, R. Narayan, L.C. Ho, Astrophys. J. **525**, L89 (1999)

C.S. Reynolds, T. Di Matteo, A.C. Fabian, U. Hwang, C.R. Canizares, Mon. Not. R. Astron. Soc. **283**, L111 (1996)

M.M. Reynoso, G.E. Romero, O.A. Sampayo, Astron. Astrophys. **454**, 11 (2006)

W.K.M. Rice, D.H. Forgan, P.J. Armitage, Mon. Not. R. Astron. Soc. **420**, 1640 (2012)

H. Riffert, H. Herold, Astrophys. J. **450**, 508 (1995)

G.E. Romero, F.L. Vieyro, G.S. Vila, Astron. Astrophys. **519**, A109 (2010)

M. Ruffert, Astron. Astrophys. Suppl. Ser. **106**, 505 (1994)
M. Ruffert, Astron. Astrophys. Suppl. Ser. **113**, 133 (1995)
M. Ruffert, Astron. Astrophys. **311**, 817 (1996)
M. Ruffert, D. Arnet, Astrophys. J. **427**, 351 (1994)
M. Ruffert, H.-T. Janka, Astron. Astrophys. **344**, 573 (1999)
G.B. Rybicki, A.P. Lightman, *Radiative Processes in Astrophysics* (Wiley, New York, 1979)
I. Sakelliou, Mon. Not. R. Astron. Soc. **318**, 1164 (2000)
M.M. Schulreich, D. Breitschwerdt, Astron. Astrophys. **531**, A13 (2011)
E. Schulz, Astrophys. J. **693**, 1310 (2009)
E. Schulz, Astrophys. J. **747**, 106 (2012)
N.I. Shakura, Astron. Ž. **49**, 921 (1972)
N.I. Shakura, R.A. Sunyaev, Astron. Astrophys. **24**, 337 (1973)
S.L. Shapiro, A.P. Lightman, Astrophys. J. **204**, 555 (1976)
S.L. Shapiro, S.A. Teukolsky, *Black Holes, White Dwarfs and Neutron Stars: The Physics of Compact Objects* (Wiley, New York, 1983)
S.L. Shapiro, A.P. Lightman, D.M. Eardly, Astrophys. J. **204**, 187 (1976)
E. Shima, T. Matsuda, H. Takeda, K. Sawada, Mon. Not. R. Astron. Soc. **217**, 367 (1985)
E. Shima, T. Matsuda, U. Anzer, G. Börner, H.M.J. Boffin, Astron. Astrophys. **337**, 311 (1998)
I. Shlosman, S. Begelman, Astrophys. J. **341**, 685 (1989)
N. Soker, Mon. Not. R. Astron. Soc. **350**, 1366 (2004)
S. Stepney, P.W. Guilbert, Mon. Not. R. Astron. Soc. **204**, 1269 (1983)
I.R. Stevens, D.M. Acreman, T.J. Ponman, Mon. Not. R. Astron. Soc. **310**, 663 (1999)
J.M. Stone, J.E. Pringle, M.C. Begelman, Mon. Not. R. Astron. Soc. **310**, 1002 (1999)
H. Störzer, Astron. Astrophys. **271**, 25 (1993)
R.E. Taam, E.L. Sandquist, Annu. Rev. Astron. Astrophys. **38**, 113 (2000)
T. Theuns, H.M.J. Boffin, A. Jorisses, Mon. Not. R. Astron. Soc. **280**, 1264 (1996)
A. Toomre, Astrophys. J. **139**, 1217 (1964)
A. Trova, J.-M. Huré, F. Hersant, Mon. Not. R. Astron. Soc. **424**, 2635 (2012)
F.L. Vieyro, G.E. Romero, Astron. Astrophys. **542**, A7 (2012)
Q. Wu, F. Yuan, X. Cao, Astrophys. J. **669**, 96 (2007)
F. Yuang, ASP Conf. Ser. **373**, 95 (2007)
F. Yuan, S. Markoff, H. Falcke, Astron. Astrophys. **383**, 854 (2002)
F. Yuan, E. Quataert, R. Narayan, Astrophys. J. **598**, 301 (2003)
F. Yuan, E. Quataert, R. Narayan, Astrophys. J. **606**, 984 (2004)
F. Yuan, R. Ma, R. Narayan, Astrophys. J. **679**, 984 (2008)
I. Zalamea, A.M. Beloborodov, Mon. Not. R. Astron. Soc. **410**, 2302 (2011)
O. Zanotti, C. Roedig, L. Rezolla, L. Del Zanna, Mon. Not. R. Astron. Soc. **417**, 2899 (2011)
D. Zhang, Z.G. Dai, Astrophys. J. **703**, 461 (2009)

Chapter 5
Jets

5.1 Phenomenology

Jets are collimated flows of particles and electromagnetic fields. They are observed in a wide variety of astrophysical systems, from protostars to active galactic nuclei. Astrophysical jets seems to be associated with accretion onto a compact, spinning, central object. One more ingredient appears to be fundamental for the formation of jets (at least for relativistic jets), and that is a large-scale magnetic field.

The most remarkable property of jets is that their length exceeds the size of the compact object by many orders of magnitude. For instance, in AGN the jets are generated in a region of no more than 100 gravitational radii of the central black hole ($\sim 10^{15}$ cm), and propagate up to distances of $\sim 10^{24}$ cm, well into the intergalactic medium. Along the jets, the specific volume of plasma increases enormously and the corresponding adiabatic losses, in combination with various radiative losses, ensure that the particles lose essentially all their "thermal" energy very quickly. Yet, high-resolution radio and X-ray observations show that the jet brightness does not decline so rapidly. This suggests that most of the jet energy is in a different form and that the observed emission is the results of its slow dissipation.

We shall present the basics of the current best models developed to explain the launching, collimation, and acceleration of jets. We shall begin by the magneto-hydrodynamic model, discussing it in some detail both in the non-relativistic and relativistic case. It is the accretion disk that plays the leading role in this model. Later we shall introduce the Blandford-Znajek mechanism, in which the main part is played by the central rotating black hole and its magnetosphere.

5.2 The Equations of Magnetohydrodynamics

The theory of magnetohydrodynamics (MHD) is one possible approach to the problem of a fluid in the presence of an electromagnetic field. For the MHD approximation to be valid two basic conditions must be fulfilled: the fluid must be electrically

G.E. Romero, G.S. Vila, *Introduction to Black Hole Astrophysics*,
Lecture Notes in Physics 876, DOI 10.1007/978-3-642-39596-3_5,
© Springer-Verlag Berlin Heidelberg 2014

quasi-neutral (i.e. with a very small charge density) and it must form a plasma. The second condition implies that the Debye length λ_D of the charged particles is much smaller than the typical length scale of the system, and at the same time the number of particles in a volume λ_D^3 is large enough to treat them as a fluid.

In *non-relativistic* MHD the relevant field is the magnetic field $\mathbf{B}$. Indeed, if the velocity of the flow is $v \ll c$ everywhere, it can be shown that $|\mathbf{E}| \ll |\mathbf{B}|$. This also means that the displacement current may be ignored, so that

$$\nabla \times \mathbf{B} = \frac{4\pi}{c} \mathbf{J} \tag{5.1}$$

where $\mathbf{J}$ is the electric current vector. Taking further into account that the charge density is, by assumption, very small, the expression for the Lorentz force simplifies to

$$\mathbf{f}_L = \frac{\mathbf{J} \times \mathbf{B}}{c} = \frac{1}{4\pi}(\nabla \times \mathbf{B}) \times \mathbf{B}. \tag{5.2}$$

Another of the hypothesis of MHD is that the flow satisfies of Ohm's law in the simple form

$$\mathbf{J} = \sigma\left(\mathbf{E} + \frac{\mathbf{v}}{c} \times \mathbf{B}\right), \tag{5.3}$$

where σ is the electric conductivity of the plasma.

To these equations we must add the rest of Maxwell's equations, namely

$$\nabla \cdot \mathbf{B} = 0, \tag{5.4}$$

$$\nabla \times \mathbf{E} = -\frac{1}{c}\frac{\partial \mathbf{B}}{\partial t}, \tag{5.5}$$

$$\nabla \cdot \mathbf{E} = 4\pi \rho_e. \tag{5.6}$$

The charge density ρ_e is not a relevant parameter; if needed, it can be calculated from Eq. (5.6) once the problem is solved and $\mathbf{E}$ is known.

Combining Eqs. (5.1), (5.3), and (5.5) we get a central equation in MHD, the so-called *induction equation* for the magnetic field

$$\frac{\partial \mathbf{B}}{\partial t} = \nabla \times (\mathbf{v} \times \mathbf{B}) + \frac{c^2}{4\pi\sigma}\nabla^2\mathbf{B}. \tag{5.7}$$

The induction equation states that the magnetic field at a given point in space varies in time because it is advected with the flow (first term on the right-hand side) and because it diffuses (second term on the right-hand side). A supposition usually made in astrophysical applications of MHD is that the conductivity of the plasma is very large. Then the diffusive term in Eq. (5.7) can be neglected compared to the convective term. This approximation is known as *ideal* MHD.

The set of equations of MHD is to be completed with the equations for the conservation of mass and momentum (including the Lorentz force but not viscosity)

$$\frac{\partial \rho}{\partial t} + \nabla \cdot (\rho \mathbf{v}) = 0, \tag{5.8}$$

$$\rho \left[\frac{\partial \mathbf{v}}{\partial t} + (\mathbf{v} \cdot \nabla)\mathbf{v} \right] = -\nabla P - \rho \nabla \Phi + \frac{1}{4\pi}(\nabla \times \mathbf{B}) \times \mathbf{B}, \tag{5.9}$$

plus an equation for the conservation of energy and an equation of state. Here, as usual, P is the pressure, ρ is the mass density, and Φ is the gravitational potential.

5.3 The Structure of Non-relativistic Ideal MHD Jets

We shall develop in some detail the theory that describes the structure of jets in steady state in the non relativistic ideal MHD approximation. The set of equations to be solved is

$$\nabla \times \mathbf{B} = \frac{4\pi}{c}\mathbf{J}, \tag{5.10}$$

$$\nabla \cdot \mathbf{B} = 0, \tag{5.11}$$

$$\nabla \times \mathbf{E} = 0, \tag{5.12}$$

$$\nabla \cdot \mathbf{E} = 4\pi \rho_{\mathrm{e}}, \tag{5.13}$$

$$\mathbf{E} + \frac{1}{c}\mathbf{v} \times \mathbf{B} = 0, \tag{5.14}$$

$$\nabla \times (\mathbf{v} \times \mathbf{B}) = 0, \tag{5.15}$$

$$\nabla \cdot (\rho \mathbf{v}) = 0, \tag{5.16}$$

$$\rho(\mathbf{v} \cdot \nabla)\mathbf{v} = -\nabla P - \rho \nabla \Phi + \frac{1}{4\pi}(\nabla \times \mathbf{B}) \times \mathbf{B}. \tag{5.17}$$

We shall work in cylindrical coordinates (r, ϕ, z) and make the key assumption that the flow is axisymmetric, i.e. $\partial_\phi = 0$. Because of this symmetry there exists a number of conserved quantities that constrain the dynamics of the flow, allowing to obtain some important results without fully solving the equations.

Let us decompose the magnetic field as

$$\mathbf{B} = \mathbf{B}_{\mathrm{p}} + B_\phi \hat{\phi}. \tag{5.18}$$

We call $\mathbf{B}_{\mathrm{p}} \equiv B_r \hat{r} + B_z \hat{z}$ the *poloidal* component of the field, whereas B_ϕ is the *toroidal* component. It follows from Eq. (5.11) and the condition of axial symmetry that $\mathbf{B}_{\mathrm{p}}$ may be written in terms of a scalar *flux function* or *stream function* Ψ as

$$\mathbf{B}_{\mathrm{p}} = \nabla \times \left(\frac{\Psi}{r}\hat{\phi} \right) = \frac{1}{r}\nabla\Psi \times \hat{\phi}, \tag{5.19}$$

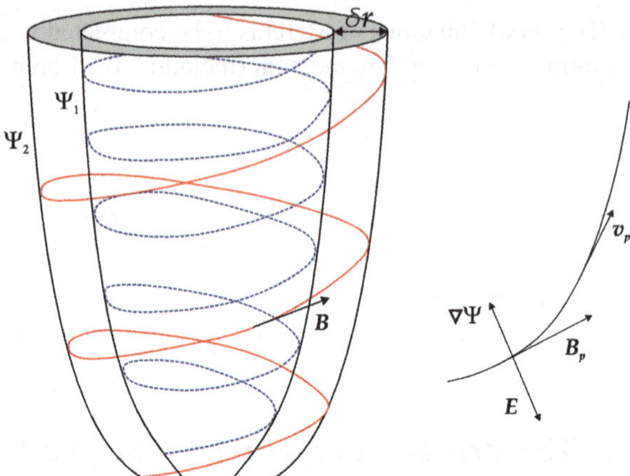

Fig. 5.1 *Left*: two different magnetic surfaces characterized by values of the flux function $\Psi = \Psi_1$ and $\Psi = \Psi_2$, separated by a radial distance δr. The magnetic field lines are indicated. *Right*: projection of a field line in the poloidal plane. The poloidal velocity $\mathbf{v}_p$ is parallel to the poloidal magnetic field $\mathbf{B}_p$, whereas the electric field $\mathbf{E}$ and the gradient of the flux function $\nabla\Psi$ are normal to it

thus

$$\mathbf{B}_r = -\frac{1}{r}\frac{\partial\Psi}{\partial z}, \qquad \mathbf{B}_z = \frac{1}{r}\frac{\partial\Psi}{\partial r}. \tag{5.20}$$

The toroidal component B_ϕ bears no direct relation to Ψ.

From Eq. (5.19) we get that the magnetic flux through a circular cross section of radius r of the jet is $2\pi\Psi(r,z)$, hence the name of flux function. It is also straightforward to check that

$$\mathbf{B}\cdot\nabla\Psi = \mathbf{B}_p\cdot\nabla\Psi = 0, \tag{5.21}$$

so Ψ is constant along magnetic field lines. Equivalently, the vectors $\mathbf{B}$ and $\mathbf{B}_p$ lie on surfaces where $\Psi = $ constant; these are called *magnetic surfaces*, see Fig. 5.1.

The induction equation (5.15) implies that

$$\mathbf{v}\times\mathbf{B} = \nabla\chi \tag{5.22}$$

for some function χ. This result has a number of interesting consequences. First, notice that since $\mathbf{E} = -(\mathbf{v}/c)\times\mathbf{B}$, then $\mathbf{E} = -\nabla\chi/c$; the function χ is therefore the scalar potential of electromagnetism. Second, since $\partial_\phi\chi = 0$ we have that $E_\phi = 0$; the electric field has no toroidal component. But also

$$\nabla\chi|_\phi = \mathbf{v}_p\times\mathbf{B}_p = 0, \tag{5.23}$$

which means that the poloidal velocity is parallel to the poloidal magnetic field. Then we can write

$$\mathbf{v}_p = \kappa(r,z)\mathbf{B}_p \tag{5.24}$$

for some scalar function $\kappa(r,z)$. Applying this result in the continuity equation we get that $\mathbf{B}_p\cdot\nabla(\rho\kappa) = 0$—the product $\rho\kappa$ is constant on magnetic surfaces and must be a function of Ψ only. Hence

$$\rho\kappa = \frac{\rho v_{\mathrm{p}}}{B_{\mathrm{p}}} = \eta(\Psi), \tag{5.25}$$

what states that the poloidal mass flux per unit of poloidal magnetic field is constant along field lines. Notice that ρv_{p} is proportional to the mass flux through a ring of area dA between r and $r + dr$ on a cross section of the jet, whereas the magnetic flux across the same annulus is proportional to B_{p}. We can then write

$$\eta(\Psi) = \frac{d\Psi_{\mathrm{m}}}{d\Psi}, \tag{5.26}$$

where $d\Psi_{\mathrm{m}} = \rho v_{\mathrm{p}} dA$ and $d\Psi = B_{\mathrm{p}} dA$. Because of this result, η is sometimes called the *mass load function*.

Let us now analyze the poloidal component of Eq. (5.22),

$$\mathbf{v}_{\mathrm{p}} \times B_{\phi}\hat{\phi} + v_{\phi}\hat{\phi} \times \mathbf{B}_{\mathrm{p}} = \nabla\chi|_{\mathrm{p}}. \tag{5.27}$$

Using that $\chi = \chi(\Psi)$ and Eqs. (5.19) and (5.24) we obtain that

$$v_{\phi} = r\Omega(\Psi) + \frac{1}{\rho}\eta(\Psi)B_{\phi}. \tag{5.28}$$

Here $\Omega(\Psi) \equiv d\chi/d\Psi$ is a function with dimensions of angular velocity. Since it depends only on Ψ, it is constant along field lines and on magnetic surfaces.[1] Notice that Ω is *not* equal to the angular velocity of matter $\Omega_{\mathrm{m}} = v_{\phi}/r$, unless ρ becomes infinite.

Putting Eqs. (5.25) and (5.28) together, we find an expression for the velocity field in terms of the magnetic field

$$\mathbf{v} = r\Omega(\Psi)\hat{\phi} + \frac{1}{\rho}\eta(\Psi)\mathbf{B}. \tag{5.29}$$

Although the streamlines are contained in the magnetic surfaces, the total velocity is *not* parallel to the total magnetic field. Equation (5.29) is usually interpreted saying that the plasma moves "like beads threaded in rotating wires", the wires being the magnetic field lines that rotate with angular velocity Ω. Indeed, in a reference frame that rotates with Ω, from Eq. (5.29) we see that the velocity is completely directed along $\mathbf{B}$. This is why the function Ω is sometimes referred to as the angular velocity of the magnetic field lines. All the lines on the same magnetic surface rotate with the same velocity, but this differs between surfaces since $\Omega = \Omega(\Psi)$.

We can determine the value of $\Omega(\Psi)$ on each field line as follows. Suppose that there is a point $\mathbf{r}_0$ (the *footpoint*) on a poloidal field line where $B_{\phi}(\mathbf{r}_0) = 0$. For

---

[1] We can now conveniently write the electric field only in terms of the flux function as $\mathbf{E} = -(\Omega/c)\nabla\psi$.

example, if the system has reflexion symmetry with respect to the equatorial plane, such point must lie on the plane $z = 0$. Then $\Omega(\Psi) = v_\phi(\mathbf{r}_0)/r_0 = \Omega_m(\mathbf{r}_0)$; at the footpoint (and only there) the angular velocity of the field lines equals the angular velocity of matter.

Let us now investigate the equation of motion (5.17) starting by the toroidal component. After applying some vector identities, this component may be written as

$$\rho \mathbf{v}_p \cdot \nabla(r v_\phi) = \frac{1}{4\pi} \mathbf{B}_p \cdot \nabla(r B_\phi). \tag{5.30}$$

Using Eq. (5.24) to eliminate the poloidal velocity we obtain

$$\mathbf{B}_p \cdot \nabla\left(r v_\phi - \frac{r B_\phi}{4\pi \rho \kappa}\right) = 0. \tag{5.31}$$

Then the quantity

$$r v_\phi - \frac{r B_\phi}{4\pi \rho \kappa} \equiv L(\Psi) \tag{5.32}$$

is conserved along poloidal field lines and is only function of Ψ. The constant L is nothing else but the angular momentum per unit mass. Both matter and toroidal magnetic field contribute to it.

Combining Eqs. (5.24), (5.28), and (5.32) we can now write B_ϕ and v_ϕ in terms of conserved quantities,

$$r B_\phi = 4\pi \eta \left(\frac{L - r^2 \Omega}{M_A^2 - 1}\right), \tag{5.33}$$

$$r v_\phi = \frac{M_A^2 L - r^2 \Omega}{M_A^2 - 1}. \tag{5.34}$$

Here we have defined the *Mach-Alfvén number*

$$M_A^2 \equiv \frac{v_p^2}{v_{Ap}^2}, \tag{5.35}$$

where v_{Ap} is the poloidal component of the *Alfvén velocity*

$$\mathbf{v}_A \equiv \frac{\mathbf{B}}{\sqrt{4\pi \rho}}. \tag{5.36}$$

The *Alfvén radius* r_A is the point on each poloidal field line where $M_A = 1$; the loci of all r_A define the *Alfvén surface*.

The numerator of Eq. (5.34) must vanish at $r = r_A$ in order to avoid a divergence. This fixes the value of the angular momentum on each field line,

$$L(\Psi) = r_A^2 \Omega(\Psi). \tag{5.37}$$

From Eq. (5.25) and the definition of the Alfvén velocity, it follows that the mass density at the Alfvén radius on each poloidal field line is $\rho_A = 4\pi\eta^2$.

We now see that up to the Alfvén surface the flow may be considered as approximately co-rotating with the field lines, since from Eq. (5.34) it results that $v_\phi \approx \Omega r$ for $M_A^2 \ll 1$. The magnetic field, on the other hand, is dominated by the poloidal component. From Eqs. (5.25) and (5.29) we get

$$\frac{B_\phi}{B_p} = \frac{v_\phi - \Omega r}{v_p}. \tag{5.38}$$

For $r \ll r_A$ the numerator is close to zero and then $B_\phi \ll B_p$.

Far from the Alfvén surface we expect that $\rho < \rho_A$ (Heyvaerts and Norman 1989). Writing Eq. (5.34) as

$$v_\phi = \Omega r \left(1 - \frac{1 - r_A^2/r^2}{1 - \rho/\rho_A} \right), \tag{5.39}$$

we find that $v_\phi \approx 0$ for $r \gg r_A$. The velocity in this region is predominantly poloidal; from Eqs. (5.25) and (5.36) we get

$$v_p^2 = \left(\frac{\rho_A}{\rho} \right) v_{Ap}^2 > v_{Ap}^2. \tag{5.40}$$

Finally, we must obtain an equation for the energy. For definiteness we shall adopt a polytropic equation of state

$$P = K\rho^\gamma, \tag{5.41}$$

where $K(\Psi)$ is constant on magnetic surfaces. The desired expression for the energy is obtained taking the dot product of $\mathbf{B}_p$ and Eq. (5.17), yielding

$$E(\Psi) = \frac{1}{2}v^2 + h + \Phi - \frac{r\Omega B_\phi}{4\pi\eta}. \tag{5.42}$$

This is the Bernoulli equation for the flow, that states the conservation of energy per unit mass along a poloidal field line. The function h is the specific enthalpy of the plasma. For an equation of state of the form (5.41) with $\gamma \neq 1$, it is given by $h = \gamma a_s^2/(\gamma - 1)$.

Like the angular momentum, the energy also has a contribution from the magnetic field. It can be shown from Eqs. (5.14) and (5.29) that the poloidal component of the Poynting vector is

$$\mathbf{S}_p = \frac{c}{4\pi}(\mathbf{E} \times \mathbf{B})_p = -\frac{r\Omega B_\phi}{4\pi}\mathbf{B}_p, \tag{5.43}$$

so the fourth term on the right-hand side of Eq. (5.42) accounts for the poloidal flux of electromagnetic energy.

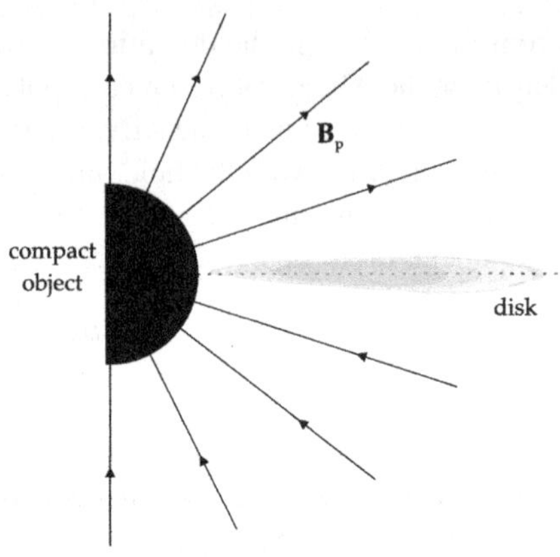

Fig. 5.2 Poloidal magnetic field lines in the split monopole configuration

The last equation to be analyzed is the poloidal component of the equation of motion (5.17). This cannot be reduced to a conservation law. However, taking the dot product with $\nabla\Psi$ we obtain a differential equation for the flux function $\Psi(r, z)$ in terms of the mass density and conserved quantities. This equation is known as the *Grad-Shafranov* or *transfield equation*. It may be written in many different ways; here we quote the result as given in Ferreira (2002)

$$\nabla \cdot \left[(M_A - 1)\frac{\nabla\Psi}{4\pi r^2} \right] - \left(B_\phi^2 + M_A^2 B_p^2 \right)\frac{\eta'}{4\pi\eta}$$

$$= \rho\left[E' - \Omega_m(\Omega r_A)' - \left(\Omega_m r^2 - \Omega r_A^2 \right)\Omega' - \frac{a_s^2}{\gamma(\gamma - 1)}K' \right]. \quad (5.44)$$

The Bernoulli and the Grad-Shafranov equations may be expressed only in terms of the flux function Ψ and another variable, say the mass density ρ, to obtain a system of two coupled, non-linear differential equations. This system of equations is, however, so complicated that analytic solutions can be obtained only under some simplifying assumption such as self-similarity (e.g Blandford and Payne 1982; Li et al. 1992; Vlahakis and Tsinganos 1998), or in some asymptotic regime (e.g. Begelman and Li 1994; Heyvaerts and Norman 2003a; Lyubarsky 2009). A different approach is to assume some functional form for the poloidal magnetic field. A possible choice is the *split monopole* configuration, in which the poloidal magnetic field is radial and has opposite polarity on both sides of the equatorial plane, see Fig. 5.2. In cylindrical coordinates a split monopole poloidal field decays as $B_p = B_0(r_0/r)^2$, where r_0 is the radius of the footpoint of the field line and $B_0 = B_p(r_0)$. The split monopole configuration was adopted in some very well-known early works on winds and jets, such as those of Weber and Davis (1967) and Michel (1969). The relative simplicity of those models allowed to obtain analytical results that helped grasp the basics of the MHD theory of outflows. Beware, however, that any model with prescribed field line shapes does not satisfy the Grad-Shafranov equation and thus is not self-consistent.

Five additional unknown functions of Ψ appear in Eqs. (5.42) and (5.44), the integrals of motion $E(\Psi)$, $L(\Psi)$, $\Omega(\Psi)$, $\eta(\Psi)$, and $K(\Psi)$. They must be prescribed or fixed from the boundary conditions. For example, if the jet is launched from an accretion disk rotating with Keplerian angular velocity, we have that $\Omega = \Omega_{\mathrm{K}}(r_0)$, where r_0 is the radius of the footpoint of the magnetic field line.

An important property of the equations of axisymmetric MHD flows is that, besides the Alfvén surface, they have other two critical surfaces. These correspond to the points where the poloidal velocity reaches the speed of each of the *magnetosonic waves* propagating along the poloidal magnetic field. These are given by the roots of the equation

$$v^4 - v^2\left(a_s^2 + v_A^2\right) + a_s^2 v_A^2 \cos^2\alpha = 0 \tag{5.45}$$

where $\cos\alpha = B/B_{\mathrm{p}}$.[2] The two solutions, v_{SM} and v_{FM}, are called the *slow* and *fast magnetosonic speed*, respectively. The corresponding critical surfaces are called the *slow* and *fast magnetosonic surfaces*. Demanding that the solution is regular at these point imposes two constraints on the integrals of motion.

The mathematical characteristics of the Grad-Shafranov equation allow to define yet another critical velocity. This is the poloidal *cusp velocity*

$$v_{\mathrm{C}}^2 = \frac{a_s^2 v_{\mathrm{Ap}}^2}{a_s^2 + v_A^2} = \frac{v_{\mathrm{SM}}^2 v_{\mathrm{FM}}^2}{v_{\mathrm{SM}}^2 + v_{\mathrm{FM}}^2}. \tag{5.46}$$

It is smaller than all the other characteristic speeds of the problem, including the speed of sound. The Grad-Shafranov equation is elliptic as long as $v_{\mathrm{p}} < v_{\mathrm{C}}$. It becomes hyperbolic for $v_{\mathrm{C}} < v_{\mathrm{p}} < v_{\mathrm{SM}}$, then again elliptic for $v_{\mathrm{SM}} < v_{\mathrm{p}} < v_{\mathrm{FM}}$, and finally again hyperbolic for $v_{\mathrm{p}} > v_{\mathrm{FM}}$. This adds much difficulty to the numerical resolution of the problem.

5.3.1 Collimation

Beyond the Alfvén surface the toroidal component of the magnetic field becomes larger than the poloidal component. The force exerted by the toroidal field helps collimate the jet, where by *collimation* we understand the degree of parallelism of the streamlines. A cylindrical jet where all the streamlines are parallel to the symmetry axis is then perfectly collimated. Notice that collimation is different from *confinement*, that refers to the "width" of the outflow.

An useful and equivalent way of writing the Grad-Shafranov equation is (e.g. Ustyugova et al. 1999; Ferreira 2002)

$$\left(v_{\mathrm{p}}^2 - v_{\mathrm{Ap}}^2\right)\frac{1}{R_{\mathrm{c}}} = -\frac{1}{8\pi\rho r^2}\frac{\partial}{\partial n}(r B_\phi)^2 + \frac{\cos\theta\, v_\phi^2}{r} - \frac{1}{\rho}\frac{\partial}{\partial n}\left(P + \frac{B_{\mathrm{p}}^2}{8\pi}\right) - \frac{\partial\Phi}{\partial n}. \tag{5.47}$$

---

[2] In the general case α is the angle between the magnetic field and the wave vector. Since we are interested in waves propagating along the poloidal magnetic field, here $\cos\alpha = B/B_{\mathrm{p}}$.

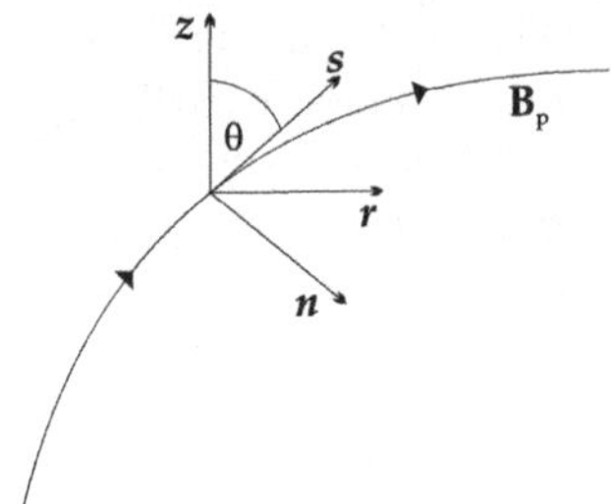

Fig. 5.3 A poloidal magnetic field line in the meridional plane. The unit vectors tangent and normal to the line at an arbitrary point are indicated. The angle θ is the inclination of the line

Here θ is the angle between the poloidal field line and the z-axis, n is a coordinate in the outwards direction perpendicular to the line, and R_{c} is the radius of curvature of the line. As seen from Fig. 5.3, $\cos\theta = \hat{s}\cdot\hat{z} = \hat{n}\cdot\hat{r}$, whereas

$$\frac{\hat{n}}{R_{\mathrm{c}}} = \frac{\partial\hat{s}}{\partial s} \quad \text{or} \quad \frac{1}{R_{\mathrm{c}}} = \frac{\partial\theta}{\partial s}. \tag{5.48}$$

Let us simplify Eq. (5.47) in the limit $r \gg r_{\mathrm{A}}$. In this region, as we have seen, the toroidal velocity is very small, the poloidal velocity is larger than the Alfvén value, and the toroidal magnetic field is much larger than the poloidal component. We can also assume that the effects of pressure and gravitational attraction are of no importance. Equation (5.47) then reduces to

$$\frac{v_{\mathrm{p}}^2}{R_{\mathrm{c}}} = -\frac{1}{8\pi\rho r^2}\frac{\partial}{\partial n}(r B_\phi)^2. \tag{5.49}$$

The curvature of the field lines is determined by the variation of the toroidal field for radius larger than the Alfvén radius. The force exerted by the toroidal field is usually called *hoop stress*. Notice that it may have a collimating or a decollimating effect depending on the shape of the surfaces $(r B_\phi)^2 = \text{constant}$.

By Ampère's law, the toroidal magnetic field is related to the poloidal current density as

$$B_\phi(r, z) = \frac{2}{cr}\int_S \mathbf{J}\cdot d\mathbf{S} = \frac{2}{cr}I(r, z), \tag{5.50}$$

where I is the net electric current through a circular cross section of radius r at height z in the jet. Then, the force exerted by the toroidal field depends on the sign of the enclosed current. Heyvaerts and Norman (1989) showed that the asymptotic shape of the field lines is cylindrical if there is a net current at infinity, $I(r, z) \neq 0$ for $r \to \infty$. If the current at infinity vanishes the jet is instead collimated to paraboloids. The outflow may also remain asymptotically uncollimated (conical with radial streamlines). It is also possible to obtain a mixed asymptotic behavior, with the core of the jet (close to the axis) collimated to cylinders and an uncollimated outer region. This occurs when there exists a return current outside the core or in a current sheet.[3]

---

[3]Recall that the current I is not constant on magnetic surfaces, and so $\mathbf{J}$ can cross them.

5.3.2 Acceleration

To investigate the acceleration of the flow we must analyze the poloidal component of the equation of motion (5.17). We project it onto a unit vector $\hat{s}$ tangent to the poloidal field lines to obtain

$$\frac{1}{2}\frac{\partial v_{\mathrm{p}}^2}{\partial s} = \Omega_{\mathrm{m}}^2 r \sin\theta - \frac{1}{\rho}\frac{\partial P}{\partial s} - \frac{\partial\Phi}{\partial s} - \frac{1}{8\pi\rho r^2}\frac{\partial}{\partial s}(r B_\phi)^2. \tag{5.51}$$

The gravitational force opposes the outwards acceleration of the flow, whereas the pressure gradient favors it because the pressure drops as the outflow expands. The centrifugal term (first term on the right-hand side) always helps to accelerate the plasma as long as the field lines are inclined outwards from the symmetry axis. Under the same conditions, however, the action of the magnetic field may accelerate or decelerate the flow depending on the behavior of B_ϕ. So the acceleration is, in general, *magnetocentrifugal.*

As we have already discussed, centrifugal acceleration is a good approximation up to the Alfvén radius, since for $r \lesssim r_{\mathrm{A}}$ the flow "co-rotates" with the field lines. Then $v_\phi/r = \Omega_{\mathrm{m}} \approx \Omega$ and the first term in Eq. (5.51) increases linearly with r. Further acceleration beyond r_{A} depends on the details of the magnetic field configuration.

5.3.3 Launching

The launching of an outflow from an accretion disk requires that this is threaded by an ordered, large-scale magnetic field. Since astrophysical black holes are not expected to have charge, this field must be generated in the disk itself by electric currents. We shall not deal with the mechanism of generation of large-scale magnetic field in accretion disks; you can see, for example, the works of Tout and Pringle (1996), Spruit and Uzdensky (2005), Lovelace et al. (2009), and Bisnovatyi-Kogan and Lovelace (2007, 2012) on this topic.

The magnetic field near the disk surface cannot have an arbitrary geometry if it is to mediate the launching of the jet. A general constraint on the inclination of the field lines can be found by considering the shape of the effective potential Φ_{eff} to which a particle near the disk surface is subject (Blandford and Payne 1982). The effective potential per unit mass is the sum of the gravitational potential and a centrifugal term,

$$\Phi_{\mathrm{eff}} = -\frac{GM_{\mathrm{BH}}}{\sqrt{r^2 + z^2}} - \frac{1}{2}\Omega_{\mathrm{m}}^2 r^2. \tag{5.52}$$

We have seen that before the Alfvén radius matter is approximately in co-rotation with the magnetic field lines at angular velocity Ω. Since the footpoints of the lines are anchored to the disk, Ω may be assumed to equal the angular velocity of the

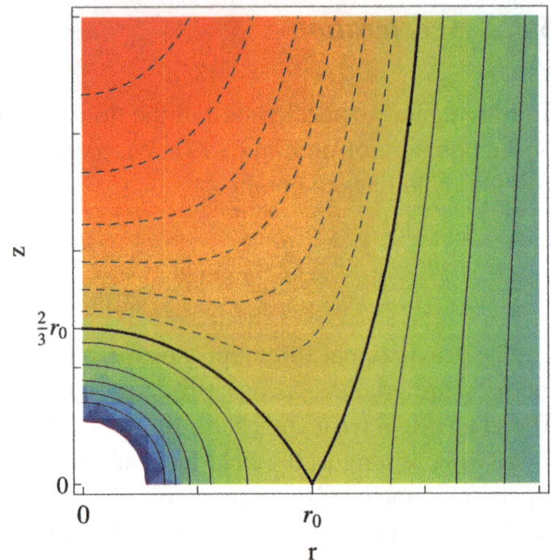

Fig. 5.4 Equipotential curves for the effective potential. The thick solid line divides the regions of increasing (*dotted lines*) and decreasing (*thin solid lines*) potential at an arbitrary point $r = r_0$ on the plane $z = 0$

disk. In a geometrically thin disk the rotation velocity is Keplerian, so $\Omega = \Omega_{\rm K}(r_0)$, where r_0 is the radius of the footpoint of the field line. We may then write Eq. (5.52) as

$$\Phi_{\rm eff} = -GM_{\rm BH}\left[\frac{r_0}{\sqrt{r^2 + z^2}} + \frac{1}{2}\left(\frac{r}{r_0}\right)^2\right]. \tag{5.53}$$

The equipotential curves of $\Phi_{\rm eff}$ are plotted in Fig. 5.4. With respect to the solid thick line, the effective potential increases towards the region of dotted lines and decreases towards the regions of thin solid lines. The first derivatives of $\Phi_{\rm eff}$ vanish at $(r_0, 0)$. For a particle at this position in the disk surface to be in unstable equilibrium with respect to a small displacement along the field line, we must demand that the second derivative of the effective potential along the field line at $(r_0, 0)$ is negative. This yields the constraint

$$\frac{\partial^2 \Phi_{\rm eff}}{\partial s^2}(r_0, 0) = -\frac{GM_{\rm BH}}{r_0^3}\left(3\sin\theta^2 - \cos\theta^2\right) < 0, \tag{5.54}$$

where θ is the angle between the field line and the z-axis at $(r_0, 0)$. The condition for unstable equilibrium is then that $\theta > 30°$; this is the minimum inclination the field lines must have in order to accelerate matter outwards from the surface of the disk.[4,5] Notice that it is independent of r_0. Since the critical angle is relatively large, jets launched by this mechanism are not well collimated near the base.

---

[4]This is the minimum angle for instability in flat and Schwarzschild space-time. It is slightly smaller for a Kerr black hole with spin $a_* = -1$, and approaches $90°$ for $a_* = 1$; a_* is taken positive (negative) if the black hole and the disk rotate in the same (opposite) sense.

[5]The critical angle is the same for a field line inclined *inwards*, but this case is of no interest for the launching of an outflow.

For inclinations larger than 30° matter has to overcome a potential barrier before reaching a region of decreasing potential. This may happen if the zone above the surface of the disk is sufficiently hot. Eventually, for $\theta \ll 30°$ the maximum of Φ_{eff} of along the field line is very far away from the disk and no significant outflow is expected to be launched from this region.

5.4 Relativistic MHD Jets

We now discuss the generalization of the theory presented in the previous section to the case of axisymmetric MHD relativistic outflows (those with bulk velocities that may get close to the speed of light) in flat space-time. Many of the results we obtained for non-relativistic outflows remain valid in the relativistic case. Perhaps the most important difference is that in relativistic MHD the electric field cannot be neglected compared to the magnetic field. Moreover, we shall see that the force exerted by electric field plays a fundamental role in the collimation of the jet.

Let $\gamma = 1/\sqrt{1 - v^2/c^2}$ be the local bulk Lorentz factor of the jet, $\mathbf{u} = \gamma \mathbf{v}$ the spatial part of the 4-velocity per unit mass, and ρ and P the mass density and the pressure in the rest frame of the outflow (also called the co-moving reference frame). The relativistic versions of the continuity equation and the equation of motion in steady state read (e.g. Li et al. 1992; Heyvaerts and Norman 2003a, 2003b; Vlahakis 2004; Lyubarsky 2009)

$$\nabla \cdot (\rho \mathbf{u}) = 0 \tag{5.55}$$

and

$$\rho(\mathbf{u} \cdot \nabla)(\xi \mathbf{u}) = -\nabla P + \rho_e \mathbf{E} + \frac{\mathbf{J} \times \mathbf{B}}{c} + \gamma \rho \nabla \left(\xi \gamma \frac{GM_{\text{BH}}}{\sqrt{r^2 + z^2}} \right). \tag{5.56}$$

The relativistic enthalpy per unit mass is $c^2 \xi$, where ξ is defined as

$$\xi = 1 + \int \frac{1}{\rho c^2} dP = 1 + \frac{\Gamma}{\Gamma - 1} \frac{KP}{\rho c^2} \tag{5.57}$$

calculated at constant entropy. The last equality is valid when a polytropic equation of state $P = K(\Psi)\rho^{\Gamma}$ is adopted. Projecting Eq. (5.56) onto a vector normal to the poloidal magnetic field yields the relativistic Grad-Shafranov equation, whereas the relativistic Bernoulli equation follows from $(v_\phi/c)^2 + (v_{\text{p}}/c)^2 + \gamma^{-2} = 1$; see for example Vlahakis (2004) and Lyubarsky (2009).

The condition of ideal MHD Eq. (5.14) takes exactly the same form in the relativistic regime. This condition amounts to demanding that in the rest frame of the flow (denoted here with primes) Ohm's law is $\mathbf{J}'/\sigma = \mathbf{E}'$, so that in the limit of very large conductivity $\mathbf{E}' = 0$. The components of the fields parallel to $\mathbf{v}$ do not change between reference frames, $\mathbf{E}_{//} = \mathbf{E}'_{//} = 0$. For the perpendicular components we must demand that $\mathbf{E}'_{\perp} = \gamma(\mathbf{E}_{\perp} + (\mathbf{v}/c) \times \mathbf{B}) = 0$. Then $\mathbf{E} + (\mathbf{v}/c) \times \mathbf{B} = 0$ also in the relativistic limit.

As in the non-relativistic case, the condition of axial symmetry allows to write the fields in terms of the flux function Ψ,

$$\mathbf{B}_{\mathrm{p}} = \frac{1}{r}\nabla\Psi \times \hat{\phi}, \tag{5.58}$$

$$\mathbf{E} = -\frac{\Omega}{c}\nabla\psi, \tag{5.59}$$

where, again, $\Omega(\Psi)$ is the angular velocity of the magnetic field lines. It defines, for each field line, the radius of the *light cylinder* $R_{\mathrm{L}} = c/\Omega$, where the velocity of the lines equals the speed of light. This is a relevant length scale in relativistic jets. Notice that in terms of R_{L} the absolute value of the electric field is

$$E = \frac{r}{R_{\mathrm{L}}}B_{\mathrm{p}}, \tag{5.60}$$

so for $r > R_{\mathrm{L}}$ the electric field becomes larger than the poloidal magnetic field. The velocity of the plasma is related to the magnetic field through Ω and the mass-load function $\eta(\Psi)$ as

$$\mathbf{u} = \gamma r\Omega\hat{\phi} + \frac{\eta}{\rho}\mathbf{B}. \tag{5.61}$$

Besides η, Ω, and K, the other two conserved quantities along poloidal magnetic field lines are the angular momentum $L(\Psi)$ and the energy $\tilde{E}(\Psi)$. The expressions in the relativistic regime are

$$L(\Psi) = \gamma\xi r v_{\phi} - \frac{r B_{\phi}}{4\pi\eta}, \tag{5.62}$$

$$\tilde{E}(\Psi) = \gamma\xi\left(c^2 - \frac{GM_{\mathrm{BH}}}{\sqrt{r^2 + z^2}}\right) - \frac{r\Omega B_{\phi}}{4\pi\eta}. \tag{5.63}$$

Notice that $\tilde{E}$ now has a contribution from the rest-mass energy.

It is useful to define from these another constant of motion as

$$\mu(\Psi) \equiv \frac{\mathcal{F}}{\mathcal{F}_{\mathrm{m}}}. \tag{5.64}$$

Here $\mathcal{F}_{\mathrm{m}} = \rho c^2 u_{\mathrm{p}}$ is the poloidal rest-mass energy flux and $\mathcal{F}$ is the total poloidal energy flux (sum of the kinetic, rest-mass, Poynting, thermal, and gravitational energy fluxes). Far away from the gravitating center, and in the case of a cold plasma for which the thermal energy can be neglected,

$$\mu(\Psi) \approx \frac{\mathcal{F}_k + \mathcal{F}_{\mathrm{m}} + \mathcal{F}_S}{\mathcal{F}_{\mathrm{m}}} = \gamma(\sigma + 1), \tag{5.65}$$

where $\mathcal{F}_S = (c/4\pi)EB_\phi$ is the Poynting flux, $\mathcal{F}_k = (\gamma - 1)\rho c^2 u_{\mathrm{p}}$ is the kinetic energy flux, and we have defined the *magnetization* parameter

$$\sigma \equiv \frac{\mathcal{F}_S}{\mathcal{F}_k + \mathcal{F}_m}. \tag{5.66}$$

The magnetization is not constant on field lines. In fact, if the outflow accelerates by conversion of electromagnetic to kinetic energy, we expect σ to decrease with the distance to the compact object. If the conversion were perfectly efficient, then asymptotically $\sigma \approx 0$. From Eq. (5.65), then, we immediately obtain that the maximum Lorentz factor the flow can achieve on a field line is $\gamma_{\mathrm{max}} = \mu$.

Having defined the relevant variables, we now discuss the most important results on the asymptotic acceleration and collimation of relativistic jets. As we shall see, the efficiency of the acceleration of the flow is closely related to the shape (i.e. the degree of collimation) of the poloidal field lines.

In the region $r \gg R_{\mathrm{L}}$ and for $\gamma \gg 1$, the velocity of the outflow is approximately given by the value of the drift velocity in a crossed electromagnetic field (e.g. Tchekhovskoy et al. 2008)

$$v \approx v_{\mathrm{dr}} = c\frac{|\mathbf{E} \times \mathbf{B}|}{B^2}. \tag{5.67}$$

The corresponding Lorentz factor is

$$\gamma^2 \approx \gamma_{\mathrm{dr}}^2 = \frac{B^2}{B^2 - E^2} \tag{5.68}$$

or, equivalently,

$$\frac{1}{\gamma^2} \approx \frac{B_{\mathrm{p}}^2}{B^2} + \frac{B_\phi^2 - E^2}{B^2}. \tag{5.69}$$

The component of the velocity along the magnetic field is negligible and does not contribute significantly to γ.

From Eqs. (5.60) and (5.68) we may obtain an estimation of the value of the toroidal magnetic field in the limit $r \gg R_{\mathrm{L}}$ and $\gamma \gg 1$,

$$B_\phi \approx \frac{\gamma E}{\sqrt{\gamma^2 - 1}} \approx B_{\mathrm{p}}\left(\frac{r}{R_{\mathrm{L}}}\right)\left(1 + \frac{1}{2\gamma^2}\right). \tag{5.70}$$

Thus we get that $B^2/B_{\mathrm{p}}^2 \approx 1 + (r/R_{\mathrm{L}})^2 \approx (r/R_{\mathrm{L}})^2$. The second term in Eq. (5.69) is related the curvature of the field lines,

$$\left(\frac{B^2}{B_\phi^2 - E^2}\right)^{1/2} \approx \mathcal{C}\left(\frac{R_{\mathrm{c}}}{r}\right)^{1/2}, \tag{5.71}$$

where $\mathcal{C}$ is a constant of the order of unity. Its exact value depends on the details of the geometry of the field lines; for a highly collimated flow with $\Omega = \mathrm{constant}$, for example, $\mathcal{C} = \sqrt{3}$ (Tchekhovskoy et al. 2008). We may then write Eq. (5.69) in a convenient form as

$$\frac{1}{\gamma^2} \equiv \frac{1}{\gamma_1^2} + \frac{1}{\gamma_2^2} \approx \frac{R_{\mathrm{L}}^2}{r^2} + \frac{r}{\mathcal{C}^2 R_{\mathrm{c}}}. \tag{5.72}$$

The Lorentz factor at a point will be effectively determined by the smallest among the values of γ_1 and γ_2.

Whether $\gamma \propto \gamma_1$ or $\gamma \propto \gamma_2$ depends on which terms asymptotically dominate the force balance in the Grad-Shafranov equation; a detailed discussion on this point is presented in Komissarov et al. (2009). The *linear* or *first acceleration regime* $\gamma \propto \gamma_1 \propto r$ corresponds to a situation where the balance between the centrifugal and the electromagnetic force keeps the outflow in equilibrium. When the curvature of the field lines becomes important, the magnetic tension dominates over the centrifugal force and the equilibrium is maintained only by electromagnetic forces. This *second acceleration regime* is characterized by $\gamma \propto \gamma_2$.

To obtain an expression for the Lorentz factor we need to find out the dependence of γ_2 on r. For a given curve $z(r)$ the curvature radius is given by

$$R_{\mathrm{c}} = \frac{(\dot{z}^2 + 1)^{3/2}}{\ddot{z}}, \tag{5.73}$$

where the dots denote the derivative with respect to r. Assume that the poloidal field lines may be parameterized as $z(r) \propto r^k$ with $k > 1$. Sufficiently far from the compact object $\dot{z} \gg 1$, so that $R_{\mathrm{c}} \propto r^{2k-1}$ and $\gamma_2 \propto r^{k-1}$. If $k = 2$ (parabolic field lines) we get that $\gamma \propto r$. If $1 < k < 2$ the flow is poorly collimated (less than parabolically) and $\gamma \sim \gamma_2 \propto r^{k-1}$. Finally, if $k > 2$ (well collimated flow) the curvature of the field lines is very small (i.e. R_{c} is very large) and $\gamma \sim \gamma_1 \propto r$. The case of a conical outflow with $k = 1$ must be analyzed separately; the result is that, asymptotically, $\gamma(r)$ grows only logarithmically (e.g. Beskin et al. 2008).

We know, however, that the outflow cannot accelerate forever: the Lorentz factor has a maximum possible value $\gamma_{\mathrm{max}} = \mu$ along each field line. How close can γ get to this maximum? In other words, is the conversion of magnetic into kinetic energy efficient? Again, this is related to the collimation of the jet as we now show.

Let us go back to the definition of $\mu(\Psi)$ in Eq. (5.65)

$$\mu = \gamma - \frac{\Omega r B_\phi B_{\mathrm{p}}}{4\pi c^2 \rho u_{\mathrm{p}}}. \tag{5.74}$$

We can use the asymptotic expression in Eq. (5.70) to eliminate the toroidal magnetic field, and the definition of η in Eq. (5.61) to eliminate ρu_{p}. This gives

$$\mu \approx \gamma - \frac{\Omega^2 r^2 B_{\mathrm{p}}}{4\pi c^3 \eta}. \tag{5.75}$$

Since μ, Ω, and η are constant on field lines, the evolution of the Lorentz factor depends on that of the product $r^2 B_{\mathrm{p}}$: a decrease in the value of this quantity results in an increase in the value of γ.

Consider now an annulus of width δr between two magnetic surfaces Ψ and $\Psi + \delta \Psi$ on a cross section of the jet at constant z as seen in Fig. 5.1. The poloidal magnetic flux through this annulus is $\delta \Psi \propto B_{\mathrm{p}} r \delta r$, then

$$r^2 B_{\mathrm{p}} \propto \left(\frac{r}{\delta r} \right) \delta \psi. \tag{5.76}$$

For $r^2 B_{\mathrm{p}}$ to decrease we must demand that $r/\delta r$ decreases too. This means that the outflow accelerates as long as the radial separation δr between two neighboring poloidal magnetic field lines increases faster than r. Such condition is usually referred to as *differential collimation* or *bunching of field lines*.

The terminal value of the Lorentz factor γ_∞ can be calculated analytically in very few cases. If the poloidal field is assumed to be a split monopole $B_{\mathrm{p}} \propto 1/r^2$, from Eq. (5.75) we get that γ is constant and there is no acceleration at all. Equation (5.75) is, however, only approximate. As shown by Michel (1969; see also Beskin et al. 2008 and Tchekhovskoy et al. 2009), the outflow actually does accelerate reaching an asymptotic Lorentz factor $\gamma_\infty \propto \mu^{1/3} \ll \mu$ and a magnetization $\sigma_\infty \propto \mu^{2/3} \gg 1$. Since the magnetization remains large the jet is asymptotically magnetically dominated, so the acceleration in an exact split magnetic monopole field is highly inefficient. Remember, nevertheless, that a parameterization of the magnetic field strength prescribed in advance does not satisfy the force balance equation, so the results obtained with such models may be misleading. We shall see that high acceleration efficiencies have been obtained in self-consistent numerical simulations of MHD jets.

The reason why the acceleration of the outflow ultimately stops is that the forces exerted by the magnetic and the electric field almost cancel each other. An argument based on the loss of causality can be formulated to understand why this is related to the collimation of the flow (e.g. Komissarov et al. 2009; Tchekhovskoy et al. 2009). To maintain the required differential collimation the field lines must be able to "communicate" with each other across the jet. Consider the emission of fast magnetosonic waves from a point in the jet. In the comoving reference frame the waves are emitted isotropically, but in the lab frame the propagation is restricted to a cone with an opening half-angle

$$\sin \theta_{\mathrm{FM}} = \frac{\gamma_{\mathrm{FM}} \upsilon_{\mathrm{FM}}}{\gamma \upsilon} \tag{5.77}$$

in the local direction of motion of the flow. Here υ_{FM} and γ_{FM} are the fast magnetosonic speed and the corresponding Lorentz factor, respectively. An efficient differential collimation is possible only if the flow can communicate with the jet axis, so the inclination angle of the field lines must satisfy

$$\sin \theta \leq \sin \theta_{\mathrm{FM}}. \tag{5.78}$$

The locus of all points where $\sin \theta = \sin \theta_{\mathrm{FM}}$ defines a "causality surface" beyond which efficient acceleration stops.

We can estimate how efficient the acceleration of the flow is from Eq. (5.78). The phase velocity of the fast magnetosonic waves for a cold flow ($a_s \sim 0$) is

$$\frac{v_{\mathrm{FM}}^2}{c^2} = \frac{B'^2}{B'^2 + 4\pi \rho c^2}, \tag{5.79}$$

where B' is value of the magnetic field in the rest frame of the jet. Then

$$\gamma_{\mathrm{FM}}^2 \frac{v_{\mathrm{FM}}^2}{c^2} = \frac{B'^2}{4\pi \rho c^2}. \tag{5.80}$$

The magnetic field in the comoving frame is related to the fields in the lab frame as $B'^2 = B^2 - E^2$. For $r \ll R_{\mathrm{L}}$, applying Eq. (5.68) and using that $B_{\mathrm{p}} \ll B_\phi \sim E$ we get

$$\gamma_{\mathrm{FM}}^2 \frac{v_{\mathrm{FM}}^2}{c^2} \approx \frac{E B_\phi}{4\pi \gamma^2 \rho c^2}. \tag{5.81}$$

Recalling the expressions for the Poynting, rest-mass, and kinetic energy flux, for $v_{\mathrm{p}} \sim c$ we can write

$$\gamma_{\mathrm{FM}}^2 \frac{v_{\mathrm{FM}}^2}{c^2} \approx \frac{\mathcal{F}_S}{\mathcal{F}_{\mathrm{k}} + \mathcal{F}_{\mathrm{m}}} = \frac{\mu}{\gamma} - 1. \tag{5.82}$$

Back to Eq. (5.78), we finally obtain the desired relation between the Lorentz factor and the inclination angle of the poloidal field lines,

$$\sin\theta^2 \le \sin\theta_{\mathrm{FM}}^2 \approx \left(\frac{\mu}{\gamma} - 1\right)\left(\frac{1}{\gamma^2 - 1}\right) \approx \frac{\mu}{\gamma^3}. \tag{5.83}$$

But $\mu = \gamma_{\max}$, so

$$\gamma \lesssim \left(\frac{\gamma_{\max}}{\sin^2\theta}\right)^{1/3}. \tag{5.84}$$

As we had previously found, poorly collimated flows are inefficient accelerators: for $\gamma_{\max} \gg 1$, the maximum possible value of the Lorentz factor is $\gamma \lesssim \gamma_{\max}^{1/3} \ll \gamma_{\max}$.

The question is, then, if relativistic jets can collimate enough as to develop very large Lorentz factors. The shape of the magnetic surfaces is dictated by the solution of the Grad-Shafranov equation for the force balance across the poloidal field lines. It turns out that the necessary degree of collimation may be achieved as the combined effect of self-collimation *and* the pressure exerted by an external medium (Tchekhovskoy et al. 2008; Lyubarsky 2009; Komissarov et al. 2009). Not any arbitrary external pressure profile, however, is appropriate. If $P_{\mathrm{ext}} \propto z^{-\alpha}$ with $\alpha < 2$, the shape of the field lines is $z \propto r^{4/\alpha}$ and from the discussion above we see that the jet accelerates with $\gamma \propto r$. For $\alpha = 2$ a profile $z \propto r^a$ with $1 < a \le 2$ is obtained, that corresponds to a poor collimation and inefficient acceleration. Finally, for $\alpha > 2$, the external pressure drops too fast and the jet eventually becomes conical and stops accelerating.

5.5   The Blandford-Znajek Mechanism

A Kerr black hole of mass M and spin parameter a_* has an amount of energy (its *reducible mass*)

$$E_{\text{rot}} = Mc^2\left(1 - \sqrt{\frac{c^2 r_{\text{h}}}{2GM}}\right) = Mc^2\left\{1 - \sqrt{\frac{1}{2}\left[1 + \left(1 - a_*^2\right)^{1/2}\right]}\right\} \tag{5.85}$$

that is available to be extracted. This energy is potentially large: for a maximally rotating hole ($a_* = 1$) it equals $\sim 0.3Mc^2$. We have already commented in Chap. 2 on the Penrose process, one possible mechanism of tapping rotational energy from black holes. The conditions that would allow efficient extraction of energy by the Penrose process, however, are not expected to be fulfilled in the environments of astrophysical black holes (Bardeen et al. 1972; Bejger et al. 2012).

If the black hole is immersed in a magnetosphere, rotational energy may be transferred to the electromagnetic field and escape to infinity as a Poynting flux. This process is a specially interesting candidate to explain the formation of astrophysical jets. It was studied in a seminal paper by Blandford and Znajek (1977), and the name "Blandford-Znajek mechanism" is nowadays broadly used to refer to any process of extraction of rotational energy from a black hole by means of an electromagnetic field.

The original model of Blandford and Znajek is based on the works on pulsar magnetospheres by Goldreich and Julian (1969) and on electrodynamics in Kerr space-time by Wald (1974). Blandford and Znajek (1977) considered a Kerr black hole in a magnetic field generated by currents flowing in an equatorial accretion disk. They assumed that the magnetosphere is *force-free*: the inertia of matter is completely neglected everywhere outside the disk, but the charge density is large enough to screen the component of the electric field parallel to the magnetic field. The charges, in the form of electron-positron pairs, would be introduced in the magnetosphere by annihilation of photons radiated by any accelerated seed particles.

In covariant notation the force-free condition reads

$$J_\mu F^{\mu\nu} = 0, \tag{5.86}$$

where $F^{\mu\nu}$ is the electromagnetic field tensor defined in Eq. (2.87) in terms of the four-potential A_μ, and the current J_μ satisfies the inhomogeneous Maxwell equations

$$F^{\mu\nu}_{;\nu} = \frac{4\pi}{c}J^\mu. \tag{5.87}$$

Blandford and Znajek (1977) looked for a solution of Eqs. (5.86) and (5.87) in Kerr space-time using Boyer-Lindquist coordinates (recall Eqs. (2.55)), demanding it to be stationary and axisymmetric ($\partial_t = \partial_\phi = 0$). In analogy to the formalism developed in the previous sections, there exists a flux function Ψ with the property that the surfaces $\Psi = \text{constant}$ are magnetic surfaces. It turns out that $\Psi = A_\phi$ is a possible choice. An equation for A_ϕ may be written down in terms of the metric and the functions

$$-\omega \equiv \frac{A_{t,r}}{A_{\phi,r}} = \frac{A_{t,\theta}}{A_{\phi,\theta}}$$

(5.88)

and

$$B_T \equiv \frac{\Delta}{\Sigma} \sin\theta \, B_\phi.$$

(5.89)

Both ω and B_T are constant along field lines, and thus only depend on A_ϕ. The scalar electrostatic potential A_t is also constant on field lines, hence Eq. (5.88) is equivalent to the function $\Omega = d\chi/d\Psi$ we introduced before. This is why ω is generally referred to as the angular velocity of the field lines.

To complete the formulation of the problem, Blandford and Znajek (1977) imposed boundary conditions at infinity and on the horizon. At infinity, they forced the fields to match known solutions valid in flat space-time. At the horizon ($r = r_\mathrm{h}$), they demanded that the electromagnetic field remained finite as seen by an observer crossing the horizon in free-fall. It was shown by Znajek (1977) that to satisfy the latter condition $A_\phi(r_\mathrm{h}, \theta)$ must be finite and

$$B_T(r_\mathrm{h}, \theta) = \frac{2r_\mathrm{h}(\omega - \Omega_\mathrm{H})}{r_\mathrm{h}^2 + (a^2/c^2)\cos^2\theta} \sin\theta \, A_{\phi,\theta}(r_\mathrm{h}, \theta),$$

(5.90)

where

$$\Omega_\mathrm{H} \equiv \frac{a}{r_\mathrm{h}^2 + a^2/c^2}$$

(5.91)

is called the angular velocity of the black hole.[6]

Calculating the general solution for A_ϕ is highly difficult, and the result found by Blandford and Znajek (1977) is a perturbative solution. They considered an unperturbed magnetic field configuration (monopole or paraboloidal) in Schwarzschild space time and, on it, they built the solution for a slowly rotating Kerr black hole as an expansion in the spin parameter $a_* \ll 1$.

The most important prediction of the Blandford-Znajek model regards the flux of electromagnetic energy from the black hole. Under the force-free approximation there is no contribution of matter to the energy-momentum tensor. The equations for the conservation of energy and momentum then reduce to

$$E^{\mu\nu}_{;\nu} = 0$$

(5.92)

where

$$E^{\mu\nu} = \frac{c}{4\pi}\left(-F^{\mu\rho}F^\nu_\rho + \frac{1}{4}g^{\mu\nu}F^{\sigma\lambda}F_{\sigma\lambda}\right)$$

(5.93)

---

[6]Notice that it equals $d\phi/dt$ in Eq. (2.59) evaluated at the horizon.

is the energy-momentum tensor of the electromagnetic field, recall Sect. 2.6.1. Let ξ_μ be a Killing vector, then from Eqs. (1.21) and (5.92) we get

$$\left(\xi_\mu E^{\mu\nu}\right)_{;\nu} = 0. \tag{5.94}$$

Since the metric is stationary $\xi^\mu = (1,0,0,0)$ is a time-like Killing vector in Boyer-Lindquist coordinates. Then from Eq. (5.94) we obtain that the four vector $\mathcal{E}^\nu = E_t^\nu$ is conserved; we identify this quantity with the electromagnetic energy flux measured by a stationary observer at infinity.[7]

The radial component evaluated at the event horizon is

$$\mathcal{E}^r(r_\mathrm{h}, \theta) = \omega(\Omega_\mathrm{H} - \omega)\left(B^r\right)^2 r_\mathrm{h} \sin^2\theta. \tag{5.95}$$

An observer at infinity will report an outgoing flux of electromagnetic energy from the black hole ($\mathcal{E}^r \geq 0$) if $0 \leq \omega \leq \Omega_\mathrm{H}$. Notice that in deducing Eq. (5.95), Blandford and Znajek (1977) applied Eq. (5.90). We shall say more about the meaning of this condition below. The total radial energy flow observed at infinity is obtained integrating Eq. (5.95),

$$L = 2\pi \int_0^\pi \sqrt{-g}\,\mathcal{E}^r\, d\theta. \tag{5.96}$$

To rigorously calculate the value of the electromagnetic energy flux we must supply an expression for ω and solve Maxwell's equations to find B^r. But if we are just interested in obtaining a quick order-of-magnitude estimate, we can still write down an approximate expression for L. Assume $\Omega_\mathrm{H} = $ constant and a monopole magnetic field $B_\mathrm{n} = B^r \sin\theta = $ constant. Integration of Eq. (5.96) then yields (Lee et al. 2000, see also Beskin 2010)

$$L = \frac{G^2}{c^3} f(x) \frac{\omega(\Omega_\mathrm{H} - \omega)}{\Omega_\mathrm{H}^2} B_\mathrm{n}^2 M^2 a_*^2, \tag{5.97}$$

where $x = a/cr_\mathrm{h}$,

$$f(x) = \frac{1+x^2}{x^2}\left[\left(x + \frac{1}{x}\right)\arctan x - 1\right], \tag{5.98}$$

and $f \sim 1$ in the allowed range $0 \leq x \leq 1$. If we further take $\omega = \Omega_\mathrm{H}/2$[8] we obtain

$$L \approx 10^{46}\left(\frac{B_\mathrm{n}}{10^4\,\mathrm{G}}\right)^2\left(\frac{M}{10^9\,M_\odot}\right)^2 a_*^2\,\mathrm{erg\,s^{-1}}. \tag{5.99}$$

---

[7]Analogous considerations show that the flux of electromagnetic angular momentum $\mathcal{L}^\mu = -E_\phi^\mu$ is also conserved.

[8]This is the value of ω calculated by Blandford and Znajek (1977) to first order in a_* in the perturbed split monopole solution.

This is a very interesting result: it shows that the rate of energy extraction through the Blandford-Znajek mechanism is, in theory, enough to account for the observed power of jets in AGN.

The initial work of Blandford and Znajek (1977) triggered an enormous amount of research, but it has been as well subject to strong criticism. From the results of recent theoretical and numerical work, however, it became apparent that the confusion and controversy arose mainly from the incorrect role ascribed to the event horizon.

The boundary condition at the horizon, Eq. (5.90), seems to be a fundamental piece in the formulation of Blandford and Znajek (1977). This contributed to put forward an analogy that could, supposedly, help to understand more easily how the Blandford-Znajek process works. In the *Membrane Paradigm* (Thorne et al. 1986) the event horizon (strictly a surface infinitesimally outside it, the so-called *stretched horizon*) is replaced by a rotating conductor of surface resistivity $R_\mathrm{H} = 4\pi/c \approx 377$ Ohm, plus other well-defined electrical, mechanical, and thermodynamical properties. The problem is then formulated in terms of the classical laws of physics without the need to resort to general relativistic electromagnetism.

In the Membrane Paradigm the stretched horizon behaves just as an unipolar inductor like the famous Faraday disk. The Lorentz force acting on (fictitious) free charges on the stretched horizon induces charge separation; this in turn generates an electric field and poloidal currents. It is also the rotation of the conductor at the stretched horizon that drives the rotation of the magnetic field lines. A model like this does, in fact, very well explain the energy loss of pulsars (see Beskin 2010 for an introduction to pulsar magnetospheres) and other magnetized rotating objects with a material surface.

Much of the criticism made about the Blandford-Znajek process comes from its strong (and erroneous) association with the Membrane Paradigm. One of such claims was formulated by Punsly and Coronity (1990a, 1990b, see also Punsly 2001) and is related to causality. Their argument goes, qualitatively, like this. The Blandford-Znajek solution describes, actually, two flows: an outflow towards infinity and an inflow towards the event horizon. Each flow has its own critical surfaces where it reaches each of the characteristic speeds of MHD winds. It turns out that the ingoing wind has to cross all its critical surfaces before reaching the horizon (Takahashi et al. 1990; Komissarov 2004). On arriving to the horizon, then, the inflow is already causally disconnected from the outgoing wind. Nothing that occurs at the horizon can affect the outflow at infinity, hence the horizon cannot be an unipolar inductor. Any condition like Eq. (5.90) imposed at the horizon lacks sense. Punsly and Coronity concluded that the solution of Blandford and Znajek (1977) must be unstable (no real system could evolve to such steady-state configuration) and therefore not physically meaningful. They argued that it should be a matter-dominated plasma that creates the necessary electric field to drive the currents. If the electric field were generated "in vacuum" it would be, according to them, screened by free pairs created in the magnetosphere. Punsly and Coronity (1990a, 1990b) developed a different model of magnetohydrodynamics in Kerr space-time that predicted energy extraction from rotating black holes.

In a series of papers that combined numerical and analytical work, Komissarov (2001, 2002, 2003, 2004, 2009, but see also McKinney 2006; McKinney and Gammie 2004; Beskin and Kuznetsova 2000) addressed the issues raised against the Blandford-Znajek mechanism. The results strongly suggest that none of them does really undermine its validity.

The first fact to be noticed is that Znajek's condition at the horizon is not really a boundary condition. From a physical point of view it just imposes a reasonable demand, namely that a free-falling observer measures a finite electromagnetic field. Blandford and Znajek (1977) had to impose it because they worked in Boyer-Lindquist coordinates, but there is no need for it when a regular set of coordinates is used (e.g. McKinney and Gammie 2004).[9] Indeed, it can be seen that Znajek's condition is actually a regularity condition (e.g. Komissarov 2001). In the force-free approximation the event horizon coincides with the fast surface of the ingoing flux; demanding the solution to cross it smoothly yields Eq. (5.90).[10] When the inertia of matter is considered, the fast surface shifts outwards and no condition on the horizon is required.

If the horizon is causally disconnected from the electromagnetic outflow, and having relieved it of the apparently special role attributed to it by Znajek's condition, a possible solution to the arguments against the Blandford-Znajek mechanism suggests itself: it is not driven by the event horizon. The "engine" must be located outside the horizon, and the natural candidate (if we think of the Penrose process) is the ergosphere. This idea is supported by analytical results and simulations.

General relativistic, time-dependent, force-free simulations of magnetospheres around Kerr black holes by Komissarov (2001, 2003, 2004) do predict an outgoing Poynting flux. For an initial split monopole magnetic field, the system tends to a steady state that matches very well the perturbative solution of Blandford and Znajek (1977) for $a_* \lesssim 0.5$. Furthermore, for all values of ω consistent with extraction of electromagnetic energy from the hole, the Alfvén surface of the ingoing flow lies inside the ergosphere (Komissarov 2002, 2003). There always exists, then, an outer region of the ergosphere that is causally connected with the outflow. These results indicate that the Blandford-Znajek solution is stable and causal.

Another interesting outcome of the simulations is that all field lines that thread the ergosphere are forced into rotation with the black hole, no matter whether they also cross the event horizon or not. This is in contradiction with the Membrane Paradigm, according to which only those lines that cross the stretched horizon should rotate.

Finally, let us comment on the last issue raised by Punsly and Coronity (1990a, 1990b) concerning the origin of the electric field. The electric field that drives poloidal currents in pulsars is generated by the charge separation induced by the rotation of the star. There is, however, no such need in a rotating black hole: even in vacuum an electric field appears inside the ergosphere (e.g. Wald 1974). The

---

[9]An example of regular coordinates in Kerr space-time are Kerr-Schild coordinates.

[10]Also in the limit of force-free, degenerate MHD, fast waves propagate at the speed of light.

important result is that this electric field cannot be completely screened by free charges—no stationary solution with completely screened electric field and zero poloidal currents is allowed inside the ergosphere (Komissarov 2004). The concept of an unipolar inductor cannot be applied to black holes magnetospheres, and there is no sense in looking for any surface that serves that purpose.

The weaknesses of the Blandford-Znajek mechanism pointed out by Punsly and Coronity (1990a, 1990b) actually demonstrate a failure of the Membrane Paradigm. Trying to build up an analogy between black hole and pulsar magnetospheres has proved misleading. The phenomena that occur in the environments of a rotating black hole are peculiar to these objects and depend fundamentally on the existence of the ergosphere and the frame dragging effect. The Blandford-Znajek mechanism, then, should be actually regarded as akin to a "Penrose-like" process. This interpretation, that we just mention loosely here, may be developed formally, see for example Komissarov (2009).

5.6 Numerical Simulations

Numerical simulations provide a way to explore regions of the parameter space for which we have none or little prior analytical knowledge;[11] well-established analytical results are useful, of course, to check the validity of the numerical codes. Simulating "real" astrophysical jets is a great challenge. The values of the parameters that characterize sources with jets (length, Lorentz factor and power of the outflow, mass of the central object, and typical magnetic field strength, to name some) span many orders of magnitude (see e.g. Levinson 2010 for some figures). Only in recent years the simulations have become complex enough as to reproduce, with a good degree of verisimilitude, the observed large-scale properties of astrophysical jets. Numerical work has also widely contributed to our comprehension of the physics that govern their evolution. Simultaneously we have significantly progressed in understanding the microphysical processes that take place in jets, such as the origin of the radiation they emit, how particles get accelerated in them, etc. The next natural step would be to couple microscopic and macroscopic phenomena, but it is a difficult task and the advances in that direction are still in their beginnings (see, however, Bosch-Ramon et al. 2011; Huarte-Espinosa et al. 2011; Drappeau et al. 2011; Polko et al. 2013).

In this section we briefly review just a few of the latest numerical works on the MHD theory of jets. The selection is mainly intended to illustrate the results discussed in the previous section, and thus largely incomplete. Many other contributions deserve to be mentioned as well, you can see for instance Koide et al. (2002), Mizuno et al. (2004), McKinney and Narayan (2007), Romanova et al. (2009), Porth and Fendt (2010), Barkov and Khangulyan (2012), Lii et al. (2012).

---

[11] The production of magnetized, supersonic jets of plasma in laboratory experiments is nowadays possible. This could be yet another way to learn about astrophysical jets, as long as the experimental conditions in the laboratory can be correctly scaled. See the articles by Remington et al. (2006) and Ciardi (2010) for reviews on this topic.

One of the goals of numerical simulations is to reproduce the conditions derived from observations in different astrophysical sources with relativistic outflows. The Crab pulsar offers an interesting case.[12] The pulsar wind is completely magnetically dominated at launch but the magnetization drops to very low levels ($\sigma_\infty \lesssim 10^{-2}$) in the terminal region (e.g. Kennel and Coroniti 1984). Classical models of pulsar winds rely on the split-monopole configuration as a good approximation to the dipolar magnetic field far from the neutron star. We have seen, however, that the efficiency of magnetic-to-kinetic energy conversion in a split-monopole magnetosphere is very poor. Can observation and theory be reconciled somehow? It appears, indeed, that the wind eventually becomes matter-dominated in the polar region.

Tchekhovskoy et al. (2009) performed ideal MHD simulations of unconfined relativistic winds launched from a rotating star, following the evolution of the outflows for 10 orders of magnitude in radius. The calculations are time-dependent with an initial pure split-monopole magnetic field. Figure 5.5 shows the steady-state magnetic field distribution and Lorentz factor in two different simulation runs. Whereas the field lines near the equator little differ from radial, they are much more collimated near the axis. In the model in the left panel the maximum Lorentz factor is $\sim$40 but $\sigma_\infty \gg 1$, implying a poor efficiency of energy conversion. In the model in the right panel the density of the plasma at the launching region was lowered near the axis to increase the value of μ, recall Eq. (5.74). Now the acceleration is very efficient near the rotation axis: along these magnetic field lines $\sigma_\infty \ll 1$ and $\gamma_{\max} \sim 200$. This region of the outflow may be identified with a relativistic jet. Towards the equator, on the contrary, the efficiency of acceleration is low. The results can be readily understood applying the concepts of Sect. 5.4. Figure 5.6 shows the evolution of μ, σ, and γ along single field lines. As expected, μ remains constant all over and σ decreases as γ increases. The vertical dotted lines indicate the position of the fast magnetosonic and causality surfaces. The acceleration efficiency is notoriously reduced after crossing the former and practically quenched after crossing the latter.

There are two important conclusions to be drawn from these results. First, that quasi-monopole magnetospheres can launch relativistic winds with good magnetic-to-kinetic energy conversion efficiency near the axis. And, second, that there is no need, a priori, for an external confining medium. A similar result was found by Komissarov et al. (2009), namely that the field lines collimate near the axis and that the Lorentz factor is higher in this region (although still moderate in their model) than towards the equator.

In sources other than pulsars the jets propagate in a dense environment whose influence cannot be ignored. Such is the situation in a collapsar—the most widely accepted scenario for the origin of long gamma-ray bursts. In the collapsar model the burst is driven by the collapse of a massive rotating star into a black hole. This is accompanied by the formation of a transient (typically lasting less than 10 s) jet that propagates through the stellar material as sketched in Fig. 5.7. Jets in GRBs (both

---

[12]This source hosts, of course, a neutron star and not a black hole.

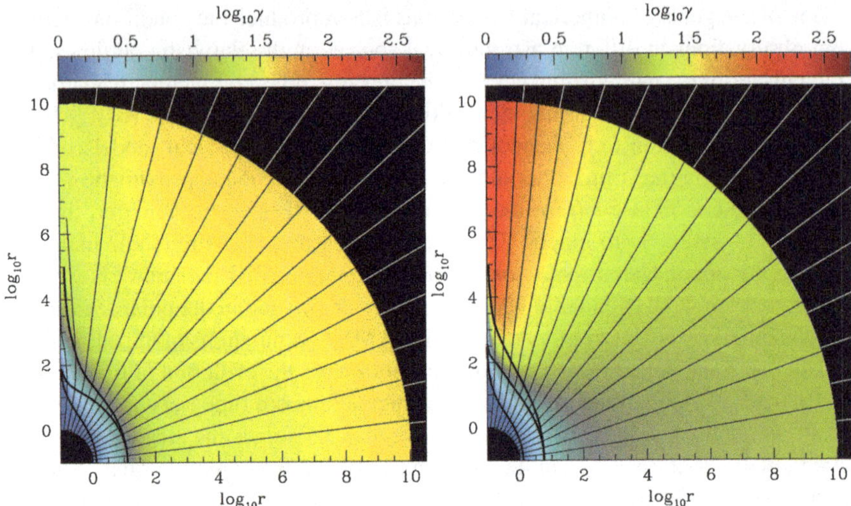

Fig. 5.5 Poloidal magnetic field lines (*thin black lines*) and Lorentz factor (*color scale*) in two simulations of ideal MHD winds from a magnetized rotating star. The *white lines* in the outer part of the figures show the initial (purely radial) magnetic field. From the star outwards, the thick black lines mark the Alfvén, fast magnetosonic, and causality surfaces, respectively. From Tchekhovskoy et al. (2009). Reproduced by permission of the AAS

long and short) are, perhaps, the most challenging when it comes to simulations: one seeks to simulate outflows with Lorentz factors up to $\sim 10^3$ with an energy output of $\sim 10^{51}$ erg, that keep very well collimated (opening angles of a few degrees) while propagating in the dense material of a collapsing star (e.g. Mészáros 2002; Piran 2004).

Tchekhovskoy et al. (2008) considered a simplified model of jet-launching source formed by a rotating black hole and an infinitesimally thin disk around it. The hole rotates with angular velocity $\Omega_0 = \Omega_{\mathrm{H}}(a_* = 1)/2$ whereas the disk is Keplerian with $\Omega_{\mathrm{d}} = \Omega_0 r^{-3/2}$. Field lines thread both components; their angular velocity Ω is determined by the angular velocity of the object to which they are anchored at the footpoint. The jet is identified with those lines that thread the black hole; the field lines that thread the disk launch a wind that supplies the external pressure to the jet. The jets are expected to be highly magnetized ($B \sim 10^{15}$ G) and relativistic, so the (time-dependent) simulations are carried out in the force-free, ideal MHD approximation in flat space-time.

Figure 5.8 shows the result of a simulation. The jet accelerates up to a Lorentz factor of $\sim 10^3$. Depending on the values of the model parameters, the maximum value of γ is attained between the jet axis and its boundary (as in the case of this figure) or at the boundary itself. The total energy carried by the jet is typically $\sim 10^{51}$–10^{52} erg, in agreement with the energetics inferred for long GRBs. The jet opening angle at the surface of the star predicted by the simulations is $\theta_{\mathrm{j}} \sim 10^{-3}$–$10^{-1}$, also consistent with observations.

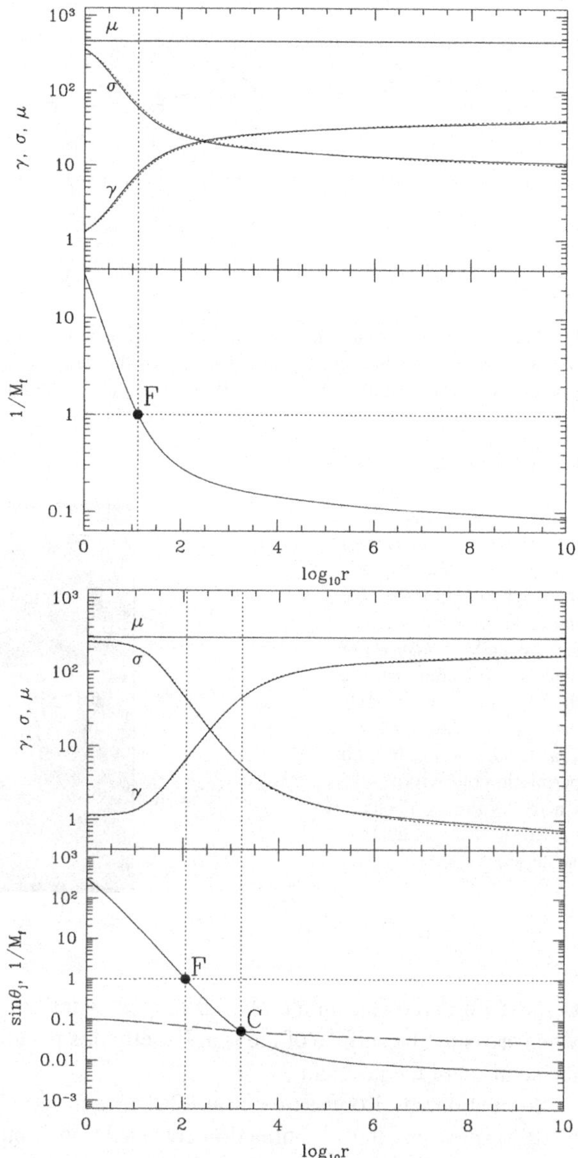

Fig. 5.6 Variation with distance along a field line of some relevant parameters in the models of Fig. 5.5. *Top*: μ, σ, γ. *Bottom*: inverse of the fast Mach number $M_{\mathrm{f}} = \gamma v / \gamma_{\mathrm{FM}} v_{\mathrm{FM}}$ (*solid line*) and inclination angle θ_{j} of the field line (*dashed line*). The values of θ_{j} at the footpoint are $90°$ (*top panels*) and $\sim 6°$ (*bottom panels*). The position of the fast and causality surfaces are indicated with F and C, respectively. For the equatorial field line they both coincide. From Tchekhovskoy et al. (2009). Reproduced by permission of the AAS

There is yet another characteristic of GRBs that simulations should comply with: observations imply that the Lorentz factor and the opening angle of the jet satisfy $\gamma \theta_{\mathrm{j}} \sim 10\text{--}30$. Confined relativistic MHD jets, however, have $\gamma \theta_{\mathrm{j}} \lesssim 1$ (Komissarov et al. 2009; Lyubarsky 2009). Unconfined jets may reach larger values, but we have already seen that such outflows eventually become conical and practically cease to accelerate. Then, if the jet leaves the star before accelerating enough it may not afterwards achieve (at least on any reasonable length scale) a value of γ comparable

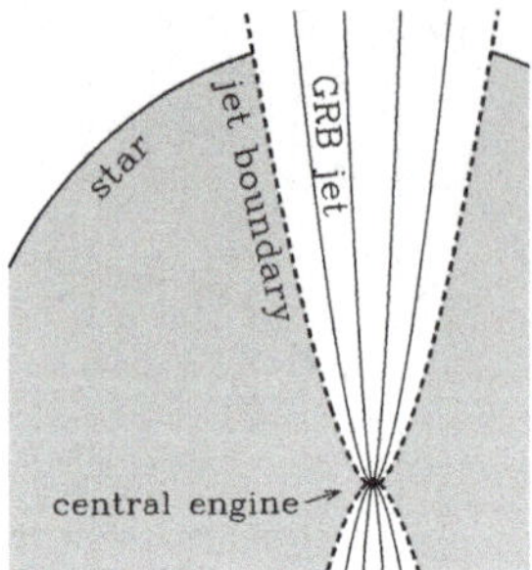

Fig. 5.7 Sketch of a collapsar. A magnetized, relativistic jet makes its way through the collapsing star, that provides a confining medium. From Tchekhovskoy et al. (2008). Reproduced by permission of Oxford University Press on behalf of the Royal Astronomical Society

Fig. 5.8 Lorentz factor of the outflow as a function of position in a simulation by Tchekhovskoy et al. (2008). The *thin solid lines* represent the magnetic field lines, the *thick solid line* the Alfvén surface, and the *dashed line* marks the boundary between the jet and the wind from the disk. From Tchekhovskoy et al. (2008). Reproduced by permission of Oxford University Press on behalf of the Royal Astronomical Society

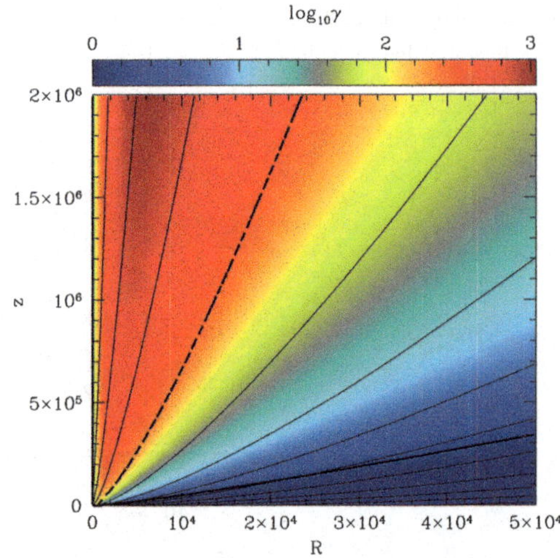

to those observed. But in a collapsar the jet switches from confined to unconfined when crossing the surface of the star. Exactly this peculiarity seems to be the key to the solution of the problem.

Simulations by Komissarov et al. (2010; see Fig. 5.9) and Tchekhovskoy et al. (2010b) show that in the confined-unconfined transition the outflow's bulk velocity is suddenly boosted, whereas the opening angle only slightly increases and afterwards remains constant. The result are values of $\gamma\theta_j \gtrsim 10$ as pursued. Komissarov et al. (2010) suggested an explanation for this effect: when the outflow becomes unconfined, a fast magnetosonic rarefaction wave propagates from the jet boundary to the axis producing a drop in magnetic pressure. The passage of the wave across the jet results in differential collimation of the field lines precisely in the manner required for the flow to accelerate. Notice that acceleration of confined jets by differential collimation can operate for very large length scales, whereas the collimation associated to the rarefaction wave ceases as soon as the wave has reached the axis;

Fig. 5.9 Simulations of a jet permanently confined by a rigid wall (*top*) and one that gets suddenly unconfined (*bottom*), both initialized with the same set of parameters. The lines represent the magnetic field lines and the color code the Lorentz factor. The latter achieves its maximum allowed value, $\gamma_{\mathrm{max}} = \mu$, in the boundary of the unconfined jet. The conversion of magnetic to kinetic energy in the confined jet is substantially more inefficient. The *red area* in the plot for the unconfined jet is vacuum. From Komissarov et al. (2010). Reproduced by permission of Oxford University Press on behalf of the Royal Astronomical Society

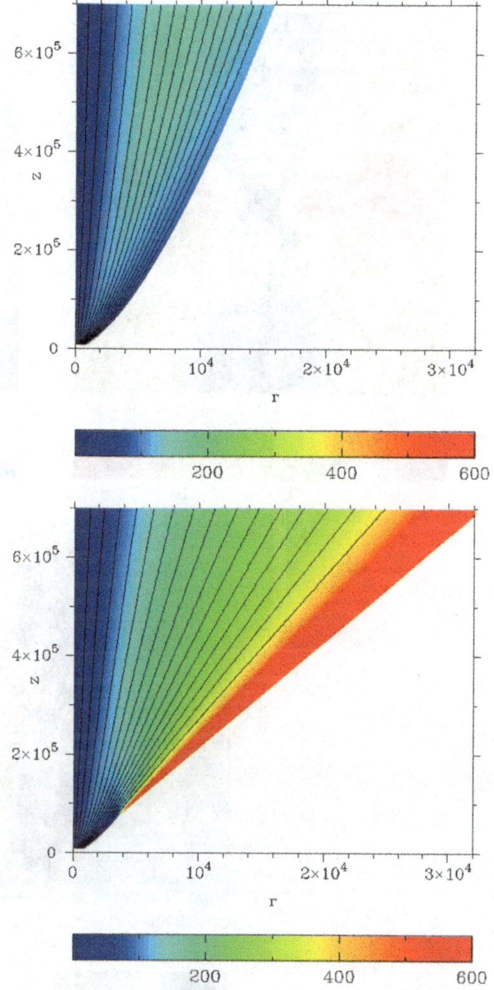

any further acceleration (at least related to collimation) is negligible. At that stage, however, the unconfined jet has already achieved a low magnetization $\sigma \lesssim 1$.

The central engine of a long gamma-ray burst is also an interesting scenario to test the efficiency of the Blandford-Znajek mechanism to launch powerful ultra relativistic jets. Figure 5.10 shows a snapshot of the core of a collapsar according to a 2D general relativistic MHD simulation by Barkov and Komissarov (2008). A pair of jets are launched from the surroundings of a black hole of mass $M_{\mathrm{BH}} = 3M_{\odot}$ and spin parameter $a_* = 0.9$, where the magnetic field reaches a huge strength of $\sim 10^{15}$ G. About 80 % of the energy carried by the jet is tapped from the black hole at a rate of 10^{51} erg s^{-1}. The value of the electromagnetic energy flux is consistent with the prediction of the Blandford-Znajek model.

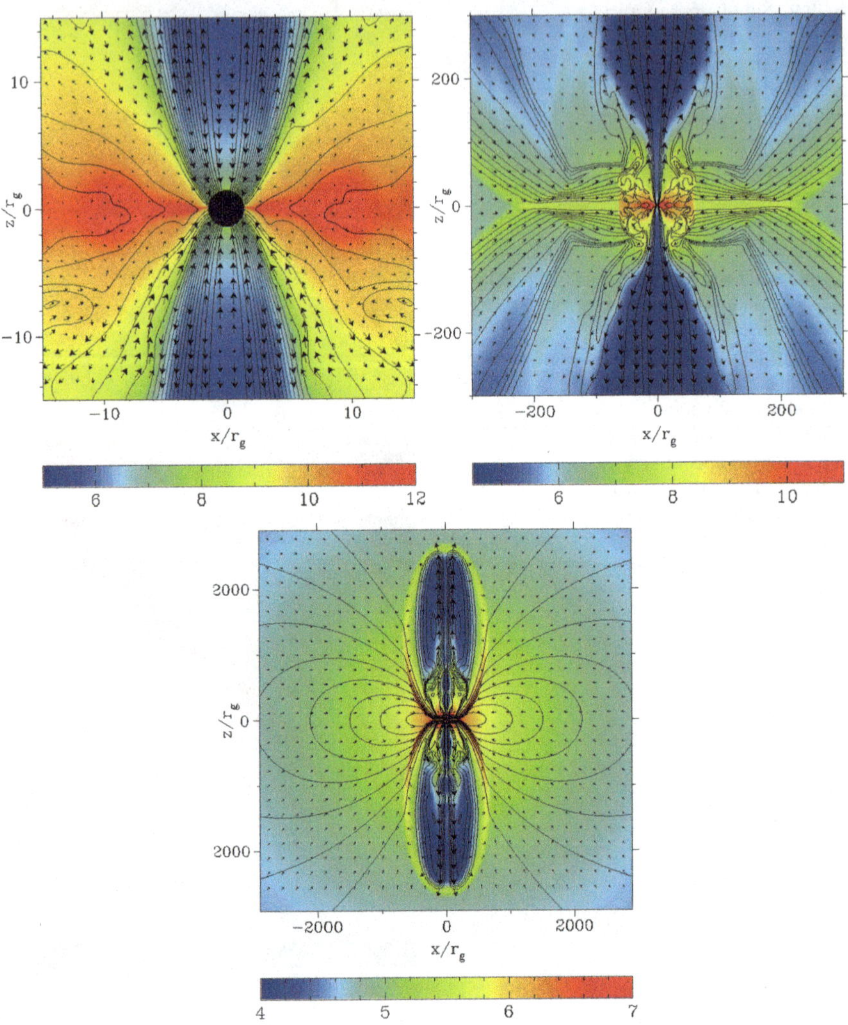

Fig. 5.10 Snapshots of the surroundings of the black hole in a collapsar at different spatial scales. The colors indicate the rest mass density, the solid contours the magnetic field lines, and the arrows the velocity field. Lengths are measured in units of the gravitational radius $r_{\mathrm{g}} = G M_{\mathrm{BH}}/c^2$. Two Poynting-dominated jets are clearly observed. Notice as well that there is a region where the direction of the wind changes from inwards (towards the black hole) to outwards (away from the black hole). From Barkov and Komissarov (2008). Reproduced by permission of Oxford University Press on behalf of the Royal Astronomical Society

5.7 Are Astrophysical Jets Disk-Driven or Black Hole-Driven?

The exact mechanism by which astrophysical jets are launched is unknown. The two strongest suspects are those we studied in this chapter, namely the magnetohydrodynamic or Blandford-Payne process (disk-driven jets) and the Blandford-Znajek process (black hole-driven jets); the quest for evidence favoring one or another is ongoing. The properties of putative black hole-driven jets are expected to depend on spin. Searching for correlations between spin and some other observational parameter should be a way to find out if the Blandford-Znajek process is at least partially involved in the launching of the outflows. Unluckily this is not so easy to do at present, mainly because accurately measuring the spin of astrophysical black holes is difficult. Two methods are usually employed (e.g. McClintock et al. 2011). One consists in fitting the spectrum of the accretion disk to the relativistic model of Novikov and Thorne (1973). In the second method the profile of the Fe Kα line at $\sim$6.4 keV is fitted as well. This fluorescence line is emitted in the inner disk and its shape is distorted by gravitational effects, thus providing information about the black hole spin.

The classical example of a phenomenon blamed on black hole spin is the "radio loud/radio quiet dichotomy" observed in AGN. The ratio $L_{\mathrm{radio}}/L_{\mathrm{opt}}$ of radio-to-optical luminosity is $\sim 10^3$–10^4 times larger in radio loud than in radio quiet sources, when compared at constant L_{opt} (Sikora et al. 2007). Since the radio emission is thought to originate in the jets, the dichotomy reveals that jets in radio loud AGN are much more powerful. One possible explanation for this is that the outflows are driven by some variation of the Blandford-Znajek mechanism, and jets in radio loud AGN are more powerful because the central black holes have larger values of spin. This model is known as the *spin paradigm* (e.g. Blandford 1990; Sikora et al. 2007).

The quadratic dependence of the power outflux on a_* predicted by Blandford and Znajek (1977), recall Eq. (5.99), is too weak to explain the disparity in $L_{\mathrm{radio}}/L_{\mathrm{opt}}$ between radio loud and radio quite AGN, at least without requiring a very large difference in the value of spin between the two populations. The Blandford-Znajek formula is, however, a first order approximation. Refinements of the same model yield a steeper dependence of the power output on a_* and Ω_{H}. Tchekhovskoy et al. (2010a, see also references therein) showed analytically and through simulations that the steepening is enhanced when the accretion flow surrounding the black hole is geometrically thick, i.e. $H/R \gtrsim 1$, like in an ADAF or a thick accretion disk. Their results are nicely summarized in Fig. 5.11. It is possible to obtain a scaling as steep as $L \propto \Omega_{\mathrm{H}}^4$ or even $\propto \Omega_{\mathrm{H}}^6$. In this scenario a plausible variation in spin from $a_* \sim 0.15$ to $a_* \sim 1$ between radio quiet and radio loud sources would be sufficient to account for the dichotomy.

Still other effects may influence the energy output from a spinning black hole. Two of them are the configuration of the magnetosphere and the sense of rotation of the black hole with respect to the accretion disk. A short discussion on this is given by Meier (2011, and references therein). An open magnetosphere connects the black hole with infinity, whereas in a closed magnetosphere the magnetic field lines bend towards the disk. Hence it is expected that open magnetospheres produce Poynting jets more easily. A closed magnetosphere, nevertheless, may turn into an open one if there is enough shear between the disk and the rotating black hole. The strongest shear develops in systems with retrograde black holes (those that rotate in opposite sense with respect to the disk) and in rapidly spinning ($0.75 < a_* < 0.99$) prograde black holes (that rotate in the same sense as the disk). Retrograde black

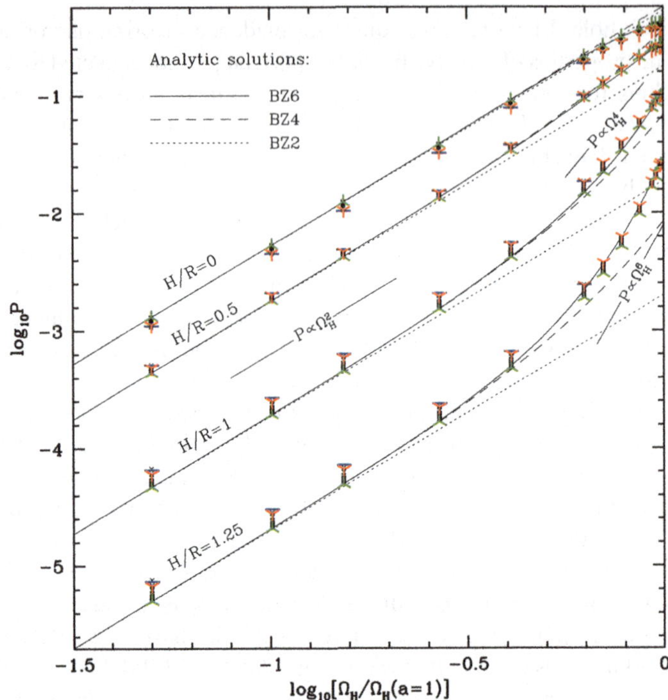

Fig. 5.11 Power output as a function of the angular velocity of the black hole for different height–to-radius ratios of the accretion flow. The *symbols* indicate the results of numerical simulations, whereas the *lines* are analytic approximations up to different orders in a_* and Ω_H. The one labeled "BZ2" corresponds to the original result of Blandford and Znajek (1977). Notice that the dependence of power on spin steepens as $\Omega_H \to 1$ (i.e. $a_* \to 1$) for thick accretion flows. From Tchekhovskoy et al. (2010a). Reproduced by permission of the AAS

holes should be then more likely to launch Blandford-Znajek jets. But, in addition, there is the prediction that jets from retrograde black holes are 10–100 times more powerful (Garofalo 2009; Garofalo et al. 2010). This result is basically founded on the "gap" model by Reynolds et al. (2006), in which the magnetic flux plunges into the black hole at the ISCO and is trapped by the hole. Retrograde black holes accrete the large-scale magnetic field of the disk more efficiently because the radius of the ISCO is larger. The combination of all these factors naturally leads to an asymmetry between the properties of jets from retrograde and prograde black holes, these being more powerful and more likely to occur in the former.

Recent 3D time-dependent simulations by Tchekhovskoy and McKinney (2012) have challenged to some extent this modified version of the spin paradigm. They considered a black hole of spin $a_* = \pm 0.9$ surrounded by a torus of inner radius $r_{in} = 15 r_{grav}$ and typical scale height $H/r \sim 0.3$–0.6. After the simulations converge to the steady state it is observed that the energy extraction from system with the prograde black hole is more efficient than that from the retrograde black hole.

In both cases the efficiency grows as the disk thickness increases.[13] Furthermore, $\sim$80 % of the energy outflow is carried by a Blandford-Znajek-powered jet (it scales quadratically with the magnetic flux on the event horizon) and the rest by a disk-powered wind. The results are independent of the disk inner radius and the initial magnetic flux of the accretion flow, and show no evident formation of a magnetic flux-free gap around the black hole. The gap model, however, strictly applies to thin disks with weak magnetic fields, conditions that are different from the setup of Tchekhovskoy and McKinney (2012).

Regarding specifically the launching of jets in X-ray binaries (XRBs) two recent studies have led to opposite conclusions. Before describing the results it is timely to notice that black hole XRBs show two types of outflows: continuous jets and discrete ejections. Continuous jets are detected during the so-called "low-hard" X-ray state (we shall say more about the spectral states of XRBs in the next chapter). These are steady, mildly relativistic outflows extending for $\gtrsim$10 AU in length. Discrete jets (or *blobs*) are transient ejections of plasma that propagate for distances of parsecs at relativistic speeds. They are observed during transitions between spectral states, a period also characterized by a climb in the system's luminosity up to near the Eddington value.

Fender et al. (2010) considered a sample of $\sim$15 XRBs with reported values of the black hole spin. They searched for correlations between spin and jet power (both for continuous and discrete outflows) or jet bulk speed (in the case of discrete ejections). They found none, expect for a dubious link between the power of transient jets and spin but only in those sources where the spin was calculated by the Fe Kα line method. To estimate the total jet power L_{jet} these authors relied on empirical laws linking it to the radio and X-ray luminosities (e.g. Gallo et al. 2003; Merloni et al. 2003).

Narayan and McClintock (2012), on the other hand, analyzed data from only four sources with transient outflows. They adopted the peak radio luminosity at 5 GHz as a proxy for the jet power. This value was found to increase with the spin parameter as $L_{\mathrm{jet}} \propto a_*^2$, in agreement with the result of Blandford and Znajek (1977) for slowly rotating black holes. Moreover, a correlation with the angular velocity of black hole of the form $L_{\mathrm{jet}} \propto \Omega_{\mathrm{H}}^2$ was also discovered. This scaling was suggested by Tchekhovskoy et al. (2010) and McKinney (2005) to be valid for values of a_* close to unity.

Narayan and McClintock (2012) argue that their results prove that the Blandford-Znajek process is most probably responsible for powering transient jets in XRBs. This affirmation is additionally supported by the fact that the values of a_*, L_{jet}, and Ω_{H} used to fit the correlations span orders of magnitude. The differences with the results of Fender et al. (2010) might be due to them choosing a less adequate proxy

---

[13]Simulations in 2D by Fragile et al. (2012) showed no correlation between the disk thickness and the jet power. These simulations probe a different regime than those by Tchekhovskoy and McKinney (2012), since the authors consider disks subject to cooling with $0.04 \lesssim H/r \lesssim 0.16$ and a flow not dominated by magnetic pressure. As a conclusion, they suggest that it is the corona and not the disk wind that provides confinement to the jet.

for the jet power. As to continuous jets, the independence of L_{jet} on the spin may be reasonably explained: these outflows are expected to be launched far from the black hole ($\gtrsim 50 r_{\text{g}}$) where relativistic effects are already negligible.

In any case, both the works of Fender et al. (2010) and Narayan and McClintock (2012) are based on data from a few sources for which there exist estimates (some very uncertain) of the black hole spin. The results, although definitely very exciting, must be better interpreted with caution and not taken as definite for the moment.

There is currently no concluding answer, then, for the question of what drives astrophysical relativistic jets. In view of the results we have gathered here it appears that both electromagnetic field and matter are fundamental pieces in the process. The most likely scenario is, perhaps, that both Blandford-Znajek and Blandford-Payne mechanisms are to some degree responsible for powering jets, one or another being relevant possibly at different spatial scales.

5.8 More on Jet Dynamics

We have seen that the "standard" axisymmetric, non-dissipative magnetohydrodynamic model provides a complete description of the process of launching, acceleration, and collimation of jets. To work efficiently, however, it relies on a number of factors that include the presence of a large-scale magnetic field and an external medium with particular properties. Up to what extent these quite finely-tuned conditions are met in real astrophysical sources we do not know.

There is yet another particularly troublesome issue with the standard model. One of its central predictions (although we have seen some ways to circumvent it) is a low efficiency of magnetic-to-kinetic energy conversion—far from the launching region relativistic jets still carry a large fraction of their energy as electromagnetic energy. It is also known that magnetized flows, say with $\sigma \gtrsim 0.1$, are not favorable environments for efficient conversion of kinetic energy into thermal energy at shocks (e.g. Mimica and Aloy 2010). The detection of non-thermal radiation from relativistic jets, on the other hand, allows to infer that these are populated with particles accelerated to very high energies. It is in this point that theory and observation clash: the mechanism of particle acceleration generally assumed to be the most relevant (first order Fermi-like acceleration, e.g. Drury 1983) involves precisely diffusion across shock fronts.[14]

There are alternative models that partially overcome these problematic features. Below we shortly examine two of them: magnetic towers and impulsive acceleration. Like the standard model both have the magnetic field as the fundamental ingredient. We shall not discuss *hydrodynamic* jets. Let us just mention that the hydrodynamic acceleration of outflows is a very robust mechanism without some of

---

[14]The reason is that shock waves are commonplace in a variety of astrophysical sources. Besides the first order Fermi process other acceleration mechanisms rely up to some extent in the presence of shocks, see for example Derishev et al. (2003). Of course, particles may also be accelerated without shocks by means of an electric field as expected to occur in pulsar magnetospheres.

the difficulties of the MHD formulation; you can find some simple calculations that demonstrate it in Komissarov (2011). A basic exposition of instabilities and the interaction of jets with the medium follows. Any mathematics is avoided here; we put special emphasis on illustrating the effects of these processes on the structure of the flow, and how they may be related to the dissipation of energy and formation of shocks.

5.8.1 *Alternative Mechanisms of Launching and Acceleration*

5.8.1.1 Magnetic Towers

The "magnetic tower" model was introduced in a series of papers by Lynden-Bell and Boily (1994) and Lynden-Bell (1996, 2003, 2006), and has been studied theoretically and numerically by many others; see for example Li et al. (2001), Kato et al. (2004a, 2004b), Nakamura et al. (2006, 2007, 2008), Uzdensky and MacFadyen (2006, 2007), and Huarte-Espinosa et al. (2012).

In its elementary formulation, the mechanism of formation of a magnetic tower is the following. Consider an accretion disk with an initial configuration of magnetic field consisting in loops whose two footpoints are anchored to the disk at different radii. An alternative configuration could be one with one footpoint of the line anchored to the disk and another to the central rotating star. The plasma immediately above (below) the disk is assumed to be in force-free state. Since the disk is in differential rotation, each field line is twisted at a rate equal to the difference between the angular velocity at the radii the footpoints are anchored. Then the field develops a toroidal component.

The magnetic pressure exerted by the toroidal field makes the loops expand in the way shown in Fig. 5.12. If there is no external pressure, however, after less than a complete turn the field stretches to infinity and the inner and outer parts of the disk initially connected by the magnetic loops become disconnected. The inclination angle of the field lines with respect to the symmetry axis is $60°$, so the process yields a very poor collimation. When a non-vanishing external pressure is added the picture changes completely. As soon as the magnetic pressure inside the loops equals the external pressure the lateral expansion of the field lines stops. Instead, these continue to expand vertically and a structure in the shape of a column develops; this is the "magnetic tower". By pressure balance Uzdensky and MacFadyen (2006) estimate the radius of the tower to be

$$R \sim \left(\frac{\Psi_0^2}{8\pi\, P_{\text{ext}}} \right)^{1/4}, \tag{5.100}$$

where Ψ_0 is the poloidal magnetic flux per unit toroidal angle in the tower and P_{ext} is the external pressure. The height of the tower increases with every turn of the disk at an approximately constant rate.

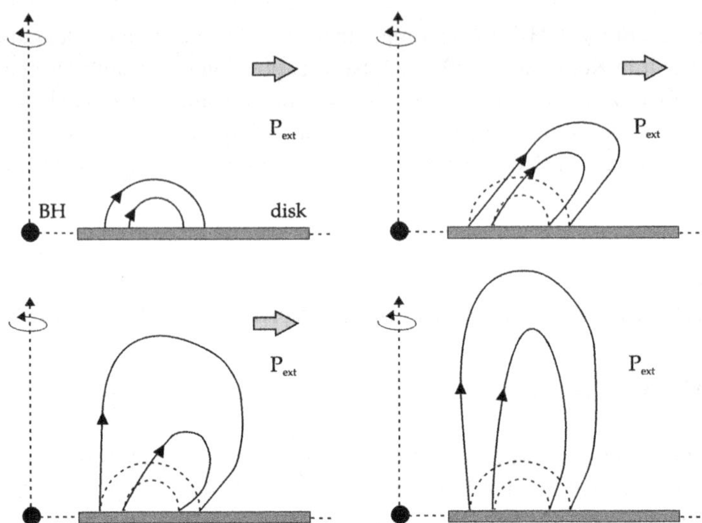

Fig. 5.12 Formation of a magnetic tower: evolution of the magnetic field lines in the poloidal plane. Loops of magnetic field anchored to an accretion disk in differential rotation around a black hole (BH) inflate, remaining confined by the ambient pressure. Adapted from Uzdensky and Mac-Fadyen (2006)

The rise of magnetic towers has been observed in numerical simulations. Kato et al. (2004a) performed time-dependent MHD simulations of a neutron star surrounded by an accretion disk connected by an initially dipolar magnetic field. The temporal evolution of the flow and the magnetic field configuration are shown in Fig. 5.13. The magnetic loops at first expand poorly collimated, but later get collimated into a cylindrical column as pressure balance between the interior and the exterior of the tower is established. The head of the jet advances at a speed $\sim 0.1c$ into the medium. Notice that matter moves away from the disk in the outside of the tower and towards the star in the inner region, thus a current sheet develops. Magnetic reconnection in the current sheet injects plasmoids in the jet.

In an additional MHD simulation Kato et al. (2004b) addressed the role of the external pressure and followed the temporal evolution of the magnetic tower until the system reached the steady state. They found that the development of the tower depends quite strongly on the value of the external pressure: when this is too large the rise of the tower is suppressed. Even for (nonzero) values of the pressure that let the jet be launched, they observe that the magnetic tower ordered structure is only transient. After some time it switches to a quasi-steady complex array of field lines, as seen in Fig. 5.14.

These simulations predict that near the axis the magnetic field has a significant poloidal component, whereas the toroidal field is dominant in the outside boundary of the jet. More recent simulations by Huarte-Espinosa et al. (2012) yield similar results. Roughly, the interior of the tower may be described as a cavity with low matter density and magnetically-dominated (with a low thermal-to-magnetic energy

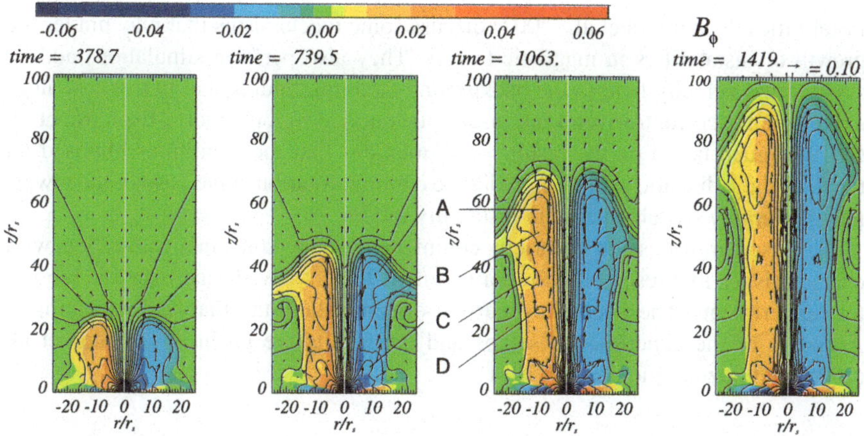

Fig. 5.13 Formation of a magnetic tower jet in a neutron star/accretion disk system. The *arrows* indicate the velocity field, the solid lines the poloidal magnetic field, and the *color code* the strength of the toroidal magnetic field. Letters A, B, C, and D mark the plasmoids injected in the jet at magnetic reconnection sites. Each plot corresponds to a different simulation time. Lengths are measured in units of the Schwarzschild radius of the star. From Kato et al. (2004a). Reproduced by permission of the AAS

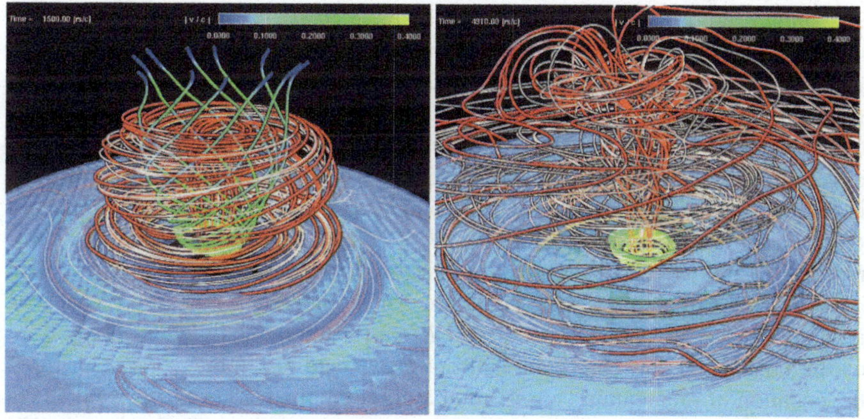

Fig. 5.14 Three dimensional plot of the magnetic field lines in a magnetic tower. *Left*: field configuration for short timescales. *Right*: field configuration in (quasi-)steady state. From Kato et al. (2004b). Reproduced by permission of the AAS

ratio). The cavity is collimated by the external medium. The axial core of the tower has higher density and higher thermal-to-magnetic energy ratio, and is collimated by the toroidal field in the cavity.

Magnetic field configurations with strong toroidal components are prone to become unstable (we shall say more on this topic below). Kato et al (2004b; see also Nakamura et al. 2006), however, found no apparent sign of disruption of the jet by instabilities. Nakamura et al. (2007) discuss some mechanisms that may hinder the growth of instabilities in magnetic towers. They also perform simulations adding some initial velocity field to the background medium, finding that in this case non-axisymmetric current-driven kink instabilities may develop. Huarte-Espinosa et al. (2012) argued based on the results of simulations that the conditions imposed on the flow (i.e. adiabatic evolution, radiative cooling, rotation at the base of the tower) may preferably switch on certain instability modes.

To close our discussion we briefly comment that supersonic magnetized jets with characteristics that resemble those of magnetic towers have been created in laboratory experiments. These jets show current-driven instabilities that lead to clumping. Accounts of the experimental setups and the results are given in Lebedev et al. (2005) and Suzuki-Vidal et al. (2011).

5.8.1.2 Impulsive Acceleration

Contopoulos (1995) proposed a MHD model of jet launching that relies only on the existence of a toroidal magnetic field. The model is motivated by the result that the differential rotation of the accretion disk is expected to amplify the toroidal, not the poloidal, component of the magnetic field. When the poloidal component is set to zero the magnetic field configuration is that of circular rings (axial symmetry is assumed) all along the outflow from the base. Since there is no poloidal field the magnetocentrifugal mechanism does not work, and the outflow is driven by a vertical gradient of magnetic pressure and not by the rotation of the disk.

In the non-relativistic version of the model the flow is described by the set of equations in Sect. 5.3. But as $\mathbf{B}_p = 0$ the concept of flux function is now useless. In its place, a stream function $\Phi(r, z)$ may be defined such that the poloidal velocity is calculated from its derivatives. The poloidal velocity is tangent to the lines of $\Phi = \text{constant}$ in the poloidal plane, and $\Phi/2$ is the mass flux through a section of radius r of the jet. As in the general case, there are functions of Φ that remain constant along streamlines. One is the angular momentum $L(\Phi) = v_\phi r$, that now is solely carried by matter. The constant $E(\Phi)$ related to the poloidal energy flow, on the other hand, has contributions from matter and electromagnetic field. From the equation of momentum conservation a differential equation for Φ is obtained. Contopoulos (1995) showed that it reduces to the general Grad-Shafranov equation in the limit when $B_p/B_\phi \to 0$.[15] The mathematical description of the flow, however, is simpler than in the general case because there is only one critical surface—the fast magnetosonic—instead of three.

A self-similar solution of these set of equations is presented in Contopoulos (1995). At large distances from the central object the solution has the same properties (a steady collimated outflow) than models with $B_p \neq 0$; this is to be expected since we have seen that asymptotically the toroidal magnetic field dominates over

---

[15]Recall that, in general, fixing the shape of the poloidal field does not yield a consistent solution of the MHD equations since such $\mathbf{B}$ does not satisfy the Grad-Shafranov equation.

the poloidal component. The two types of models differ in the characteristics imposed on the launching region. In the magnetocentrifugal model, there must exist a poloidal magnetic field with a certain inclination with respect to the disk to initially accelerate the plasma. In the model with $B_p = 0$, the necessary condition is a continuous supply of toroidal magnetic flux by the disk. As we have already said, and as Contopoulos (1995) notices, a purely toroidal magnetic field configuration is prone to strong instabilities. The acceleration of the flow by vertical gradient of magnetic pressure is then not expected to operate over large length scales. But if transient events of toroidal field expulsion by the disk occurred, this may be an efficient mechanism to launch discrete jets (blobs of plasma) instead of continuous outflows.

Lately, the idea of impulsive acceleration of relativistic outflows has been taken up in the works of Lyutikov (2010a, 2010b), Lyutikov and Lister (2010), Granot et al. (2011), and Granot (2012a, 2012b). A basic version of the mechanism that illustrates its potential is discussed in Granot et al. (2011). They consider a one dimensional planar shell[16] of cold, highly magnetized plasma that rests on one side on a conducting wall and faces vacuum on the other. At $t = 0$ the length of the shell is l_0, the magnetic field is B_0 (uniform and parallel to the wall), and the magnetization $\sigma_0 \gg 1$. The evolution of the shell for $t > 0$ consists of three stages. Initially, as the shell is left to expand against vacuum driven by magnetic pressure gradient, a rarefaction wave propagates towards the wall. The front of the wave travels almost at the speed of light and reaches the wall at $t_0 \sim l_0/c$, signaling the end of the first or "impulsive" phase. By this time the shell has effectively separated from the wall; it propagates keeping a constant thickness $\sim 2l_0$ with a low density and low energy tail behind. Around the end of the impulsive stage, the mean Lorentz factor and the magnetization of the plasma are $\langle \Gamma \rangle \sim \sigma_0^{1/3}$ and $\langle \sigma \rangle \sim \sigma_0^{2/3}$, respectively. During the second phase, until $t_c \sim t_0 \sigma_0^2$, the flow accelerates as $\langle \Gamma \rangle \propto t^{1/3}$ eventually reaching $\langle \Gamma \rangle \sim \sigma_0$ and $\sigma \sim 1$. There is no further increase of the Lorentz factor during the third phase, whereas the magnetization continues to decrease as $\sigma \sim t^{-1}$ and the flow eventually becomes matter-dominated. Granot et al. (2011) nicely explain how the working principle of the process is related to the conservation of momentum, and resort to familiar phenomena such as rocket launching and a pair of masses attached to a spring for analogies.

The attractive feature of the impulsive acceleration is that, under a variety of conditions, it leads to a more rapid and efficient conversion of magnetic-to-kinetic energy compared to the standard MHD steady acceleration. The impulsive mechanism is likely to operate more efficiently in variable astrophysical sources that produce discrete ejecta.[17] Granot et al. (2011) argue that to maximize the advantages of the mechanism the time between ejections should be relatively long, enough as to leave large (almost) empty regions between shells. In this manner, when the shells collide they would be already in the kinetic energy-dominated phase thus leading to an efficient energy dissipation at shocks.

---

[16]The results apply as well to the case of a spherical shell.

[17]Such events are observed for instance in microquasars, see next chapter.

Granot et al. (2011) discuss the plausibility that a situation similar to those of their simplified model arises in jet launching sources. They also develop applications of the model using values of the characteristic parameters of the order expected to reproduce the conditions in GRBs and AGN. The conclusion is that the combination of steady MHD acceleration/collimation followed by the onset of impulsive acceleration may explain the large Lorentz factors inferred for GRBs, and help to mitigate the problem of the efficiency of energy dissipation in shocks in AGN jets.

5.8.2 *Instabilities and Interaction with the Medium*

Astrophysical jets propagate keeping well collimated for extremely long distances (up to millions of light years in AGN) until they brake by interaction with the medium. This is indeed a striking fact, considering that jets are subject to instabilities that manifest as distortions in their structure such as bends, knots, and twists. Eventually, the growth of instabilities may lead to serious decollimation and even the complete disruption of jets. Since observations make it clear that this extreme situation is not the most usual, some stabilizing mechanisms must operate to prevent it. We shall restrict our discussion on instabilities in jets to the very basics; comprehensive and updated reviews on theoretical and numerical work in this topic are presented in Hardee (2011) and Perucho (2012).

The analytical study of the stability of a magnetized flow is performed linearizing the equations of MHD. In the simplest case of a cylindrical axisymmetric jet one seeks for an expansion of the solution in Fourier components of the form $\exp[i(\mathbf{k}\cdot\mathbf{r}+m\phi-\omega t)]$. The integer $m=0,\pm1,\ldots$ specifies the azimuthal order of the normal mode. A mode is unstable if ω or $\mathbf{k}$ are complex so that the amplitude of the perturbation grows. When looking for solutions of the linearized equations it is assumed that the perturbations remain of small amplitude. If this is not satisfied a full non-linear stability analysis must be carried out.

The two types of instabilities most relevant to astrophysical jets are the current-driven instability (CDI) and the Kelvin-Helmholtz instability (KHI). Both are expected to occur in Poynting-dominated regions of the jet, but only the KHI in the kinetically-dominated zones. There is a second type of instability that, like the CDI, occurs even in the absence of bulk motion. This is the so-called pressure-driven instability, triggered by the lack of balance between gas and magnetic pressure. We shall not deal with the pressure-driven instability here; you can see the works of Begelman (1998), Kersalé et al. (2000), and Longaretti (2003) for more on it.

Current-driven instabilities are triggered by gradients of the force exerted by a toroidal magnetic field, and thus take place in jets that carry axial currents. Two well-known modes of the CDI are the pinch or sausage ($m=0$) mode and the helical kink ($m=\pm1$) mode; their effects are sketched in Fig. 5.15 in the simplified case of a column of plasma with toroidal magnetic field only. If some section of the column narrows, the field strength locally increases. So does the magnetic tension, leading to further compression. This is the onset of the pinch mode that causes the column

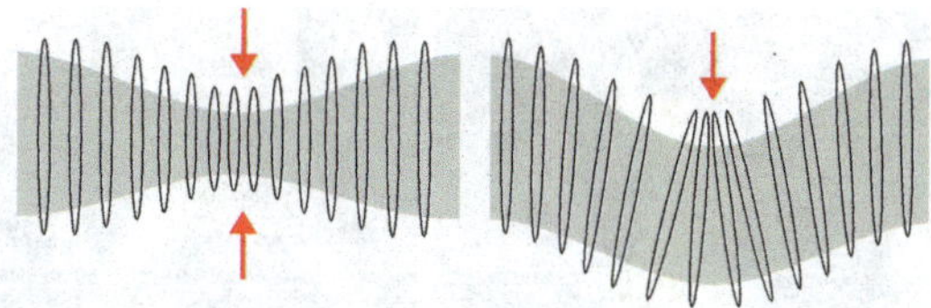

Fig. 5.15 Instabilities in a cylindrical column of plasma driven by a toroidal magnetic field (rings). *Left*: pinch mode. *Right*: kink mode. The *arrows* indicate the direction of the net force on the column

to strangle. The kink mode switches on if the column is somehow perturbed and bends. The toroidal field in the inner part of the kink becomes larger than in the outer part, the magnetic pressure in the inner part rises, and the column continues to bend more and more. Kink modes are generally regarded as the most threatening for the stability of the jets; these excite large-scale helical motions that can seriously distort the structure of the flow or even cause its disruption.

Both pinch and kink modes are stabilized by the presence of a poloidal magnetic field. Indeed, stability criteria[18] show that what determines the onset of the CDI is a large value of the ratio B_ϕ/B_p, called the *helicity* of the magnetic field.[19] The growth of CDIs may be affected as well by the expansion and rotation of the jet. It has been suggested that non-linear CDIs may help to convert magnetic energy into kinetic energy (Hardee 2011).

The Kelvin-Helmholtz instability develops at the separation surface between two flows with different velocities—the jet and the surrounding medium, for example. The result is the mixing of both media. In jets, the mass entrainment driven by the KHI may lead to decceleration, severe loss of collimation, and disruption. The spatial growth rate of the KHI, however, is slowed down by a number of factors. These include a large Lorentz factor, a large Mach number, low temperature or high density of the plasma, and the radial expansion of the jet. A wind that surrounds the jet or a thick shear layer at the boundary operate in the same sense. The magnetic field may also inhibit KHIs in several ways. A most interesting results is that magnetohydrodynamic jets are KH stable as long as they remain sub-Alfvénic (e.g. Romero 1995, Hardee and Rosen 1999, 2002).

The latest numerical works on the stability of jets are based on complex simulations. A throughout numerical study of CDIs, for instance, is presented in a series of papers by Mizuno et al. (2007, 2009, 2011, 2012). The simulations were designed to study the effects of rotation, external winds, velocity shears, etc, on the development of the CDI in relativistic jets. The initial equilibrium configuration is that of force-free plasma with low gas pressure, threaded by a helical magnetic field characterized by a "pitch" $P \equiv r B_z/B_\phi$. A small amplitude velocity perturbation is imposed on this configuration.

The upper panels of Fig. 5.16 show the three-dimensional structure of a constant density surface after the transition to the non-linear phase of the instability in three

---

[18] In a static force-free plasma, for example, the Kruskal-Shafranov criterion predicts instability when $-B_\phi/B_\mathrm{p} > 2\pi r/L$, where r is the radius and L the length of the plasma column.

[19] In the frame of the standard MHD model, then, we expect that jets become unstable beyond the Alfvén surface.

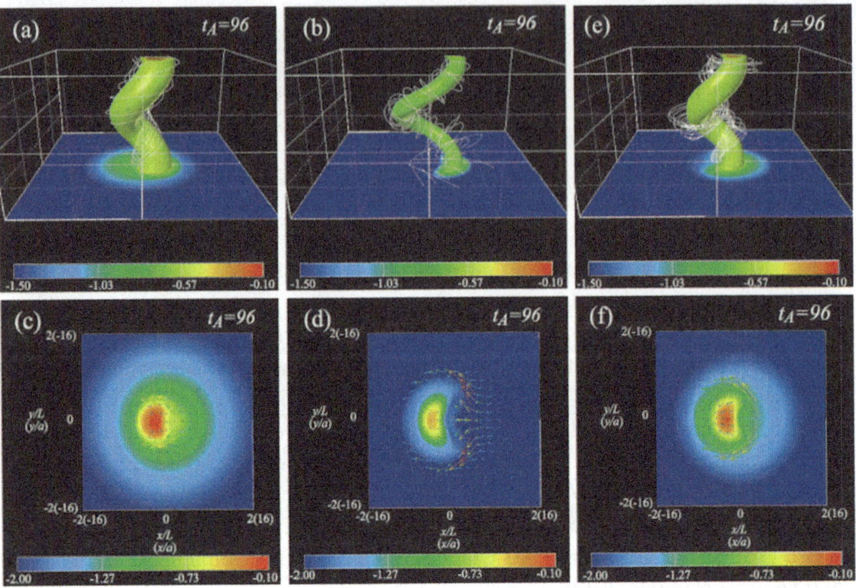

Fig. 5.16 *Top row*: 3D structure of a constant density surface. *Bottom row*: cross section of the jet showing the radial distribution of the logarithm of the density. The *white lines* are the magnetic field lines. From left to right, the plots correspond to simulations with radially increasing, radially decreasing, and constant value of the magnetic pitch. From Mizuno et al. (2009). Reproduced by permission of the AAS

cases: from left to right, radially increasing, radially decreasing, and constant value of the pitch. The column is twisted and wrapped by magnetic field lines although in no case disrupted. The growth of the CDI kink instability triggers radial motions of the plasma, more or less pronounced depending on the case. This can be clearly seen from the density distributions on a cross section of the jet in the bottom panels. Notice that the high-density core is displaced from its initial position on the axis. The growth rate of the instability is found to depend on the behavior of the magnetic pitch. Increasing the pitch with radius has a stabilizing effect: it slows down the growth during the linear phase and retards the transition to the non-linear regime. The opposite is true for a column with radially decreasing magnetic pitch.

Adding differential rotation to the column or a radially decreasing density and gas pressure profile completely alters the outcome. Compare the previous results with those in Fig. 5.17; the jet is now completely disrupted. These simulations by Mizuno et al. (2012) also confirm the analytical result that the radial distribution of the poloidal magnetic field strongly conditions the growth of the instability. More or less homogeneous poloidal fields produce a slow linear growth and saturation of the amplitude of the perturbations in the non-linear phase, with no significant disruption of the jet. An increase of $\mathbf{B}_\mathrm{p}$ towards the jet axis, on the other hand, leads to the growth of the perturbations until the disruption of the structure of the flow. Recall from Sect. 5.6 that an accumulation of poloidal magnetic flux in the

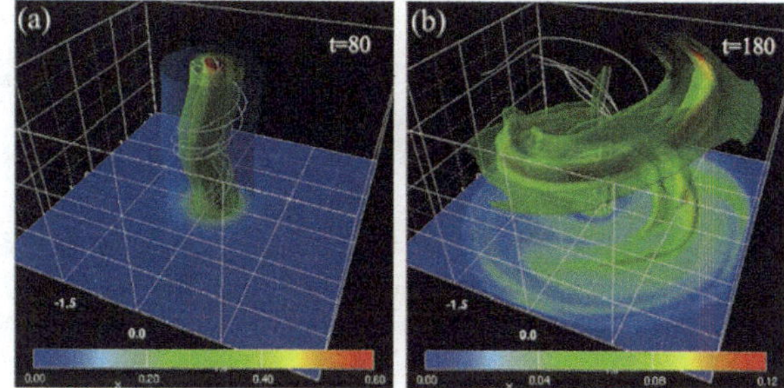

Fig. 5.17 Same as in the top row of Fig. 5.16, but for a differentially rotating jet with constant magnetic pitch at two time instants. From Mizuno et al. (2012). Reproduced by permission of the AAS

core of relativistic jets is predicted by the results of MHD simulations. Based on this Mizuno et al. (2012) argue that Poynting-dominated jets may become strongly kink-unstable when the magnetization falls to $\sigma \sim 1$. The complete disruption of the ordered field structure should dissipate magnetic energy, perhaps powering episodes of flaring emission.

Instabilities may be also triggered by the interaction of the jet with the medium it traverses. One possible cause is the development of recollimation (or reconfinement) shocks that occur when the jet becomes supersonic and underpressured with respect to the external medium (e.g. Sanders 1983; Falle 1991; Komissarov and Falle 1997). Recollimation shocks prompt instabilities by exciting large amplitude non-linear pinches or plasma motions that couple to the KHI (Perucho 2012). Figure 5.18 shows a snapshot of the simulated mass density distribution in a recollimated jet. The outflow propagates in a medium of density that decreases with the distance to the injection point. When the linear size of the jet is of the order of ~ 1.5 kpc it becomes underpressured and a recollimation shock appears. Successive phases of expansion and recollimation follow as the flow becomes over and underpressured with respect to its surroundings. The shocks set in oscillations and mass entrainment that loads the jet and slows it down. Eventually the matter from the external medium makes its way to the jet axis and disrupts it completely.

As we anticipated, shocks in general are expected to be sites where particles get accelerated to relativistic energies and radiate. In this context, the bright stationary or moving knotty patterns observed in many extragalactic jets have been linked to reconfinement shocks (e.g. Stawarz et al. 2006). Notice that there are other possible explanations for the origin of internal structures in jets, among them Kelvin-Helmholtz kink modes (e.g. Hardee and Eilek 2011; Perucho et al. 2012), jet precession, and pinches excited by the injection of fast and dense discrete components into the flow (e.g. Agudo et al. 2001).

Analogously, the hot spots observed in some jets from active galaxies (those historically classified as Fanaroff-Riley II) are associated with radiation emitted by accelerated particles at the termination shocks. These are the shocks that form in the boundary between the "head" of the jet and the external medium. Hydrodynamical simulations of the jet/medium interaction in the termination region have been carried out, for example, by Bordas et al. (2009) and Bosch-Ramon et al. (2011), see Fig. 5.18.

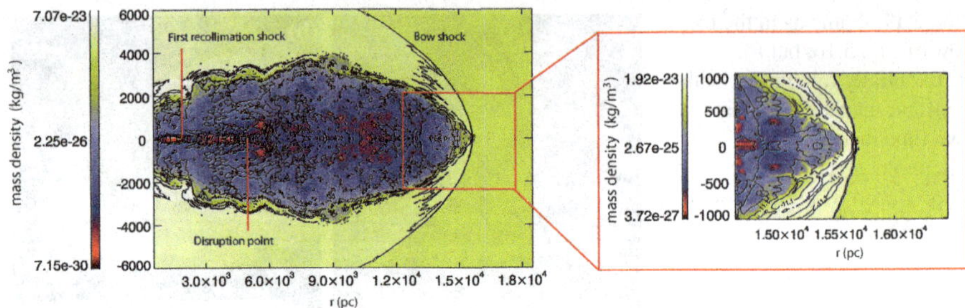

Fig. 5.18 *Left*: Logarithm of the rest mass density in a simulation of the temporal evolution of an extragactic jet. The jet is eventually disrupted by mass entrainment following a recollimation shock. A bow shock develops at the boundary between the jet and the medium. The *right panel* is a detail of the "head" of the jet. Simulation by Perucho and Martí (2007). Figure from Bordas et al. (2011), reproduced by permission of Oxford University Press on behalf of the Royal Astronomical Society

Let us finally mention that jets may be disrupted by encounters with inhomogeneities in the medium like clouds, uniform or clumpy stellar winds, and stars (e.g. Bosch-Ramon et al. 2012; Perucho and Bosch-Ramon 2012). You can see the simulated effect of the impact of a clumpy wind on the jet in a X-ray binary in Fig. 5.19.

5.9 Content and Radiation from Jets

The matter content of jets is largely unknown. Jets may be energetically dominated by a cold (thermal) proton-electron plasma plus a small contribution of relativistic particles mixed with the outflow, or by a pure relativistic leptonic electron-positron plasma. The former is expected to occur in hydrodynamic and magnetohydrodynamic jets (such jets are called "hadronic" or "heavy" jets), whereas the latter seems more likely to happen in black hole driven jets. Accretion loaded or black hole rotation powered jets with medium entrainment can produce hadronic jets, whereas black hole driven jets within a diluted medium may lead to leptonic jets. These leptonic jets would consist basically of a powerful, collimated electromagnetic wave, carrying just some electrons and positrons injected by pair creation at the base of the jet.

In hydrodynamic and low magnetization MHD jets, the kinetic energy of a particle of mass m associated with the bulk motion in the direction of propagation of the jet is $e_b = (\gamma - 1)mc^2$, where γ is the bulk Lorentz factor of the outflow. It is expected to be much larger than the energy associated with jet expansion (i.e. $e_{\exp} \approx mv_{\exp}^2/2$) and temperature (i.e. $e_T = kT$). This implies strong collimation and a low speed of sound or Alfvén speed compared to the jet velocity—the jet will be strongly supersonic/superalfvenic. If this were not the case, we learned that jets may still be collimated and accelerated with the help of external pressure, although it appears difficult that the right degree of collimation could be sustained

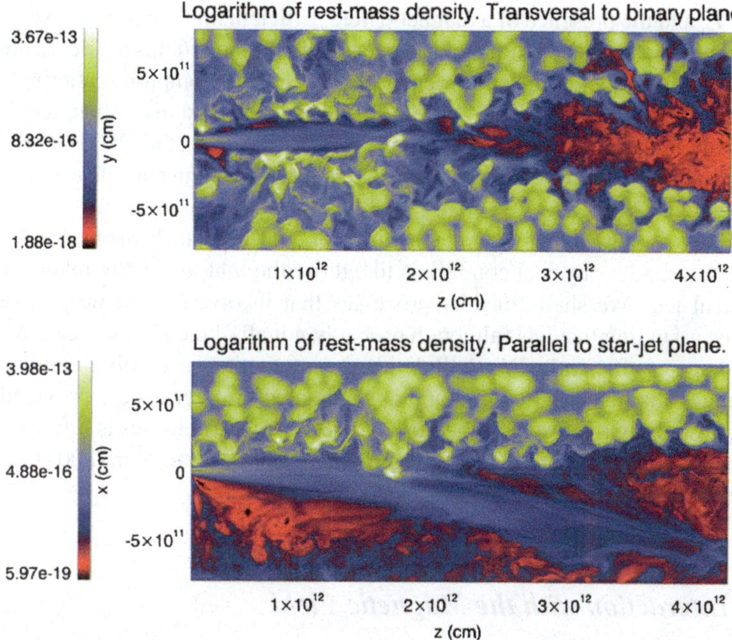

Fig. 5.19 Two spatial views of the rest mass density distribution in a hydrodynamic simulation of the interaction of a jet with the clumpy wind of the donor star in a X-ray binary. The clumps are the high-density regions shown in *yellow*. The impact of the wind produces a recollimation shock in the jet (*top panel*). A clump that penetrates the jet immediately after the shock induces a large deviation in the direction of the flow (*bottom panel*). The matter in the clumps mixes with the material of the jet, forming complex structures and producing turbulence. The entrainment slows down the jet far from the launching region. From Perucho and Bosch-Ramon (2012), reproduced with permission ©ESO

over astrophysical distances. For jets with dynamics dominated by the electromagnetic field, the temperature of the plasma could be very large as long as collimation is satisfied, i.e. the external and magnetic pressure should be well above the thermal pressure. Otherwise, the flow would be uncollimated. The realization of the different situations may take place in the same jet at different stages of its evolution. The jet could be electromagnetic/leptonic at its very base, MHD/hadronic through external medium entrainment at intermediate scales, and hydrodynamic at the largest scales after magnetic energy has been used up to accelerate the jet.

As long as jets are radiatively efficient, they will shine across the electromagnetic spectrum and may be detected, but when the conditions in the plasma are not suitable for the production of significant radiation, they may keep dark all the way to their termination regions. The radiation from jets can be thermal (continuum and lines), although the detection of thermal jets is rare since the required densities are high and the temperature and bulk velocity moderate. The detection of non-thermal synchrotron radiation from jets, mainly in the radio band, is more common. For that, it is required that some particles are accelerated up to relativistic energies, well above kT, and the presence of a magnetic field. Efficient high-energy and very high-energy emission is possible if the radiation and/or the matter fields are dense enough. This radiation can be produced, for example, by inverse Compton scattering, when accelerated leptons are present, or by proton-proton collisions, when accelerated hadrons are present (e.g. Bosch-Ramon et al. 2006; Romero et al. 2003). For a comprehensive description of radiative processes in jets, including particle transport, see Reynoso et al. (2011) and Vila et al. (2012).

Below, we review some of the mechanisms of non-thermal electromagnetic emission relevant under physical conditions like those thought to exist in relativistic astrophysical jets. We shall consider processes that involve relativistic protons and electrons, assuming that jets (although probably not all of them) can accelerate both particle species efficiently. We shall divide the interactions according to the type of target: magnetic field, matter, and photons. Abundant references are provided. The reader particularly interested in radiation processes in astrophysics is referred to the books by Rybicki and Lightman (1979), Longair (1994), Aharonian (2004), Dermer and Menon (2009), and Romero and Paredes (2011).

5.9.1 Interaction with the Magnetic Field

5.9.1.1 Synchrotron Radiation

Any accelerated charged particle will radiate in the presence of a magnetic field. When the particle is relativistic this radiation is a continuum and is called *synchrotron*. In the classical limit, the power per unit energy emitted by a single particle of mass m, energy $E = \gamma mc^2$, and charge e, whose velocity vector forms an angle α (the *pitch angle*) with the magnetic field $\mathbf{B}$ is (e.g. Ginzburg and Syrovatskii 1965; Blumenthal and Gould 1970)

$$P_{\text{synchr}} = \frac{\sqrt{3}e^3 B}{4\pi mc^2 h} \frac{E_\gamma}{E_c} \int_{E_\gamma/E_c}^{\infty} K_{5/3}(\zeta)d\xi. \qquad (5.101)$$

Here E_γ is the energy of the radiated photon, $K_{5/3}$ is a modified Bessel function, and

$$E_c = \frac{3heB\sin\alpha}{4\pi mc}\gamma^2. \qquad (5.102)$$

The synchrotron power peaks sharply around $E_\gamma \approx 0.29 E_c$. In general, this energy is much lower than the energy of the parent particle. Notice as well the dependence of formulas (5.101) and (5.102) on the mass; clearly, light particles cool more efficiently via synchrotron radiation than heavy ones. The synchrotron *energy loss rate* for a single particle is

$$\left(\frac{dE}{dt}\right)_{\text{synchr}} = -\frac{4}{3}\left(\frac{m_e}{m}\right)^2 c\sigma_T U_B \gamma^2, \qquad (5.103)$$

where $U_B = B^2/8\pi$ is the magnetic energy density. Comparing a proton and an electron with the same Lorentz factor, for example, we find that the electron loses energy $\sim 10^6$ times faster. Most of the photons are emitted in the direction of motion of the particle within a cone of semi aperture $\sim 1/\gamma$, so when the particle is relativistic the radiation is highly collimated. Another important property of the synchrotron radiation is that it is polarized.

To calculate the synchrotron spectrum emitted not by a single but by a distribution of particles, we must integrate Eq. (5.101) over particle energy and solid angle in the pitch angle space. If N is the energy distribution of relativistic particles (in units of $\mathrm{erg}^{-1}\,\mathrm{cm}^{-3}$), the total synchrotron power per unit energy per unit volume is

$$q_{\mathrm{synchr}}(E_\gamma) = \int_{E_{\min}}^{E_{\min}} \int_{\Omega_\alpha} N(E,\alpha) P_{\mathrm{synchr}}(E_\gamma, E, \alpha) \sin\alpha \, dE \, d\Omega_\alpha. \quad (5.104)$$

The particle distribution and the magnetic field may of course be inhomogeneous; the spatial dependence is implicit in Eqs. (5.101) and (5.104) (and henceforth). An important case is that of an isotropic particle distribution that is a power-law in energy, i.e. $N(E,\alpha) = (K/4\pi)E^{-p}$. In the limit $E_{\max} \to \infty$ and $E_{\min} \to 0$,[20] the emissivity can be calculated analytically and is given by

$$q_{\mathrm{synchr}}(E_\gamma) = a(p)\frac{4\pi K e^3 B^{\frac{p+1}{2}}}{hmc^2}\left(\frac{3he}{4\pi m^3 c^5}\right)^{\frac{p-1}{2}} E_\gamma^{-\frac{p-1}{2}}. \quad (5.105)$$

The constant $a(p)$ equals

$$a(p) = \frac{2^{\frac{p-1}{2}}\sqrt{3}}{8\sqrt{\pi}(p+1)}\frac{\Gamma[(3p-1)/12]\Gamma[(3p+19)/12]\Gamma[(p+5)/4]}{\Gamma[(p+7)/4]} \quad (5.106)$$

where $\Gamma[x]$ is the usual Gamma function. The key result from Eq. (5.105) is that the synchrotron spectrum of a power-law distribution of particles with index p, is another power-law in the energy of the emitted photons with index $(p-1)/2$. Thus, the characteristics of the synchrotron emission may provide valuable information on the distribution of parent particles and the mechanisms that accelerated them.

The classical formalism of synchrotron radiation is valid as long as

$$\frac{E}{m_e c^2}\frac{B}{B_{\mathrm{cr}}} \ll 1. \quad (5.107)$$

The critical value of the magnetic field is

$$B_{\mathrm{cr}} = \frac{m_e^2 c^3}{e\hbar} \approx 4.4 \times 10^{13}\,\mathrm{G}, \quad (5.108)$$

---

[20]The expression in Eq. (5.105) is then valid for $E_\gamma \gg E_{\mathrm{c}}(E_{\min})$ and $E_\gamma \ll E_{\mathrm{c}}(E_{\max})$, i.e. far from the low-energy and the high-energy cutoffs of the synchrotron spectrum.

above which quantum effects turn relevant. In particular, photons can create electron-positron pairs in the strong magnetic field

$$\gamma + B \longrightarrow e^+ + e^-. \tag{5.109}$$

The processes of pair creation and synchrotron radiation become tightly coupled in the quantum regime and can sustain an electromagnetic cascade. This phenomenon is expected to occur in the magnetosphere of pulsars. Relevant formulas for synchrotron radiation and pair creation in strong magnetic fields can be found, for instance, in Baring (1988, 1989) and Anguelov and Vankov (1999).

For most applications to astrophysical jets, however, the classical formalism is enough. The typical magnetic field strength and the maximum energy of particles in jets in active galactic nuclei and galactic X-ray binaries are expected to be satisfy condition (5.107). In these systems the emission of synchrotron radiation at radio wavelengths is the signature of the presence of jets. The shape of the radio spectrum is very characteristic: the flux density depends on the frequency as $S_\nu \propto \nu^{-l}$ with $l \sim 0$, so the spectrum is approximately flat. This is very well reproduced assuming that the synchrotron radiation is emitted by a power-law distribution of relativistic electrons in an outflow that expands as it propagates (e.g. Hjellming and Johnston 1988; Blandford and Königl 1979). In some galactic jets the radio and X-ray emission appear to be correlated (Corbel et al. 2003; Gallo et al. 2003), suggesting that the relativistic electrons in the jets are energetic enough to radiate synchrotron photons up to the X-ray band.

5.9.2 Interaction with Radiation

5.9.2.1 Inverse Compton Scattering

A photon can gain energy in the elastic scattering off an electron (or positron),[21]

$$e^- + \gamma \longrightarrow e^- + \gamma. \tag{5.110}$$

This process is called *inverse* Compton (IC) scattering.

Figure 5.20 shows a sketch of the interaction as seen in the rest frame of the electron (primed variables) and in the lab frame (unprimed variables). From the conservation of energy and momentum it is easy to calculate the energy of the scattered photon in the rest frame of the electron:

$$E'_\gamma = \frac{E'_{\mathrm{ph}}}{1 + (E'_{\mathrm{ph}}/m_e c^2)(1 - \cos\theta'_2)}, \tag{5.111}$$

---

[21] Inverse Compton scattering off protons and other particles much heavier than the electron is very inefficient and therefore generally neglected. Protons have other much more efficient channels of interaction with radiation as we shall see.

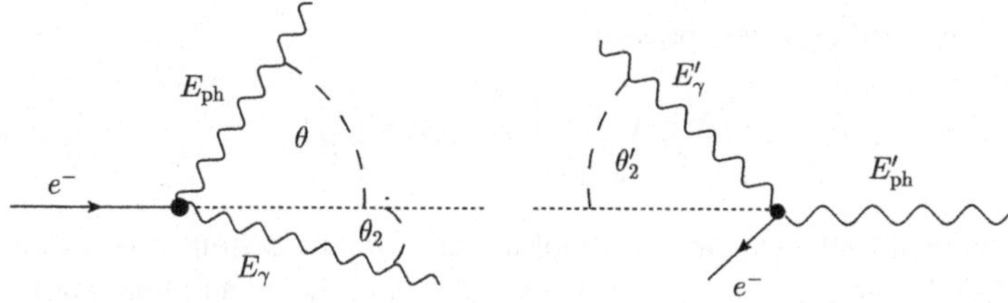

Fig. 5.20 Inverse Compton scattering in the lab frame (*left*) and in the rest frame of the electron (*right*)

where θ'_2 is the scattering angle. Transforming to the lab frame,

$$E_\gamma = \gamma_e E'_\gamma \left(1 - \beta \cos\theta'_2\right) \tag{5.112}$$

where $\gamma_e = E_e/m_e c^2$ is the Lorentz factor of the electron before the collision and $\beta = (1 - \gamma_e^{-2})^{1/2}$. When $E'_{\rm ph} \ll m_e c^2$ the interaction takes place in the *Thomson limit*; this amounts to $E_e E_{\rm ph} \ll m_e^2 c^4$ in the lab frame. In the Thomson regime the scattering is almost elastic in the rest frame of the electron and $E'_\gamma \approx E'_{\rm ph}$. The maximum energy of the scattered photon in the lab frame is then

$$E_{\gamma,\rm max} \approx 4\gamma_e^2 E_{\rm ph}, \tag{5.113}$$

that corresponds to a head-on collision. The energy of the scattered photons is large although still much smaller than that of the electron, that loses just a modest fraction of its energy in the interaction. In the opposite limit, $E_e E_{\rm ph} \gg m_e^2 c^4$, the interaction occurs in the *Klein-Nishina* regime. In this regime the electron transfers almost all its energy to the photon in a single collision; the energy loss is said to be "catastrophic".[22] However, the probability of interaction decreases drastically and so, overall, electron cooling in the Klein-Nishina regime is quite inefficient. This may be readily demonstrated analyzing the cross section, as we shall show below.

To calculate the energy loss rate for an electron in an isotropic photon distribution $n(E_{\rm ph})$, the IC spectrum must be integrated over the initial and final energy of the photons:

$$\left(\frac{dE}{dt}\right)_{\rm IC} = -\int_{E_{\rm ph}^{\rm min}}^{E_{\rm ph}^{\rm max}} dE_{\rm ph} \int_{E_{\rm ph}}^{E_\gamma^{\rm max}} dE_\gamma (E_\gamma - E_{\rm ph}) P_{\rm IC}. \tag{5.114}$$

The function $P_{\rm IC}$ and the maximum energy of the scattered photons $E_{\rm ph}^{\rm max}$ are given below in Eqs. (5.120) and (5.124), respectively. In the Thomson limit Eq. (5.114)

---

reduces to the well known expression

$$\left(\frac{dE}{dt}\right)_{\mathrm{IC,Th}} = -\frac{4}{3}c\sigma_T U_{\mathrm{ph}}\gamma^2. \tag{5.115}$$

This is identical to the synchrotron cooling rate for an electron if the magnetic energy density is replaced by the energy density U_{ph} of the target photon field.

The exact differential cross section for Compton interaction is given by the Klein-Nishina formula (e.g. Blumenthal and Gould 1970)

$$\frac{d\sigma_{\mathrm{KN}}}{d\Omega'_2 dE'_\gamma} = \frac{r_e^2}{2}\left(\frac{E'_\gamma}{E'_{\mathrm{ph}}}\right)^2\left(\frac{E'_{\mathrm{ph}}}{E'_\gamma} + \frac{E'_\gamma}{E'_{\mathrm{ph}}} - \sin^2\theta'_2\right)$$

$$\times \delta\left(E'_\gamma - \frac{E'_{\mathrm{ph}}}{1 + (E'_{\mathrm{ph}}/m_e c^2)(1 - \cos\theta'_2)}\right), \tag{5.116}$$

where r_e is the classical electron radius. The total cross section is an invariant, so it can be calculated integrating Eq. (5.116). The angle-averaged[23] result is

$$\sigma_{\mathrm{IC}} = \frac{3\sigma_T}{8x}\left[\left(1 - \frac{2}{x} - \frac{2}{x^2}\right)\ln(1+2x) + \frac{1}{2} + \frac{4}{x} - \frac{1}{2(1+2x)^2}\right], \tag{5.117}$$

where we have defined $x = E_e E_{\mathrm{ph}}/m_e^2 c^4$. In the Thomson limit ($x \ll 1$) it reduces to

$$\sigma_{\mathrm{IC}} \approx \sigma_T(1 - 2x), \tag{5.118}$$

whereas in the Klein-Nishina regime ($x \gg 1$) it indeed falls drastically as

$$\sigma_{\mathrm{IC}} \approx \frac{3}{8}\sigma_T x^{-1}\ln(4x). \tag{5.119}$$

To calculate the inverse Compton spectrum from the interaction of an electron distribution $N(E_e)$ with an isotropic target photon distribution $n(E_{\mathrm{ph}})$ we can resort to the formulas given in Blumenthal and Gould (1970). We define

$$P_{\mathrm{IC}} = \frac{3\sigma_T c}{4\gamma_e^2}\frac{n(E_{\mathrm{ph}})}{E_{\mathrm{ph}}}F(q) \tag{5.120}$$

and

$$F(q) = 2q\ln q + (1 + 2q)(1 - q) + \frac{1}{2}(1 - q)\frac{(q\Gamma_e)^2}{(1 + \Gamma_e q)}, \tag{5.121}$$

---

[23] In certain astrophysical systems the IC scattering cannot be approximated as isotropic. An example is the interaction of electrons in the jets in X-ray binaries with the radiation field of the companion star. A full treatment taking into account the collision angle is then required, see e.g. Dermer and Schlickeiser (1993) and Khangulyan et al. (2008).

where

$$\Gamma_e = \frac{4 E_{ph} E_e}{m_e^2 c^4} \tag{5.122}$$

and

$$q = \frac{E_\gamma}{\Gamma_e E_e (1 - E_\gamma / E_e)}. \tag{5.123}$$

Equation (5.120) is valid both in the Thomson and Klein-Nishina regimes as long as $\gamma_e \gg 1$. The energy of the scattered photons is constrained to be within the range

$$E_{ph} \leq E_\gamma \leq \frac{\Gamma_e}{1 + \Gamma_e} E_e, \tag{5.124}$$

hence $E_\gamma \to E_e$ for $\Gamma_e \gg 1$ in the Klein-Nishina limit. The IC emissivity is calculated integrating over the energies of the electron and the target photons,

$$q_{IC}(E_\gamma) = \int_{E_e^{min}}^{E_e^{max}} \int N(E_e) P_{IC}(E_\gamma, E_e, E_{ph}) dE_{ph} dE_e. \tag{5.125}$$

The limits in the integral over E_{ph} at fixed E_e must be taken from Eq. (5.124). In the Thomson regime and if the electron energy distribution if a power-law $N(E_e) = K E_e^{-p}$, the integral over E_e can be solved analytically. The result is that, far from the low and high-energy cutoffs, the IC emissivity follows a power-law $q_{IC} \propto E_\gamma^{-l}$ with $l = (p + 1)/2$. You can find the complete expression for the emissivity and some further calculations for the special case of a blackbody target photon field in Blumenthal and Gould (1970).

5.9.2.2 Proton-Photon Inelastic Collisions

The cross sections for interactions of high-energy hadrons with photons are small compared to those of matter-matter interactions. However, in some astrophysical environments, radiation density is larger than matter density and photohadronic processes may become relevant.[24]

Inelastic collisions of high-energy protons with photons proceed along two main channels. For a proton of energy $E_p = \gamma_p m_p c^2$, the direct creation of an electron-positron pair (*photopair* production)

$$p + \gamma \to p + e^- + e^- \tag{5.126}$$

---

[24]Photohadronic interactions of cosmic rays protons with the cosmic microwave background set an upper limit to the energy of cosmic rays arriving to Earth of $\sim 5 \times 10^{19}$ eV. This is the famous Greisen-Zatsepin-Kuzmin (GZK) cutoff.

is possible above the threshold $E_p E_{\mathrm{ph}} > 2m_e^2 c^4$. When the energy of the target photon in the rest frame of the proton exceeds

$$E_{\mathrm{ph(thr)}}^{\prime(\pi)} = m_{\pi^0} c^2 \left(1 + \frac{m_{\pi^0}}{2m_p}\right) \approx 145 \text{ MeV} \tag{5.127}$$

the creation of pions (*photopion* or *photomeson* production) takes over

$$p + \gamma \rightarrow p + a\pi^0 + b\left(\pi^+ + \pi^-\right), \tag{5.128}$$

$$p + \gamma \rightarrow n + \pi^+ + a\pi^0 + b\left(\pi^+ + \pi^-\right). \tag{5.129}$$

The integers a and b are the pion multiplicities. Near the threshold in (5.127) the production of a single pion dominates; multiple pions are created at higher energies.

Neutral pions have a mean life $\tau_{\pi^0} = 8.52 \times 10^{-17}$ s. They decay with a branching ratio ≥ 98.8 into two gamma-rays,

$$\pi^0 \rightarrow \gamma + \gamma. \tag{5.130}$$

Charged pions decay into muons and neutrinos, and muons into electrons/positrons and more neutrinos,

$$\pi^+ \rightarrow \mu^+ + \nu_\mu, \qquad \mu^+ \rightarrow e^+ + \nu_e + \overline{\nu}_\mu, \tag{5.131}$$

$$\pi^- \rightarrow \mu^- + \overline{\nu}_\mu, \qquad \mu^- \rightarrow e^- + \overline{\nu}_e + \nu_\mu. \tag{5.132}$$

The decay of charged pions and muons is a weak process. The mean lifetime of these particles—$\tau_{\pi^\pm} \approx 2.6 \times 10^{-8}$ s and $\tau_\mu \approx 2.2 \times 10^{-6}$ s—is several orders of magnitude larger than that of the neutral pion, whose decay is purely electromagnetic.

The energy loss rate of a proton in an isotropic photon distribution $n(E_{\mathrm{ph}})$ can be conveniently parameterized as (e.g. Begelman et al. 1990)

$$\left(\frac{dE}{dt}\right)_{p\gamma} = -\frac{m_p c^3}{2\gamma_p} \int_{E_{\mathrm{ph(thr)}}^{\prime(i)}/2\gamma_p}^{\infty} dE_{\mathrm{ph}} \frac{n(E_{\mathrm{ph}})}{E_{\mathrm{ph}}^2}$$

$$\times \int_{E_{\mathrm{ph(thr)}}^{\prime(i)}}^{2\gamma_p E_{\mathrm{ph}}} dE_{\mathrm{ph}}' \sigma_{p\gamma}^{(i)}\left(E_{\mathrm{ph}}'\right) K_{p\gamma}^{(i)}\left(E_{\mathrm{ph}}'\right) E_{\mathrm{ph}}', \tag{5.133}$$

where $i = e, \pi$ denotes the interaction channel, $\sigma_{p\gamma}^{(i)}$ is the cross section, and $K_{p\gamma}^{(i)}$ is the inelasticity, defined as the fractional energy loss of the proton.

The study of photohadronic interactions from very first principles is obviously complex. Data on the cross sections and inelasticities are obtained experimentally and from numerical simulations like the Monte Carlo code SOPHIA (Mücke et al. 2000). Fortunately, some simple parameterizations are available in Atoyan and Dermer (2003), Begelman et al. (1990), and Maximon (1968). The photopion cross section and inelasticity are of the order of 0.1–0.3 mb and 0.2–0.6, respectively. The cross section for pair production (often referred to as Bethe-Heitler cross section) is about two orders of magnitude larger but the inelasticity is very low, $K^{(e)}_{p\gamma} \sim 2m_e/m_p$. As a result, the cooling is completely dominated by pion production as soon as the energy of the photons is larger than the threshold in Eq. (5.127).

Kelner and Aharonian (2008, 2010) introduced convenient analytical expressions for the gamma-ray spectrum from photohadronic interactions (see also Atoyan and Dermer 2003). Given the distributions of relativistic protons $N_p(E_p)$ and target photons $n(E_{\mathrm{ph}})$, the gamma-ray emissivity can be written as

$$q_\gamma(E_\gamma) = \int_{E_p^{\min}}^{E_p^{\max}} dE_p \int_{E'^{(\pi)}_{\mathrm{ph(thr)}}/2\gamma_p}^{\infty} dE_{\mathrm{ph}} \frac{1}{E_p} N_p(E_p) n(E_{\mathrm{ph}}) \Phi(\eta, x). \tag{5.134}$$

Here $\eta = 4E_{\mathrm{ph}}E_p/m_p^2c^4$ and $x = E_\gamma/E_p$. From numerical results obtained with the code SOPHIA, the function $\Phi(\eta, x)$ can be approximated with an accuracy better than 10 %. Let us define $x_\pm$ as

$$x_\pm = \frac{1}{2(1+\eta)}\left[\eta + r^2 \pm \sqrt{(\eta - r^2 - 2r)(\eta - r^2 + 2r)}\right]. \tag{5.135}$$

Then, in the range $x_- < x < x_+$,

$$\Phi_\gamma(\eta, x) = B_\gamma \exp\left\{-s_\gamma\left[\ln\left(\frac{x}{x_-}\right)\right]^{\delta_\gamma}\right\}\left[\ln\left(\frac{2}{1+y^2}\right)\right]^{2.5+0.4\ln(\eta/\eta_0)}, \tag{5.136}$$

where

$$y = \frac{x - x_-}{x_+ - x_-} \tag{5.137}$$

and

$$\eta_0 = 2\frac{m_\pi}{m_p} + \frac{m_\pi^2}{m_p^2} \approx 0.313. \tag{5.138}$$

For $x < x_-$, the spectrum is independent of x,

$$\Phi_\gamma(\eta, x) = B_\gamma[\ln 2]^{2.5+0.4\ln(\eta/\eta_0)}, \tag{5.139}$$

and finally $\Phi_\gamma(\eta, x) = 0$ for $x > x_+$. The parameters B_γ, s_γ and δ_γ are functions of η. For values of $1.1\eta_0 < \eta < 100\eta_0$, these functions are tabulated in Kelner and Aharonian (2008).

We shall not reproduce them here, but if one wishes to calculate the injection spectra of pions, muons, electron/positrons, and neutrinos from proton-photon collisions, all the necessary formulas under different approximations are given in Schlickeiser (2002), Atoyan and Dermer (2003), Mastichiadis et al. (2005), Lipari et al. (2007), and Kelner and Aharonian (2008). Although generally disregarded, the cooling of pions and muons before decay may be considerable if these particles are created energetic and in a medium with a strong magnetic field, for example. The energy losses of pions and muons affect the spectrum of neutrinos injected at their decay. Neglecting this effect can lead to an overprediction of the neutrino emissivity, see Reynoso and Romero (2009) for a detailed application to the production of neutrinos in galactic jets. Notice that the emission of neutrinos is the imprint of hadronic interactions. The detection of neutrinos from any given type of astrophysical source (with jets, for example) would definitely prove it can accelerate protons up to relativistic energies. The two largest and most sensitive active neutrino detectors, ANTARES and IceCube, up to now have failed to pinpoint any galactic or extragalactic source of high-energy neutrinos. The search is ongoing. It is expected that the chances of detection will improve significantly with the next-generation neutrino telescope KM3NeT, and future upgrades of IceCube.

5.9.3 Interaction with Matter

5.9.3.1 Proton-Proton Collisions

The inelastic collision of a relativistic proton with a low-energy proton creates mesons. The reactions with the lowest energy thresholds correspond to the creation of pions

$$p + p \rightarrow p + p + a\pi^0 + b(\pi^+ + \pi^-), \tag{5.140}$$

$$p + p \rightarrow p + n + \pi^+ + a\pi^0 + b(\pi^+ + \pi^-), \tag{5.141}$$

$$p + p \rightarrow n + n + 2\pi^+ + a\pi^0 + b(\pi^+ + \pi^-). \tag{5.142}$$

The integers a and b are, as before, the pion multiplicities. The energy threshold of the proton for the production of a single neutral pion is

$$E_{p(\text{thr})} = m_p c^2 + 2m_{\pi^0} c^2 \left(1 + \frac{m_{\pi^0}}{4m_p}\right) \approx 1.22 \text{ GeV}. \tag{5.143}$$

The energy loss rate for a proton of energy E_p due to inelastic collisions with a target field of low-energy protons with number density n_p is (e.g. Begelman et al. 1990; Aharonian and Atoyan 2000)

$$\left(\frac{dE}{dt}\right)_{pp} \approx -cn_p K_{pp}\sigma_{pp}(E_p)E_p. \tag{5.144}$$

Here σ_{pp} is the total inelastic cross section and $K_{pp} \approx 0.5$ is the total inelasticity of the interaction. Most of the energy lost by the relativistic proton is transferred to only one or two "leading" pions. The cross section $\sigma_{pp}(E_p)$ can be accurately

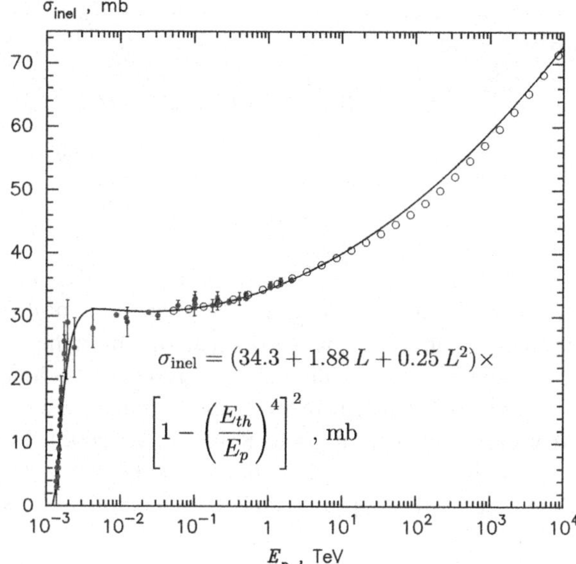

Fig. 5.21 Inelastic cross section for proton-proton collisions. The *filled circles* are experimental data and the *empty circles* the results of simulations with the code SIBYLL. Reprinted figure with permission from Kelner et al. (2006). Copyright (2006) by the American Physical Society

approximated as (Kelner et al. 2006)

$$\sigma_{\text{inel}}(E_p) = \left(34.3 + 1.88L + 0.25L^2\right)\left\{1 - \left[\frac{E_{p(\text{thr})}}{E_p}\right]^4\right\}^2 \text{ mb}, \qquad (5.145)$$

where $L = \ln(E_p/1\text{ TeV})$. As seen in Fig. 5.21, this expression correctly describes the cross section near the threshold and fits data from experiments and simulations up to at least $E_p \sim 10^4$ TeV.

Kelner et al. (2006, 2009) introduced an useful parameterization for the spectrum of gamma rays due to the decay of neutral pions created in proton-proton collisions. The formulas were obtained fitting the results of the code SIBYLL, used to study atmospheric cascades at ultra-high energies (Fletcher et al. 1994). Defining $x = E_\gamma/E_p$, the gamma-ray emissivity (in units of $\text{erg}^{-1}\,\text{cm}^{-3}\,\text{s}^{-1}$) is given by

$$q_\gamma^{(pp)}(E_\gamma) = cn_p \int_{E_\gamma}^{E_p^{\max}} \frac{1}{E_p}\sigma_{pp}(E_p)N_p(E_p)F_\gamma(x, E_p)dE_p. \qquad (5.146)$$

The function $F_\gamma(x, E_p)$ is the number of photons per unit energy created per proton-proton collision; its full expression is

$$F_\gamma(x, E_p) = B_\gamma \frac{\ln x}{x}\left[\frac{1 - x^{\beta_\gamma}}{1 + k_\gamma x^{\beta_\gamma}(1 - x^{\beta_\gamma})}\right]^4$$

$$\times \left[\frac{1}{\ln x} - \frac{4\beta_\gamma x^{\beta_\gamma}}{1 - x^{\beta_\gamma}} - \frac{4k_\gamma \beta_\gamma x^{\beta_\gamma}(1 - 2x^{\beta_\gamma})}{1 + k_\gamma x^{\beta_\gamma}(1 - x^{\beta_\gamma})}\right]. \qquad (5.147)$$

For proton energies in the range $0.1\,\text{TeV} \leq E_p \leq 10^5\,\text{TeV}$, fits to the results of SIBYLL yield

$$B_\gamma = 1.30 + 0.14L + 0.011L^2, \tag{5.148}$$

$$\beta_\gamma = \left(1.79 + 0.11L + 0.008L^2\right)^{-1}, \tag{5.149}$$

$$k_\gamma = \left(0.801 + 0.049L + 0.014L^2\right)^{-1}. \tag{5.150}$$

The function $F_\gamma(x, E_p)$ includes, along with the contribution to the gamma-ray spectrum from the π^0 decay, that from the decay of the η-meson. Around $x \sim 0.1$, the contribution to the gamma-ray spectrum from the η-mesons is about 25 %.

Equation (5.146) is valid for $E_p \gtrsim 100\,\text{GeV}$. At lower energies, the gamma-ray emissivity can be calculated to a good accuracy using the δ-*functional formalism* (Aharonian and Atoyan 2000; Kelner et al. 2006). In this approximation all the neutral pions carry a fixed fraction of the kinetic energy of the relativistic proton, $E_\pi \approx K_\pi E_p^{\text{kin}}$. The injection function of neutral pions is then

$$Q_{\pi^0}^{(pp)}(E_\pi) = \tilde{n} c n_p \int \delta(E_\pi - K_\pi E_{\text{kin}})\sigma_{pp}(E_p)N_p(E_p)dE_p$$

$$= \frac{\tilde{n}}{K_\pi} c n_p \sigma_{pp}\left(m_p c^2 + \frac{E_\pi}{K_\pi}\right)N_p\left(m_p c^2 + \frac{E_\pi}{K_\pi}\right), \tag{5.151}$$

where $\tilde{n}$ is the number of neutral pions created per proton-proton collision. The gamma-ray emissivity is directly calculated from $Q_{\pi^0}^{(pp)}$ as

$$q_\gamma^{(pp)}(E_\gamma) = 2 \int_{E_{\text{min}}}^{E_p^{\text{max}}} \frac{Q_{\pi^0}^{(pp)}(E_\pi)}{\sqrt{E_\pi^2 - m_{\pi^0}^2 c^4}} dE_\pi, \tag{5.152}$$

with

$$E_{\text{min}} = E_\gamma + \frac{m_{\pi^0}^2 c^4}{4E_\gamma}. \tag{5.153}$$

For a given value of K_π, the value of $\tilde{n}$ is fixed demanding continuity between Eqs. (5.146) and (5.152) at $E_p = 100\,\text{GeV}$.[25] Taking $K_\pi = 0.17$ provides a good agreement with the results of simulations (Aharonian and Atoyan 2000). Since for $E_p \lesssim 100\,\text{GeV}$ the cross section is essentially constant, the shape of the photon spectrum in the δ-functional formalism is similar to the shape of the parent proton spectrum, shifted in energy by a factor K_π.

An expression for the injection function of charged pions in proton-proton collisions is also given in Kelner et al. (2006).

---

[25] It is assumed that $\tilde{n}$ and K_π depend only weakly on the energy of the proton.

5.9.3.2 Relativistic Bremsstrahlung

Bremsstrahlung radiation is produced when a relativistic charged particle is accelerated in an electrostatic field. Bremsstrahlung losses are essentially catastrophic: the particle loses almost all its energy in one interaction, and most of the emitted radiation is in the form of high-energy photons. We can, however, introduce an average continuous cooling rate. For an electron of energy E_e in a plasma of fully ionized nuclei of charge eZ and number density n_p,

$$\left(\frac{dE}{dt}\right)_{\mathrm{Br}} = -4\alpha_{\mathrm{FS}} r_e^2 Z^2 c n_p E_e \left[\ln\left(\frac{2E_e}{m_e c^2}\right) - \frac{1}{3}\right], \tag{5.154}$$

where α_{FS} is the fine structure constant and r_e the classical electron radius.

The differential cross section for the emission of a photon with energy E_γ by an electron of energy $E_e \gg m_e c^2$ in the presence of a nucleus of charge eZ is (e.g. Berezinskii et al. 1990)

$$\frac{d\sigma_{\mathrm{Br}}}{dE_\gamma}(E_\gamma, E_e) = \frac{4\alpha_{\mathrm{FS}} r_e^2 Z^2}{E_\gamma} \left[1 + \left(1 - \frac{E_\gamma}{E_e}\right)^2 - \frac{2}{3}\left(1 - \frac{E_\gamma}{E_e}\right)\right]$$
$$\times \left\{\ln\left[\frac{2E_e(E_e - E_\gamma)}{m_e c^2 E_\gamma}\right] - \frac{1}{2}\right\}. \tag{5.155}$$

The total Bremsstrahlung emissivity from a distribution of electrons $N_e(E_e)$ can be directly calculated from the differential cross section and the distribution of electrons as

$$q_{\mathrm{Br}}(E_\gamma) = c n_p \int_{E_e^{\min}}^{E_e^{\max}} dE_e \frac{d\sigma_{\mathrm{Br}}}{dE_\gamma}(E_\gamma, E_e) N_e(E_e). \tag{5.156}$$

5.9.4 Absorption

5.9.4.1 Photon-Photon Annihilation

The production of electron-positron pairs by annihilation of two photons

$$\gamma + \gamma \to e^+ + e^-, \tag{5.157}$$

is a sink for photons but also a source of pairs. The process is possible only above the kinematic energy threshold

$$E_{\mathrm{ph}} E_\gamma (1 - \cos\theta) \geq 2 m_e^2 c^4, \tag{5.158}$$

where E_γ and E_{ph} are the energies of the photons, and θ is the collision angle in the observer frame. The annihilation cross section is (e.g. Gould and Schréder 1967)

$$\sigma_{\gamma\gamma}(\beta) = \frac{3}{16}\sigma_{\mathrm{T}}(1 - \beta^2)\left[(3 - \beta^4)\ln\left(\frac{1 + \beta}{1 - \beta}\right) - 2\beta(2 - \beta^2)\right]. \tag{5.159}$$

Here $\beta = (1 - \gamma_e^{-2})^{1/2}$ and γ_e is the Lorentz factor of the electron (positron) in the center of momentum frame. It is related to the energy of the photons and the collision angle as

$$\left(1 - \beta^2\right) = \frac{2m_e^2 c^4}{(1 - \cos\theta) E_{\text{ph}} E_\gamma}, \quad 0 \le \beta < 1. \tag{5.160}$$

We define the optical depth $\tau_{\gamma\gamma}(E_\gamma, R)$ as the probability that a photon of energy E_γ annihilates against another photon of the target radiation field $n(E_{\text{ph}})$, after traversing a distance R. It is given by (e.g. Gould and Schréder 1967)

$$\tau_{\gamma\gamma}(E_\gamma, R) = \frac{1}{2} \int_0^R d\ell \int_{E_{\text{ph(thr)}}}^\infty dE_{\text{ph}} \int_{-1}^{u_{\max}} du (1 - u) \sigma_{\gamma\gamma}(E_\gamma, E_{\text{ph}}, u) n(E_{\text{ph}}), \tag{5.161}$$

where $u = \cos\theta$ and ℓ is a spatial variable along the path of the photon. The integration limits are

$$E_{\text{ph(thr)}} = \frac{m_e^2 c^4}{E_\gamma}, \tag{5.162}$$

and

$$u_{\max} = 1 - \frac{2m_e^2 c^4}{E_{\text{ph}} E_\gamma}. \tag{5.163}$$

Because of the narrowness of the pair production cross section, a gamma ray of energy E_γ can effectively be absorbed by photons with energy in a narrow band centered at $E_{\text{ph}} \approx 4m_e^2 c^4 / E_\gamma$.

Notice that the absorption of radiation by interaction with matter may also be of importance, specially in extragalactic sources. At low energies ($E_\gamma \lesssim 1$ keV) the dominant mechanisms of absorption are scattering off dust and, for $E_\gamma > 13.6$ eV, photoionization. Direct Compton scattering and pair creation in photon-nuclei collisions become relevant above $E_\gamma \approx 1$ keV.

Under the conditions $\epsilon \ll m_e c^2 \lesssim E_\gamma$, the pair emissivity that results from the interaction of two isotropic photon distributions n_γ and n_{ph} can be approximated by the following expression (Böttcher and Schlickeiser 1997; Aharonian et al. 1983)

$$Q_{e^\pm}^{(\gamma\gamma)}(E_e) = \frac{3}{32} \frac{c\sigma_{\text{T}}}{m_e c^2} \int_{\gamma_e}^\infty d\epsilon_\gamma \int_{\frac{\epsilon_\gamma}{4\gamma_e(\epsilon_\gamma - \gamma_e)}}^\infty d\omega \frac{n_\gamma(\epsilon_\gamma)}{\epsilon_\gamma^3} \frac{n_{\text{ph}}(\omega)}{\omega^2} \left\{ \frac{4\epsilon_\gamma^2}{\gamma_e(\epsilon_\gamma - \gamma_e)} \right.$$

$$\times \ln\left[\frac{4\gamma_e \omega(\epsilon_\gamma - \gamma_e)}{\epsilon_\gamma}\right] - 8\epsilon_\gamma\omega + \frac{2(2\epsilon_\gamma\omega - 1)\epsilon_\gamma^2}{\gamma_e(\epsilon_\gamma - \gamma_e)}$$

$$\left. - \left(1 - \frac{1}{\epsilon_\gamma\omega}\right) \frac{\epsilon_\gamma^4}{\gamma_e^2(\epsilon_\gamma - \gamma_e)^2} \right\}. \tag{5.164}$$

Here $\epsilon_\gamma = E_\gamma / m_e c^2$ and $\omega = E_{\text{ph}}/m_e c^2$ are the adimensional photon energies. The spectrum is symmetric around $E_e = E_\gamma/2$. For $E_{\text{ph}} E_\gamma \gg m_e^2 c^4$ the interaction is catastrophic: one of the produced particles takes most of the energy of the gamma ray.

References

F.A. Aharonian, *Very High-Energy Cosmic Gamma Radiation* (World Scientific, New Jersey, 2004)

F.A. Aharonian, A.M. Atoyan, Astron. Astrophys. **362**, 937 (2000)

F.A. Aharonian, A.M. Atoyan, A.M. Nagapetian, Astrofizika **19**, 323 (1983)

I. Agudo, J.L. Gómez, J.M. Martí, J.M. Ibáñez, A.P. Marscher, A. Alberdi, M.A. Aloy, P.E. Hardee, Astrophys. J. **549**, L183 (2001)

V. Anguelov, H. Vankov, J. Phys. G, Nucl. Part. Phys. **25**, 1755 (1999)

A.M. Atoyan, C.D. Dermer, Astrophys. J. **586**, 79 (2003)

J.M. Bardeen, W.H. Press, S.A. Teukolsky, Astrophys. J. **178**, 347 (1972)

M.G. Baring, Mon. Not. R. Astron. Soc. **235**, 51 (1988)

M.G. Baring, Astrophys. J. **225**, 260 (1989)

M.V. Barkov, D.V. Khangulyan, Mon. Not. R. Astron. Soc. **421**, 1351 (2012)

M.V. Barkov, S.S. Komissarov, Mon. Not. R. Astron. Soc. **385**, L28 (2008)

M.C. Begelman, Astrophys. J. **493**, 291 (1998)

M.C. Begelman, Z.-H. Li, Astrophys. J. **426**, 269 (1994)

M.C. Begelman, B. Rudak, M. Sikora, Astrophys. J. **362**, 38 (1990)

M. Bejger, T. Piran, M. Abramowicz, F. Hakanson, Phys. Rev. Lett. **109**, 121101 (2012)

V.S. Berezinskii, S.V. Bulanov, V.A. Dogiel, V.S. Ptuskin, *Astrophysics of Cosmic Rays* (North-Holland, Amsterdam, 1990)

V.S. Beskin, *MHD Flows in Compact Astrophysical Objects* (Springer, Heidelberg, 2010)

V.S. Beskin, I.V. Kuznetsova, Riv. Nuovo Cimento **115**, 179 (2000)

V.S. Beskin, I.V. Kuznetsova, R.R. Rafikov, Mon. Not. R. Astron. Soc. **299**, 341 (2008)

G.S. Bisnovatyi-Kogan, R.V.E. Lovelace, Astrophys. J. **667**, L167 (2007)

G.S. Bisnovatyi-Kogan, R.V.E. Lovelace, Astrophys. J. **750**, 109 (2012)

R.D. Blandford, in *Active Galactic Nuclei*, ed. by T.J.-L. Courvoisier, M. Mayor (Springer, Berlin, 1990), p. 161

R.D. Blandford, A. Königl, Astrophys. J. **232**, 34 (1979)

R.D. Blandford, D.G. Payne, Mon. Not. R. Astron. Soc. **199**, 883 (1982)

R.D. Blandford, R.L. Znajek, Mon. Not. R. Astron. Soc. **179**, 433 (1977)

G.R. Blumenthal, R.J. Gould, Rev. Mod. Phys. **42**, 237 (1970)

P. Bordas, V. Bosch-Ramon, J.M. Paredes, M. Perucho, Astron. Astrophys. **497**, 325 (2009)

P. Bordas, V. Bosch-Ramon, M. Perucho, Mon. Not. R. Astron. Soc. **412**, 1229 (2011)

V. Bosch-Ramon, G.E. Romero, J.M. Paredes, Astron. Astrophys. **447**, 263 (2006)

V. Bosch-Ramon, M. Perucho, P. Bordas, Astron. Astrophys. **528**, A89 (2011)

V. Bosch-Ramon, M. Perucho, M.V. Barkov, Astron. Astrophys. **539**, A69 (2012)

M. Böttcher, R. Schlickeiser, Astron. Astrophys. **325**, 866 (1997)

A. Ciardi, Jets from young stars IV. Lect. Notes Phys. **793**, 31 (2010)

J. Contopoulos, Astrophys. J. **450**, 616 (1995)

S. Corbel, M.A. Nowak, R.P. Fender, A.K. Tzioumis, S. Markoff, Astron. Astrophys. **400**, 1007 (2003)

E.V. Derishev, F.A. Aharonian, V.V. Kocharovsky, Vl.V. Kocharovsky, Phys. Rev. D **68**, 043003 (2003)

C.D. Dermer, G. Menon, *High-Energy Radiation from Black Holes* (Princeton University Press, Princeton, 2009)

C.D. Dermer, R. Schlickeiser, Astrophys. J. **416**, 458 (1993)

S. Drappeau, S.D. Markoff, P.C. Fragile, in *Jets at All Scales*, ed. by G.E. Romero, R.A. Sunyaev, T. Belloni. Proceedings of the IAU Symposium, vol. 275 (Cambridge University Press, Cambridge, 2011), p. 104

L.O.C. Drury, Rep. Prog. Phys. **46**, 973 (1983)

S.A.E.G. Falle, Mon. Not. R. Astron. Soc. **250**, 581 (1991)

R.P. Fender, E. Gallo, D. Russell, Mon. Not. R. Astron. Soc. **406**, 1425 (2010)

J. Ferreira, in *Star Formation and the Physics of Young Stars*, ed. by J. Bouvier, J.-P. Zhan (EDP Sciences, Cambridge, 2002), p. 229

R.S. Fletcher, T.K. Gaisser, P. Lipari, T. Stanev, Phys. Rev. D **50**, 5710 (1994)

P.C. Fragile, J. Wilson, M. Rodriguez, Mon. Not. R. Astron. Soc. **424**, 524 (2012)

E. Gallo, R.P. Fender, G.G. Pooley, Mon. Not. R. Astron. Soc. **344**, 60 (2003)

D. Garofalo, Astrophys. J. **699**, 400 (2009)

D. Garofalo, D.A. Evans, R.M. Sambruna, Mon. Not. R. Astron. Soc. **406**, 975 (2010)

V.L. Ginzburg, S.I. Syrovatskii, Annu. Rev. Astron. Astrophys. **3**, 297 (1965)

P. Goldreich, W.H. Julian, Astrophys. J. **157**, 869 (1969)

R.J. Gould, G.P. Schréder, Phys. Rev. **155**, 1404 (1967)

J. Granot, Mon. Not. R. Astron. Soc. **421**, 2442 (2012a)

J. Granot, Mon. Not. R. Astron. Soc. **421**, 2467 (2012b)

J. Granot, S.S. Komissarov, A. Spitkovsky, Mon. Not. R. Astron. Soc. **411**, 1323 (2011)

P.E. Hardee, in *Jets at All Scales*, ed. by G.E. Romero, R.A. Sunyaev, T. Belloni. Proceedings of the IAU Symposium, vol. 275 (Cambridge University Press, Cambridge, 2011), p. 41

P.E. Hardee, J.A. Eilek, Astrophys. J. **735**, 61 (2011)

P.E. Hardee, A. Rosen, Astrophys. J. **524**, 650 (1999)

P.E. Hardee, A. Rosen, Astrophys. J. **576**, 204 (2002)

J. Heyvaerts, C.A. Norman, Astrophys. J. **347**, 1055 (1989)

J. Heyvaerts, C.A. Norman, Astrophys. J. **596**, 1240 (2003a)

J. Heyvaerts, C.A. Norman, Astrophys. J. **596**, 1256 (2003b)

R.M. Hjellming, K.J. Johnston, Astrophys. J. **328**, 600 (1988)

M. Huarte-Espinosa, A. Frank, E. Blackman, in *Jets at All Scales*, ed. by G.E. Romero, R.A. Sunyaev, T. Belloni. Proceedings of the IAU Symposium, vol. 275 (Cambridge University Press, Cambridge, 2011), p. 87

M. Huarte-Espinosa, A. Frank, E.G. Blackman, A. Ciardi, P. Hartigan, S.V. Lebedev, J.P. Chittenden, Astrophys. J. **757**, 66 (2012)

Y. Kato, M.R. Hayashi, R. Matsumoto, Astrophys. J. **600**, 338 (2004a)

Y. Kato, S. Mineshige, K. Shibata, Astrophys. J. **605**, 307 (2004b)

S.R. Kelner, F.A. Aharonian, Phys. Rev. D **78**, 034013 (2008)

S.R. Kelner, F.A. Aharonian, Phys. Rev. D **82**, 099901 (2010)

S.R. Kelner, F.A. Aharonian, V.V. Bugayov, Phys. Rev. D **74**, id. 034018 (2006)

S.R. Kelner, F.A. Aharonian, V.V. Bugayov, Phys. Rev. D **79**, 039901 (2009)

C.F. Kennel, F.V. Coroniti, Astrophys. J. **283**, 694 (1984)

E. Kersalé, P.-Y. Longaretti, G. Pelletier, Astron. Astrophys. **363**, 1166 (2000)

D. Khangulyan, F.A. Aharonian, AIP Conf. Proc. **745**, 359 (2005)

D. Khangulyan, F. Aharonian, V. Bosch-Ramon, Mon. Not. R. Astron. Soc. **383**, 467 (2008)

S. Koide, K. Shibata, T. Kudoh, D.L. Meier, Science **5560**, 1688 (2002)

S.S. Komissarov, Mon. Not. R. Astron. Soc. **326**, L41 (2001)

S.S. Komissarov, arXiv:astro-ph/0211141 (2002)

S.S. Komissarov, in *Proceedings of the Third International Sakharov Conference on Physics*, ed. by A. Semikhatov, V. Zaikin, M. Vasiliev (World Scientific, Moscow, 2003), p. 392

S.S. Komissarov, Mon. Not. R. Astron. Soc. **350**, 427 (2004)

S.S. Komissarov, J. Korean Phys. Soc. **54**, 2503 (2009)

S.S. Komissarov, Mem. Soc. Astron. Ital. **82**, 95 (2011)

S.S. Komissarov, S.A.E.G. Falle, Mon. Not. R. Astron. Soc. **288**, 833 (1997)

S.S. Komissarov, N. Vlahakis, A. Königl, M.V. Barkov, Mon. Not. R. Astron. Soc. **394**, 1182 (2009)

S.S. Komissarov, N. Vlahakis, A. Königl, Mon. Not. R. Astron. Soc. **407**, 17 (2010)
S.V. Lebedev, A. Ciardi, D. Ampleford, S.N. Bland, S.C. Bott, J.P. Chittenden, G. Hall, J. Rapley, A. Frank, E.G. Blackman, T. Lery, Mon. Not. R. Astron. Soc. **361**, 97 (2005)
H.K. Lee, R.A.M.J. Wijers, G.E. Brown, Phys. Rep. **325**, 83 (2000)
A. Levinson, Int. J. Mod. Phys. D **19**, 649 (2010)
Z.-H. Li, T. Chiueh, M.C. Begelman, Astrophys. J. **394**, 459 (1992)
H. Li, R.V.E. Lovelace, J.M. Finn, S.A. Colgate, Astrophys. J. **561**, 915 (2001)
P. Lii, T.M.M. Romanova, R.V.E. Lovelace, Mon. Not. R. Astron. Soc. **420**, 2020 (2012)
P. Lipari, M. Lusignoli, D. Meloni, Phys. Rev. D **75**, 123005 (2007)
M.S. Longair, *High-Energy Astrophysics*, vols. I & II (Cambridge University Press, Cambridge, 1994)
P.-Y. Longaretti, Phys. Lett. A **320**, 215 (2003)
R.V.E. Lovelace, D.M. Rothstein, G.S. Bisnovatyi-Kogan, Astrophys. J. **701**, 885 (2009)
D. Lynden-Bell, Mon. Not. R. Astron. Soc. **279**, 389 (1996)
D. Lynden-Bell, Mon. Not. R. Astron. Soc. **341**, 1360 (2003)
D. Lynden-Bell, Mon. Not. R. Astron. Soc. **369**, 1167 (2006)
D. Lynden-Bell, C. Boily, Mon. Not. R. Astron. Soc. **267**, 146 (1994)
Y. Lyubarsky, Astrophys. J. **698**, 1570 (2009)
M. Lyutikov, Phys. Rev. E **82**, 056305 (2010a)
M. Lyutikov, Mon. Not. R. Astron. Soc. **411**, 422 (2010b)
M. Lyutikov, M. Lister, Astrophys. J. **722**, 197 (2010)
A. Mastichiadis, R.J. Protheroe, J.G. Kirk, Astrophys. J. **433**, 765 (2005)
L. Maximon, J. Res. Natl. Bur. Stand. **72B**, 79 (1968)
J.E. McClintock, R. Narayan, S.W. Davis, L. Gou, A. Kullkarni, J.A. Orosz, R.F. Penna, R.A. Remillard, J.F. Steiner, Class. Quantum Gravity **28**, 114009 (2011)
J.C. McKinney, Astrophys. J. **630**, L5 (2005)
J.C. McKinney, Mon. Not. R. Astron. Soc. **367**, 1797 (2006)
J.C. McKinney, C.F. Gammie, Astrophys. J. **611**, 977 (2004)
J.C. McKinney, R. Narayan, Mon. Not. R. Astron. Soc. **375**, 531 (2007)
D.L. Meier, in *Jets at All Scales*, ed. by G.E. Romero, R.A. Sunyaev, T.M. Belloni. Proceedings of the IAU Symposium, vol. 275 (Cambridge University Press, Cambridge, 2011), p. 13
A. Merloni, S. Heinz, T. Di Matteo, Mon. Not. R. Astron. Soc. **345**, 1057 (2003)
P. Mészáros, Annu. Rev. Astron. Astrophys. **40**, 137 (2002)
F.C. Michel, Astrophys. J. **158**, 727 (1969)
P. Mimica, M.A. Aloy, Mon. Not. R. Astron. Soc. **401**, 525 (2010)
Y. Mizuno, S. Yamada, S. Koide, K. Shibata, Astrophys. J. **606**, 395 (2004)
Y. Mizuno, P.E. Hardee, K. Nishikawa, Astrophys. J. **662**, 835 (2007)
Y. Mizuno, Y. Lyubarsky, K. Nishikawa, P.E. Hardee, Astrophys. J. **700**, 684 (2009)
Y. Mizuno, P.E. Hardee, K. Nishikawa, Astrophys. J. **734**, 19 (2011)
Y. Mizuno, Y. Lyubarsky, K. Nishikawa, P.E. Hardee, Astrophys. J. **757**, 16 (2012)
R. Moderski, M. Sikora, P.S. Coppi, F.A. Aharonian, Mon. Not. R. Astron. Soc. **363**, 954 (2005)
A. Mücke, R. Engel, J.P. Rachen, R.J. Protheroe, T. Stanev, Comput. Phys. Commun. **124**, 290 (2000)
M. Nakamura, H. Li, S. Li, Astrophys. J. **652**, 1059 (2006)
M. Nakamura, H. Li, S. Li, Astrophys. J. **656**, 721 (2007)
M. Nakamura, I.L. Tregillis, H. Li, S. Li, Astrophys. J. **686**, 843 (2008)
R. Narayan, J.E. McClintock, Mon. Not. R. Astron. Soc. **419**, L69 (2012)
I.D. Novikov, K.S. Thorne, in *Black Holes*, ed. by C. DeWitt, B. DeWitt (Gordon and Breach, Paris, 1973), p. 343
M. Perucho, Int. J. Mod. Phys. Conf. Ser. **8**, 241 (2012)
M. Perucho, V. Bosch-Ramon, Astron. Astrophys. **539**, A57 (2012)
M. Perucho, J.M. Martí, Mon. Not. R. Astron. Soc. **382**, 526 (2007)
M. Perucho, I. Martí-Vidal, A.P. Lobanov, P.E. Hardee, Astron. Astrophys. **545**, A65 (2012)
T. Piran, Rev. Mod. Phys. **76**, 1143 (2004)

P. Polko, D.L. Meier, S. Markoff, Mon. Not. R. Astron. Soc. **428**, 587 (2013)

O. Porth, C. Fendt, Astrophys. J. **709**, 1100 (2010)

B. Punsly, *Black Hole Gravitohydromagnetics* (Springer, Berlin, 2001)

B. Punsly, F.V. Coronity, Astrophys. J. **350**, 518 (1990a)

B. Punsly, F.V. Coronity, Astrophys. J. **354**, 583 (1990b)

B.A. Remington, R.P. Drake, D.D. Ryutov, Rev. Mod. Phys. **78**, 755 (2006)

C.S. Reynolds, D. Garofalo, M.C. Begelman, Astrophys. J. **651**, 1023 (2006)

M.M. Reynoso, G.E. Romero, Astron. Astrophys. **493**, 1 (2009)

M.M. Reynoso, M.C. Medina, G.E. Romero, Astron. Astrophys. **531**, A30 (2011)

G.B. Rybicki, A.P. Lightman, *Radiative Processes in Astrophysics* (J. Wiley, New York, 1979)

M.M. Romanova, G.V. Ustyugova, A.V. Koldoba, R.V.E. Lovelace, Mon. Not. R. Astron. Soc. **399**, 1802 (2009)

G.E. Romero, Astrophys. Space Sci. **234**, 49 (1995)

G.E. Romero, J.M. Paredes, *Introducción a la Astrofísica Relativista*, 1st edn. (Publicacions i Edicions de la Universitat de Barcelona, Barcelona, 2011)

G.E. Romero, D.F. Torres, M.M. Kaufman-Bernadó, I.F. Mirabel, Astron. Astrophys. **410**, L1 (2003)

R.H. Sanders, Astrophys. Space Sci. **266**, 73 (1983)

R. Schlickeiser, *Cosmic Ray Astrophysics* (Springer, Berlin, 2002)

M. Sikora, L. Stawarz, J.-P. Lasota, Astrophys. J. **658**, 815 (2007)

H.C. Spruit, D.A. Uzdensky, Astrophys. J. **629**, 960 (2005)

L. Stawarz, F. Aharonian, J. Kataoka, M. Ostrowski, A. Siemiginowska, M. Sikora, Mon. Not. R. Astron. Soc. **370**, 981 (2006)

F. Suzuki-Vidal, S.V. Lebedev, S.N. Bland, G.N. Hall, G. Swadling, A.J. Harvey-Thompson, G. Burdiak, P. de Grouchy, J.P. Chittenden, A. Marocchino, M. Bocchi, A. Ciardi, A. Frank, S.C. Bott, Astrophys. Space Sci. **336**, 41 (2011)

M. Takahashi, S. Nitta, Y. Tatematsu, A. Tomimatsu, Astrophys. J. **363**, 206 (1990)

A. Tchekhovskoy, J.C. McKinney, Mon. Not. R. Astron. Soc. **423**, L55 (2012)

A. Tchekhovskoy, J.C. McKinney, R. Narayan, Mon. Not. R. Astron. Soc. **388**, 551 (2008)

A. Tchekhovskoy, J.C. McKinney, R. Narayan, Astrophys. J. **699**, 1789 (2009)

A. Tchekhovskoy, R. Narayan, J.C. McKinney, Astrophys. J. **711**, 50 (2010a)

A. Tchekhovskoy, R. Narayan, J.C. McKinney, New Astron. **15**, 749 (2010b)

K.S. Thorne, R.H. Price, D.A. Macdonald, *Black Holes: The Membrane Paradigm* (Yale University Press, New Haven, 1986)

C.A. Tout, J.E. Pringle, Mon. Not. R. Astron. Soc. **281**, 219 (1996)

G. Ustyugova, A.V. Koldoba, M.M. Romanova, V.M. Chechetkin, R.V.E. Lovelace, Astrophys. J. **516**, 221 (1999)

D.A. Uzdensky, A.I. MacFadyen, Astrophys. J. **647**, 1192 (2006)

D.A. Uzdensky, A.I. MacFadyen, Astrophys. J. **669**, 546 (2007)

G.S. Vila, G.E. Romero, N.A. Casco, Astron. Astrophys. **538**, A97 (2012)

N. Vlahakis, Astrophys. J. **600**, 324 (2004)

N. Vlahakis, K. Tsinganos, Mon. Not. R. Astron. Soc. **298**, 777 (1998)

R.M. Wald, Phys. Rev. D **10**, 1680 (1974)

E.J. Weber, L. Davis Jr., Astrophys. J. **148**, 217 (1967)

R.L. Znajek, Mon. Not. R. Astron. Soc. **179**, 457 (1977)

Chapter 6
Evidence for Black Holes

6.1 Black Holes in Action

Some of the most energetic phenomena in the Universe take place in accreting sources that host (or are suspected to host) black holes. Naturally, black holes that span almost ten orders of magnitude in mass, placed in a variety of environments, and fed with matter from different sources, bring about an incredibly rich phenomenology. By and large, though, theory and observation point to an unified description of the dynamics and radiation of inflows/outflows in sources powered by accretion onto black holes at all scales (e.g. Mirabel and Rodríguez 1998). The fundamental variables in this description are the accretion rate and the mass (and possibly the spin) of the black hole (e.g. Fender et al. 2007; Heinz and Sunyaev 2003; Sams et al. 1996). The following sections present a brief account of X-ray binaries, active galactic nuclei, and gamma-ray bursts. Ultra-luminous X-ray sources are presented later on. The theory of accretion and jets (often with examples and applications to these types of sources) and some relevant non-thermal radiative processes have been extensively developed in previous chapters; here the discussion shall be restricted to the most fundamental phenomenological aspects.

6.1.1 X-Ray Binaries and Microquasars

X-ray binaries are formed by a compact object (a neutron star[1] or a stellar-mass black hole, also called the primary) in accretion from a non-collapsed star (referred to as the donor or companion star, or just the secondary). According to the mass of the donor star XRBs are classified into *high-mass* and *low-mass*. In high-mass XRBs the donor star is an O, B, or Wolf-Rayet star of mass $M_* \sim 8$–$20 M_\odot$. These

---

[1] We shall not deal with neutron star XRBs here. See for example Migliari and Fender (2006), also Migliari et al. (2011) for a comprehensive analysis of the differences and similarities between neutron star and black hole XRBs.

G.E. Romero, G.S. Vila, *Introduction to Black Hole Astrophysics*,
Lecture Notes in Physics 876, DOI 10.1007/978-3-642-39596-3_6,
© Springer-Verlag Berlin Heidelberg 2014

Fig. 6.1 A simplified sketch of a microquasar

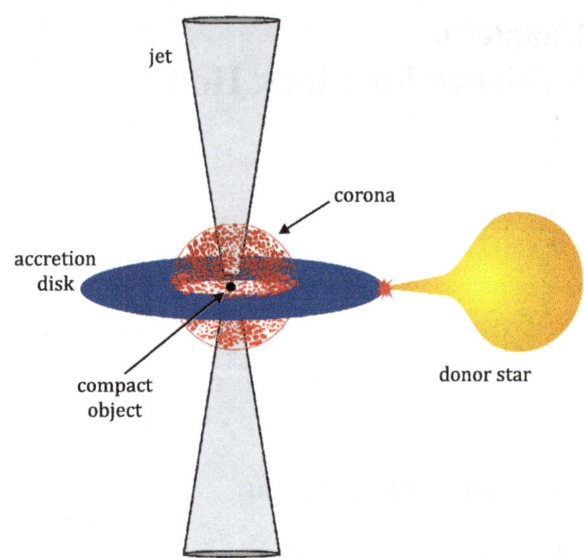

stars lose mass mainly through strong winds. Donor stars in low-mass XRBs have $M_* \lesssim 2M_\odot$. They are old stars of spectral type B or later, that transfer mass to the compact object through the overflow of the Roche lobe. Out of the 299 sources cataloged by Liu et al. (2006, 2007), 185 are low-mass and 114 are high-mass XRBs. The spatial distribution of high-mass XRBs in our galaxy traces the star forming regions in the spiral arms. This is expected, since the companion star is relatively young ($\lesssim 10^7$ yr) and should not have departed significantly from its birthplace. Low-mass X-ray binaries are concentrated towards the center of the galaxy, especially in the bulge. These systems are old ($\sim 10^9$ yr) and some have migrated from the galactic plane towards higher latitudes.

Figure 6.1 is a simplified sketch of a X-ray binary. The basic components of the accretion flow are pictured: the accretion disk, the corona, and the jets. Not all XRBs launch jets—or perhaps they do and we have not detected them yet. Those X-ray binaries with jets were named *microquasars* by Mirabel et al. (1992), alluding to their striking morphological resemblance to the extragalactic *quasars* we shall speak about in the next section. There are about 20 confirmed (with jets imaged at radio wavelengths) or firm microquasar candidates in our galaxy and one recently discovered in the galaxy NGC 7793 (Soria et al. 2010).

Microquasars produce two type of outflows: continuous steady jets and discrete ejections. Continuous jets are observed during the so-called low-hard state (see below), whereas the ejection of blobs occurs during the transition between spectral states. Jets in microquasars are mildly relativistic, with bulk Lorentz factors $\lesssim 10$. Apparent superluminal motion of the discrete ejections is observed in some sources. The typical power of microquasars jets is $L_{\mathrm{jet}} \sim 10^{37}$–$10^{40}$ erg s^{-1}. The most powerful inject enough energy in the surrounding medium as to distort it significantly. Strong signatures of jet-medium interaction are seen in the microquasars SS 433 and Cygnus X-1.

Black hole X-ray binaries go through different spectral states, classified according to the timing and spectral characteristics of the X-ray emission. The four canonical states are (e.g. McClintock and Remillard 2006; Belloni et al. 2011) the *low-hard*, *high-soft*, *very high*, and *quiescence* states. *Intermediate* states with mixed properties are also observed. These five spectral states can be briefly characterized as follows:

- **Low-hard (LH) state:** the X-ray flux follows a hard power-law $F_X \propto E^{-\alpha}$ with $\alpha \sim 0.4$–0.9 that cuts off exponentially at ~ 100 keV. A fluorescence Fe Kα line around 6.4 keV is detected in many sources. The thermal emission from the accretion disk is sometimes observed, but peaking at lower energies and less luminous than in the high-soft state. The presence of steady jets in the LH state is inferred from the flat/slightly inverted shape of the radio spectrum. The characteristic luminosity of jets in the LH state is 10^{36}–10^{37} erg s^{-1}.

  The canonical model of XRBs in low-hard state postulates that the hard X-rays are emitted in the hot corona (likely to be accreting in a radiatively inefficient regime like an ADAF) by Comptonization of photons from an outer accretion disk. Part of the radiation from the corona is reprocessed in the disk, including the excitation of Fe atoms that gives rise to the fluorescence line. In some microquasars the radio and X-ray luminosities display a tight correlation of the form (e.g. Corbel et al. 2003)

$$L_{\text{radio}} \propto L_X^{0.7}, \tag{6.1}$$

  what suggests that the jets may contribute or even dominate the X-ray emission.[2]
- **Quiescence state:** similar to a faint LH state. The X-ray spectrum is dominated by a hard power-law of very low luminosity $\sim 10^{30}$–10^{35} erg s^{-1}.
- **Very high (VH) state:** the X-ray spectrum is a power-law without indication of a cutoff up to at least 100 keV. The power-law is steeper than in LH state. Some sources in VH state show quasi-periodic oscillations (see Sect. 6.5.1). The role of jets during the VH state is not clear. In some cases the onset of the VH state coincides with the quenching of the radio emission; in other sources discrete ejections are observed during the VH state or the transition to it.
- **High-soft (HS) state:** the spectrum below ~ 10 keV is dominated by the thermal emission of an accretion disk of temperature $kT_d \sim 0.5$–1 keV. A steep power-law tail extends into the hard X-rays. The Fe Kα line is broadened, probably because the disk extends closer to the compact object than in the LH state. The radio emission is strongly suppressed, suggesting the absence of jets.
- **Intermediate states:** XRBs spend most time in the four spectral states described above, but there are also epochs when the characteristics of the X-ray emission cannot be accounted for purely in terms of one spectral state. In such cases the source is said to be in an intermediate state. Intermediate states occur, for example, during the transition between two of the main spectral states.

---

[2]Intending to extend it to AGN, a modified version of Eq. (6.1) that includes the mass of the black hole was introduced by Merloni et al. (2003) and Falcke et al. (2004). Its validity is not definitely established.

The spectral and timing properties of black hole XRBs vary in response to changes in the mass accretion rate induced by instabilities in the disk, recall Sect. 4.5.3 and Fig. 4.13. In quiescence the accretion rate is very low, of the order of $\dot{m} \sim 10^{-2}$ in Eddington units. The surroundings of the black hole are filled by an extended, radiatively very inefficient corona and the disk is quite detached. The LH state is similar but at slightly large accretion rates, $10^{-2} \lesssim \dot{m} \lesssim 0.1$. Continuous jets are ejected during quiescence and LH state. But the ADAF corona only exists for accretion rates below a critical value, so as $\dot{m}$ increases the inner radius of the disk approaches the black hole. Finally the corona disappears, perhaps expelled as a relativistic discrete jet, and the system enters the HS state. The radiative spectrum is dominated by the thermal emission of the disk plus a power-law from a very tenuous corona. The typical accretion rate in the HS state is $\dot{m} \approx 0.1$, but values as large as $\dot{m} \approx 1$ can be achieved during the LH-to-HS transition. This process is cyclic; eventually the source goes back to the LH state through a low-luminosity path.

Within the last decade a small number of XRBs have been detected at high (30 MeV–50 GeV) or very high (>50 GeV) energies, and there are a few other gamma-ray sources suspected to be XRBs. All the positively identified *gamma-ray binaries* are also radio sources and contain a massive companion star of type O or B. Two of them, Cygnus X-1 (McConnell et al. 2000; Albert et al. 2007; Sabatini et al. 2010) and Cygnus X-3 (Abdo et al. 2009a; Tavani et al. 2009), are high-mass microquasars and the others are binaries formed by a Be star and a non-accreting pulsar. There are two sources, LS 5039 (Paredes et al. 2000; Aharonian et al. 2006; Abdo et al. 2009c) and LS I +61° 303 (Albert et al. 2006; Acciari et al. 2008; Abdo et al. 2009b), whose nature is not definitely established. They could be pulsar-driven binaries or microquasars. You can see Dubus (2006) and Romero et al. (2007) for arguments in favor or each scenario in LS I +61° 303.

In pulsar-driven binaries the gamma rays are emitted by particles accelerated at shocks that develop where the pulsar wind and the decretion disk of the Be star collide. In high-mass microquasars the gamma-ray emission originates in the interaction of relativistic particles in the jets with the wind of the companion star via proton-proton collisions and/or inverse Compton scattering of stellar photons (e.g. Kaufman-Bernadó et al. 2002; Romero and Orellana 2005; Bosch-Ramon et al. 2006). To date, no gamma-ray emission has been observed from a low-mass X-ray binary although there are a couple of suspects (e.g. Falanga et al. 2010; de Martino et al. 2013). Their detection is further complicated because in general they are transient sources. Since donor stars in low-mass XRBs are old and dim, jet-star interactions are not expected to be efficient. Gamma rays in low-mass microquasars might be created in the interaction of relativistic particles with the matter, radiation, and magnetic field carried by the jet itself (e.g. Romero and Vila 2008; Vila and Romero 2010; Vila et al. 2012). In sources both with high-mass and low-mass companions, non-thermal particles in the corona may also inject high-energy photons (e.g. Romero et al. 2010; Vieyro and Romero 2012).

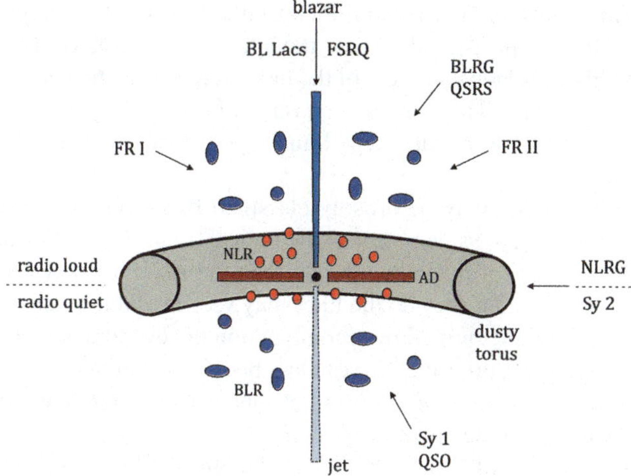

Fig. 6.2 The unification scheme of active galactic nuclei. AD, NLR, and BRL stand for accretion disk, narrow line region, and broad line region, respectively; the rest of the abbreviations are explained in the text. On the radio quiet side the jet is fainter to indicate it is absent

6.1.2 Active Galactic Nuclei

Active galaxies are those that display non-stellar emission in the nucleus; they represent about 10 % of all galaxies. The electromagnetic radiation from AGN covers the whole spectrum from radio to gamma rays. The observed fluxes and the distances determined from the cosmological redshift of spectral lines, allow to infer that AGN have huge intrinsic luminosities of the order of 10^{44} erg s^{-1} and higher. Variability on short timescales implies that the emission region is compact, whereas the presence of high-velocity clouds requires that the mass enclosed in that region is large. The estimated lifetime of some AGN indicates that this mass is most likely contained in a single object and not a in collection of them (a cluster of stars, for example). The detection of radio jets in about 10 % of AGN lends support to this hypothesis.

The current model of active galactic nuclei is conceptually quite simple; a basic scheme is pictured in Fig. 6.2. They are powered by accretion onto a supermassive black hole (10^6–$10^{10} M_\odot$) at the center of the galaxy. The black hole is surrounded by a corona and an accretion disk that radiate from the optical through the soft X-rays. A thick ring of dust and gas (a "dusty torus") obscures the central region and emits mostly in the infrared. Clouds orbit the black hole at different distances. A pair of collimated jets are launched by action of the (rotating) black hole and the disk. The outflows propagate for distances of tens of kpc and even Mpc.

The strength of this picture is that it explains the great variety of observational manifestations of AGN in terms of the orientation of the system with respect to the line of sight. The orientation determines the degree of obscuration and beaming that affect the radiation in its path to the observer. Before this "unification scheme" was settled (e.g. Antonucci 1993; Urry and Padovani 1995), what we now understand are AGN seen from different sides were thought to be objects of distinct nature. Because of the historical development of the field there was a proliferation of names still largely in use today. These include quasars, blazars, BL Lacs, radio galaxies, type 1 and type 2 Seyfert galaxies, type I and type II Fanaroff-Riley galaxies, and the list goes on ...

Active galactic nuclei may be broadly classified based on their emission lines and radio loudness. At the same power in the optical, *radio loud* AGN have a radio power about 3–4 orders of magnitude larger than the *radio quiet*. They may show emission lines or not, and if they do the lines may vary in width because of Doppler broadening. The lines are emitted by recombination of several elements (H, He, C, O, Fe) in the clouds that orbit the black hole—broad lines in the inner, high-speed clouds ($\sim 10^3$ km s^{-1}) of the *broad-line region*, and narrow lines in the outer, slower clouds ($\sim 10^2$ km s^{-1}) of the *narrow-line region*.

In sources where the disk is seen nearly edge-on the dusty torus obscures the central nucleus and only narrow lines are detected. Seyfert 2 (Sy 2) galaxies belong to this group. Broad lines and the optical-to-X-rays continuum from the nucleus is observed in sources oriented more face-on. In radio quiet quasars (called as well quasi-stellar objects, QSO) and Seyfert 1 (Sy 1) galaxies the symmetry axis of the system makes an angle $\sim 30°$ with the line of sight, permitting the detection of the emission from both the narrow and the broad line regions. Radio loud AGN (only about 10 % of all) have relativistic jets. In radio galaxies the line of sight is basically perpendicular to the jet axis so the outflows can be appreciated in their whole extension. If the torus hides the broad line region, it is a narrow line radio galaxy (NLRG). The synchrotron radio emitting regions in jets of Fanaroff-Ryley (FR) I and II galaxies display different morphologies. Jets in FR I galaxies are very bright in the inner regions and faint with the distance to the nucleus; jets in FR II galaxies remain practically dark up to the termination region, where the develop huge radio lobes and hot spots by interaction with the medium. Radio-loud AGN with broad lines are broad line radio galaxies (BLRG) and radio quasars (also called quasi-stellar radio sources, QSRS, or quasi-stellar sources). When the jet is pointed almost along the line of sight we call it a blazar. Blazars are sources of beamed, highly variable, polarized non-thermal continuum emission. Among blazars are the BL Lacertae (BL Lac) objects, with very weak or no emission lines at all. Jets in flat spectrum radio quasars (FSRQ) have radiative properties similar to blazars and show emission lines. BL Lacs and FSRQ are though to be FR I and FR II galaxies, respectively, with the jet axis close to the line of sight ($\lesssim 15°$). It is worth mentioning here that orientation alone cannot account for the radio-loud/radio-quiet dichotomy, that appears to be a true manifestation of different physical processes going on in the central engines of AGN that impact on the properties of jets.

Blazars are very interesting objects because of their extreme properties: fast and high-amplitude variability and polarization, beaming and amplification of the radiation and apparent superluminal motion induced by the relativistic speed of the jets (Lorentz factors ~ 50). The spectral energy distribution of blazars is non-thermal and covers up to 20 orders of magnitude in frequency from radio to gamma rays.

Actually, the largest fraction of the identified gamma-ray sources are blazars. They represent, for instance, 57 % (894 sources) of all entries in the second release of the catalog of sources detected with the Large Area Telescope (LAT) on board the satellite *Fermi* (Nolan et al. 2012).[3]

The SED of blazars typically presents two peaks—a synchrotron peak at low energies (up to the infrared/optical) and an inverse Compton peak at MeV-GeV (and even TeV) energies from the up-scattering of synchrotron or external photons (e.g. Böttcher 2007). It is clear, then, that jets in AGN are efficient accelerators of electrons. Relativistic protons, if present, may contribute as well to the gamma-ray spectrum of blazars and other types of AGN through synchrotron radiation, proton-proton and proton-photon collisions; see for example the models of Aharonian (2002), Mücke et al. (2003), and Reynoso et al. (2011).

Active galaxies play a fundamental role in modern astrophysics since they are thought to be the sites of acceleration of the most energetic cosmic rays that arrive to Earth. In 2007, the scientists of the Pierre Auger Observatory released a renowned work where it was announced that the arrival direction of 18 out of the 27 cosmic rays with energy >55 EeV[4] detected so far was within $\sim 3°$ from the position in the sky of the nearby ($\lesssim 75$ Mpc) AGN listed in certain catalogs (Abraham et al. 2007). It must be noticed that reconstructing the origin of cosmic rays from their direction of arrival is not an easy task; it implies the modeling of the propagation of charged energetic particles in the intergalactic and galactic magnetic fields and their interactions. What correlation studies really test is the degree of anisotropy in the arrival directions. As more events were collected, however, the significance of the correlation claimed by the Auger Collaboration washed out; the latest analysis reveal only a weak deviation from isotropy in the direction of arrival of cosmic rays at the highest energies. The anisotropy hypothesis is also rejected from the results of another cosmic ray detector, HiRes, in the northern hemisphere.[5] However, both correlation and no correlation claims have low significance levels and await to be confirmed.

Any anisotropy in the arrival direction of extragalactic cosmic rays may be due to a nearby source. An interesting candidate in this regard is the nearest AGN—the radio galaxy Centaurus A. A clustering of energetic cosmic ray events are detected from a region of $24°$ around this galaxy. Another suggestive fact is that there are several pairs of events (separated by less than $5°$) near Cen A, some of them consisting of detections made by different experiments. Of course, one should be able to explain how the jets of Cen A could accelerated protons or heavier nuclei to the required energies. Furthermore, Cen A is located in one of the most populated regions in the nearby Universe. It is up to some extent natural to expect that cosmic rays arrive from that direction although not necessarily from Cen A itself.

---

[3] A few radio galaxies, Seyfert galaxies, and other AGN have been detected by *Fermi* as well.

[4] EeV stands for "exa electron volt"; 1 EeV $= 10^{18}$ eV.

[5] But Auger and HiRes disagree on the composition of cosmic rays.

Despite the lack of conclusive experimental results and the doubts about the role Cen A, active galactic nuclei still hold as the most likely accelerators of extragalactic cosmic rays. Perhaps an enlarged version of the Auger Observatory and other forthcoming cosmic ray detectors (such as the planned JEM-EUSO experiment on board the International Space Station) will gather enough statistics to unveil their origin.

6.1.3 Gamma-Ray Bursts

Gamma-ray bursts were discovered in the late 1960s–early 1970s by the Vela series of satellites, commissioned to spot clandestine detonations of nuclear weapons on Earth. In the 1990s, the Burst and Transient Source Experiment (BATSE) of the *Compton Gamma-Ray Satellite* discovered thousands of GRBs. They were isotropically distributed in the sky, what suggested a cosmological origin. This was later confirmed with the first redshift determinations from observations in the optical. On average, one gamma-ray burst occurs per day in the Universe.

From the distance estimates and accounting for beaming, the photon fluxes measured from GRBs imply huge luminosities of $\sim 10^{50}$ erg s^{-1} and a total energy release of $\gtrsim 10^{51}$ erg. Moreover, all this energy is liberated in one flash lasting typically a few seconds. Gamma-ray bursts are one-time events, there is no evidence that they are recurrent. The bursts have two phases. During the initial *prompt* stage hard X-rays and gamma rays up to tens of GeV (typically peaking between ~ 100 keV and ~ 1 MeV) are emitted. The prompt emission is well described by a broken power-law. Then follows the *afterglow*, a phase of smoother fading emission in the X-rays, optical, and radio. The afterglow fades away on timescales of hours to weeks.

The distribution of GRBs according to the duration of the prompt phase is bimodal and shows a minimum at about $T_{90} = 2$ s, where T_{90} is defined as the time in which 90 % of the photons arrive to the detector. The spectrum of *short* gamma-ray bursts is harder than that of *long* gamma-ray bursts, what indicates that the bimodality is most probably real and not induced by an observational bias. It is thought that long GRBs are emitted during the collapse of a massive rotating star into a black hole (e.g. Paczyński 1998; MacFadyen and Woosley 1999). This idea is supported by the association of long GBRs with star forming regions and in a few cases with supernovae. Short GRBs, on the other hand, may originate as a product of the merger of two compact objects to form a stellar-mass black hole (e.g. Eichler et al. 1989; Nakar 2007). Short GRBs take place in host galaxies with varied star formation properties, including low star-forming rates. Although there are few short GRBs with known redshift, these appear to be nearer than long GRBs and also less powerful.

Both the merger and the collapse (or collapsar) models involve the release of a huge amount of gravitational energy by accretion onto the central core. A small part of this energy is eventually converted into radiation in a "fireball", an extremely relativistic collimated outflow expanding outwards with bulk Lorentz factors up to

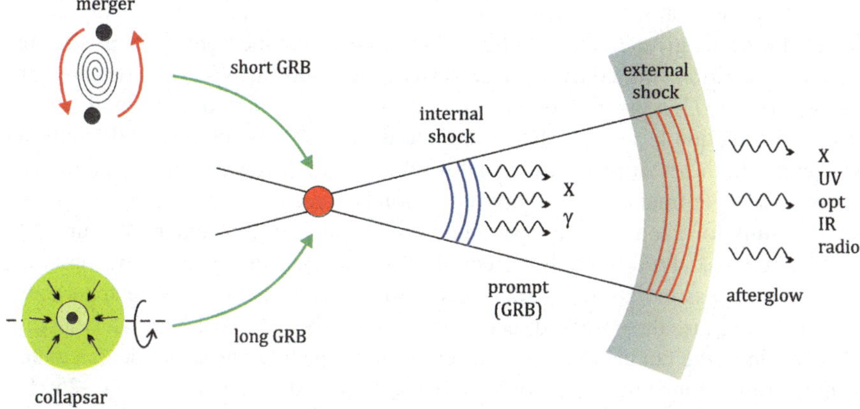

Fig. 6.3 A gamma-ray burst in the fireball model

10^3.[6] The jets might be powered by the rotation of the newborn black hole, annihilation of neutrinos emitted in the accretion disk (recall Sect. 4.5.5),[7] and/or other mechanisms. The kinetic energy of the fireball is dissipated as radiation in shocks—internal shocks (that develop within the outflow as faster shells of material catch up with slower ones) for the prompt emission and external shocks (collisions with the external medium) for the afterglow (see Fig. 6.3). Electrons and perhaps protons are accelerated up to relativistic energies at the sites of the shocks and these emit the non-thermal radiation.

Neither the central engines nor the jets in GBRs can be resolved observationally. There are, however, some theoretical and observational arguments in favor of relativistic outflows and beaming in GRBs. The first and immediate argument is that beaming reduces the energetic budget[8]—compared to an isotropically emitting source—required to account for the measured electromagnetic fluxes. The second is the "compactness" problem. The photon density at fixed energy in the source may be estimated from the observed flux, variability timescale, and distance. For values of the parameters typical of GRBs, the photon field turns out to be very dense and the optical depth for two-photon annihilation as large as $\tau_{\gamma\gamma} \sim 10^{15} \gg 1$. If this were so the source would be absolutely self-absorbed and unobservable. The compactness problem disappears if one assumes that the plasma is moving towards the observer with a large Lorentz factor Γ. Then, in the rest frame of the jet the energy of the photons and the photon density are lower than we infer. Typically $\Gamma > 100$ are required for the source to be optically thin to its own high-energy radiation.

---

[6] We only detect the gamma-ray burst (i.e. the prompt emission) when the jet is pointing to us. If the jet axis is slightly shifted from the line of sight, "orphan" afterglow emission should be detectable, i.e. radio and optical emission without a preceding gamma-ray burst.

[7] This process injects electron-positron pairs, $\bar{\nu} + \nu \rightarrow e^+ + e^-$.

[8] Approximately by a factor $\theta_{jet}^2/2$, where θ_{jet} is the opening angle of the jet.

Beaming and collimation is also compatible with the observation of "achromatic" breaks, the name given to the simultaneous steepening in the light curve at all energies. The radiation emitted by a plasma moving at relativistic speeds is beamed into an angle $\theta \sim 1/\Gamma$ in the direction of motion. If Γ is very high only a small portion of the jet's section is visible. But in the internal/external shock model the outflow decelerates in the external shocks. As the bulk Lorentz factor decreases and the jets advances a larger fraction of the emitting region becomes visible. As long as $\theta < \theta_{\text{jet}}$ beamed emission cannot be observationally distinguished form spherical. But eventually when $\Gamma < 1/\theta_{\text{jet}}$ the whole section of the jet becomes visible. An observer now receives less radiation than she/he would do if the radiation were spherically symmetric because the observed section increases only because the jet is advancing. So the light curve begins to decays faster. The jet opening angle can be calculated from the time of the break. For long GRBs $\theta_{\text{jet}} \sim 5°–10°$ whereas $\theta_{\text{jet}} \sim 5°–25°$ for short GRBs, although the latter estimate is more uncertain due to poor statistics. Gamma-ray bursts with chromatics breaks and very late breaks (which could imply large jet opening angles) have been observed.

In the classical fireball model the outflows are matter (not magnetically) dominated; recall from Chap. 5 that a low magnetization is essential for shocks to develop. Electrons accelerated at the internal shocks radiate the prompt emission through synchrotron radiation in the locally amplified magnetic field, and perhaps also synchrotron-self Compton and inverse Compton scattering of thermal photons. The afterglow radiation is synchrotron emission from electrons accelerated at the external shocks. If the jet is powered by the rotational energy of the black hole it is likely that the magnetization remains high and magnetic energy is dissipated by reconnection, instabilities, or other mechanisms different from shocks. Although the baryon load (if any) of GRB jets is unknown radiation of hadronic origin has been also considered, recently in particular to explain the GeV emission from some GRBs detected with *Fermi*-LAT. Gamma-ray bursts jets with a content of relativistic protons should be sources of neutrinos created in proton-photon interactions. Interestingly, neutrino bursts could be expected even in the absence of a burst of radiation if the jets cannot make it through the stellar envelope in a collapsar. These are called "choked" gamma-ray bursts. Short gamma-ray bursts and their progenitors must be as well sources of gravitational radiation.

6.2 Evidence for Stellar-Mass Black Holes

6.2.1 Dynamical Arguments

It is possible to derive from Kepler's laws an expression that relates the masses of the two components of a binary system and the inclination angle of the orbital plane. This is known as the *mass function* and is given by

$$f(M) = \frac{V_*^3 \, P_{\text{orb}}}{2\pi G} = \frac{M^3 \sin^3 i}{(M + M_*)^2} = \frac{M \sin^3 i}{(1 + q)^2}. \tag{6.2}$$

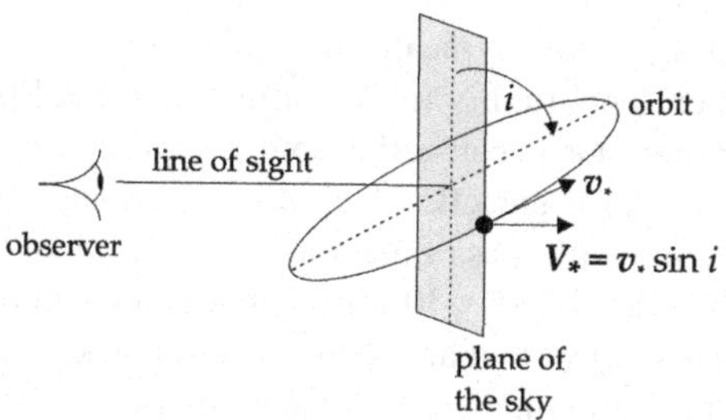

Fig. 6.4 Definition of the parameters in the binary mass function. The angle i gives the inclination of the plane of the orbit with respect to the plane of the sky (perpendicular to the line of sight). The velocity V_* is the component along the line of sight of the orbital velocity of M_* measured where indicated

Here M and M_* are the masses of the compact object and the donor star, respectively, $q = M/M_*$ is the mass ratio, i is the inclination angle of the orbit, P_{orb} is the orbital period, and V_* is the semi-amplitude of the donor star's line-of-sight velocity. The definitions of i and V_* are made clear in Fig. 6.4.

The mass function is an interesting quantity because its value can be calculated from parameters (P_{orb} and V_*) measurable from the light curve of the donor star. If i and M_* are independently known the value of M follows directly from that of f. For our purpose, however, it is enough to notice that the mass function sets an absolute lower limit to the mass of the compact object:

$$f(M) \leq M, \tag{6.3}$$

where the equal corresponds to $i = 90°$ and $q \to 0$. The upper limit for the mass of a neutron star is uncertain, mainly because of the relatively poor knowledge about the equation of state of matter at extremely high densities. It is presently thought to be somewhere in the range $M_{\mathrm{NS}} \sim 2.9$–$3.2\,M_\odot$ (e.g. Rhoades and Ruffini 1974; Kalogera and Baym 1996), and up to a $\sim$20–25 % larger if the star is rotating rapidly (Friedman and Ipser 1987; Cook et al. 1994). A value of f larger than this constitutes a piece of *dynamical* evidence for the presence of a black hole.

There are more than 20 stellar-mass black hole candidates in binary systems identified on dynamical grounds (e.g. Remillard and McClintock 2006; Casares 2010; Özel et al. 2010). In some cases $f < M_{\mathrm{NS}}$, but independent constraints on i or M_* allow to assure that $M > M_{\mathrm{NS}}$. The first to be discovered and perhaps the most examined galactic black hole candidate is Cygnus X-1. This is a binary system formed by a compact object and a massive O-type donor star. According to the latest estimates by Orosz et al. (2011) the mass of the compact object in Cygnus X-1 is $M = (14.81 \pm 0.98)\,M_\odot$, far beyond any upper limit for the mass of a neutron star. Aside a few other exceptions, however, most galactic black hole candidates have been found in low-mass binaries—particularly in those classified as "X-ray transients" or "X-ray novae". These systems spend most of their lives in quiescence (X-ray luminosity $L_X \lesssim 10^{33}$ erg s^{-1}) until some type of instability in the accretion disk triggers a sudden increase in the accretion rate. Then they enter in outburst, radiating intensely in the X rays typically for a few months. The periods of quiescence offer an excellent opportunity to determine the orbital parameters from the observation of the donor star because the optical emission is uncontaminated by the accretion flow. Very firm galactic black hole candidates have been identified in X-ray transients, some with mass functions $f > 6M_\odot$ like GRS 1915+105 and XTE J1118+480. There are other ~ 30 transient sources without determination of the mass of the compact object that are candidates to black hole binaries (e.g. Özel et al. 2010). The identification is based on the similarities of their X-ray spectrum with that of dynamically established candidates.

6.2.2 *The Search for the Signature of the Event Horizon*

The key fact that distinguishes black holes from neutron stars (and from any other type of collapsed object) is that they do not have a solid surface. Indeed, the definite proof of the existence of black holes would be the observation of some phenomenon that unequivocally revealed the presence of the horizon. X-ray binaries harbor either neutron stars or black holes,[9] so a possible way to search for the signature of the event horizon is to compare the observational characteristics of both groups of sources. Below we discuss two proposals based on the study of the X-ray emission of X-ray transients. The results are appealing but also controversial; you can find a critical assessment in Abramowicz et al. (2002).

6.2.2.1 The Luminosity of X-Ray Transients in Quiescence

Figure 6.5 shows the X-ray luminosity versus the orbital period for a sample of X-ray transients in quiescence.[10] With the exception of a few outliers, black hole binaries are systematically much dimmer (10–100 times) than neutron star binaries.

We have already discussed in Chap. 4 that the spectrum of X-ray transients in low-luminosity states is well described by the ADAF model. Whereas disk accretion is characterized by a luminosity proportional to the accretion rate $\dot{M}$, recall for example Eq. (4.100), two-temperature ADAFs exist only below a certain value $\dot{m}_{\rm crit} \sim 0.01$–$0.1$ (in Eddington units) and display a luminosity $L \propto \dot{m}^2$. Most of the gravitational energy of the infalling matter is stored as thermal energy instead of being radiated. If the accretor is a black hole all this amount of energy is advected with the flow as it crosses the event horizon and simply disappears for an external

---

[9]For simplicity, throughout this chapter we shall use the expression "black hole(s)" implying "black hole *candidate(s)*".

[10]The orbital period is chosen as a variable because in most systems the accretion rate is expected to be determined by its value. Black holes and neutron star binaries with similar $P_{\rm orb}$ should have similar accretion rates (in Eddington units), allowing proper comparison. See the discussion in Narayan et al. (2002).

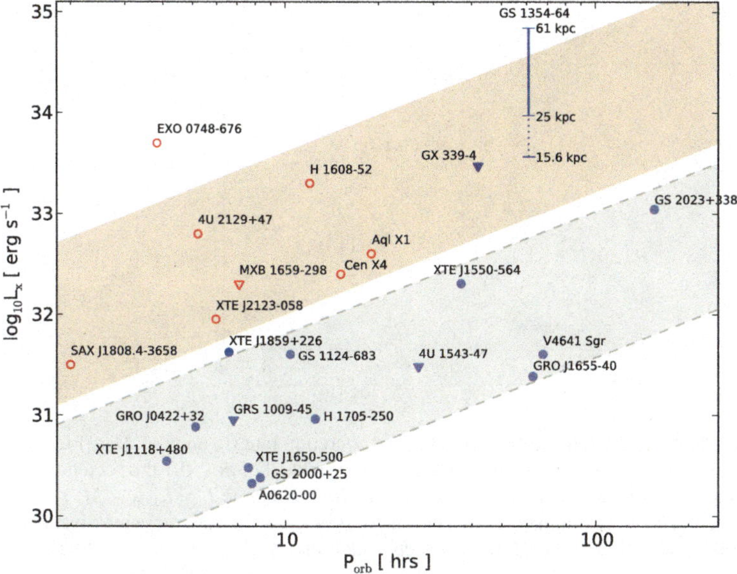

Fig. 6.5 X-ray luminosity as a function of orbital period for black hole (*filled symbols*) and neutron star (*empty symbols*) X-ray transients in quiescence. The *triangles* indicate upper limits. Two black hole candidates display luminosities above the expected. However, for one source (GX 339-4) it is not clear whether it was actually in quiescence. The second case, GS 1354-64, is discussed in the text. This figure is an updated version of that published in Reynolds and Miller (2011) that includes the latest data on the source H 1705-250 (Yang et al. 2012). Figure granted by the authors and reproduced by their permission

observer. It appears reasonable to expect that neutron star X-ray transients in quiescence also accrete in the ADAF regime with values of $\dot{m}$ of the same order. In this case, however, the thermal energy of the accretion flow does not "disappear" but must be liberated as matter comes to rest by impacting on the surface of the neutron star. The total luminosity (the emission of the ADAF plus the energy radiated at the surface of the star) in such systems would be approximately equal to the blackbody luminosity of a thin disk. Then, if the ADAF model applies both to neutron star and black hole binaries *and* black holes have an event horizon, neutron star X-ray transients should be more luminous than black hole candidates in quiescence, and the latter should show a greater difference in luminosity between outburst and quiescence.

The plot of luminosity vs. accretion rate in Fig. 6.6 summarizes the theoretical expectations from this scenario. Narayan et al. (1997, 2002) and García et al. (2001) suggested to confront them with observations as a way to test the presence of a horizon in black hole candidates. They analyzed ∼15 sources and found the data are qualitatively consistent with the model. Figure 6.5 is an updated version of their original result with more sources included. The authors note, nevertheless, that if the accretion proceeded purely in the advection-dominated regime the observed luminosity gap between neutron stars and black holes should be larger. Narayan

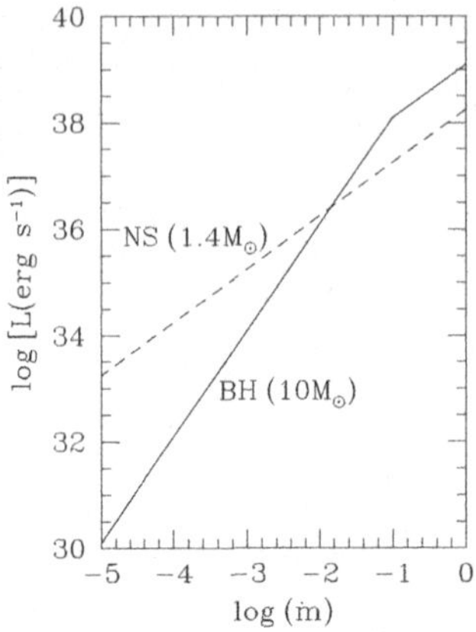

Fig. 6.6 Theoretical luminosity as a function of accretion rate (in units of $\dot{M}_{\rm Edd}$) for representative models of black hole and neutron star X-ray transients. For $\dot{m} \gtrsim 0.1$ the accretion flow forms a thin disk that radiates efficiently as a blackbody. In this regime $L \sim L_{\rm Edd} \propto M$, so black holes are expected to be more luminous than neutrons stars because they are more massive. For $\dot{m} \lesssim 0.1$ (quiescence state) the accretion flow is advection-dominated. Neutron stars should approximately display the same radiative efficiency (scaling as $L \propto \dot{m}$), but black holes should be very underluminous ($L \propto \dot{m}^2$) if they have a horizon. From Narayan et al. (1997). Reproduced by permission of the AAS

et al. (2002) argued that some refinements of the ADAF model (e.g. convection and magnetic field, propeller effect in neutron stars) may improve the quantitative accord. The interesting point to stress is that the main implication of these studies (that neutron star binaries should be much brighter than black holes binaries and that this hints at the existence of an event horizon in the latter) does not depend on the detailed characteristics of the accretion flow as long as it is radiatively inefficient.

This interpretation of the gap in X-ray luminosity between neutron star and black hole binaries in quiescence has been questioned; some alternative models are examined in Narayan et al. (2002). The criticism mainly focus on the assumption that the X rays are emitted in an radiatively inefficient accretion flow, and not with the possible connection to the existence of an event horizon in black hole candidates. Although the data from concrete sources may be explained by other means, Narayan et al. (2002) argue that the ADAF/event horizon explanation is, among all, the most general and straightforward. Recently, Reynolds and Miller (2011) published X-ray data from the black hole binary GS 1354-64 obtained with the satellite *Chandra*. The spectral shape is consistent with that of other black hole binaries but the X-ray luminosity is at least one order of magnitude larger than the expected (see Fig. 6.5), even when allowing for considerable uncertainties in the distance to the source. Reynolds and Miller (2011) suggest two explanations for the anomaly. One is that a temporal increase in the accretion rate triggers an also temporal rise in the X-ray emission. Variability or flares have been observed in other sources, although the luminosities always remained within the expected levels. The second explanation is that the actual quiescence X-ray luminosity of GS 1354-64 is steady and as high as observed. Considering that other black hole binaries with similar orbital period, mass of the black hole, and mass of the donor star are $\sim 10^{-3}$ times fainter, the confirmation of the results could indicate that the compact object in GS 1354-64 is accreting in an unknown low-luminosity regime. If this were the case, the evidence supporting the existence of an event horizon in black hole X-ray binaries gathered by the methods discussed in this section might turn debatable.

6.2.2.2 Type I X-Ray Bursts

There are yet other possible means to obtain evidence for the event horizon in black hole candidates. One phenomenon in particular ought to occur only in binaries where the compact object has a solid surface—the so-called Type I X-ray bursts (e.g. Lewin et al. 1993; Cumming 2004). The origin of the bursts is very well understood: the runaway thermonuclear burning of the accreted matter (mainly Hydrogen and Helium) accumulated on the surface of the compact object. Type I bursts last for $\gtrsim 10$ s and are recurrent; once one burst is over, matter starts to pile up on the surface again until the temperature and pressure are high enough to trigger nuclear fusion. The typical time between bursts ranges from hours to some days.

No Type I X-ray bursts have ever been detected in black hole XRBs. Although suggestive, this fact does not at all confirm the absence of a surface: it could well happen that black hole candidates do have surfaces but do not burst for some reason. Not all accreting neutron stars show Type I bursts, for example. Narayan and Heyl (2002) devised a scheme to prove the presence of an event horizon in black hole candidates based on the study of Type I X-ray bursts. The idea is to develop a model for the bursts, apply it to black hole candidates, and show that if they had a surface they should have bursts. All other phenomena that could inhibit the bursts must be discarded before concluding that the lack of bursts implies the absence of surface.

Narayan and Heyl (2002, also 2003) studied the evolution of a layer of matter (Hydrogen, Helium, and some heavier elements) on the surface of a compact object. The steady-state equations for the composition, temperature, density of the layer, and the outgoing energy flux are solved, and the stability of the solution under small perturbations is investigated. If any of the perturbed solutions is unstable it is assumed that a Type I burst effectively develops.

The model was applied to two types of accreting objects: a typical neutron star and a putative black hole candidate with a surface. Both differ in their mass and radius, and also in the value adopted for the temperature T_{in} at the bottom of the layer. The latter is related to the accretion luminosity, and therefore lower in black hole candidates. Several values for the surface mass density at the base of the layer $\Sigma_{\text{max}} = 10^{9}$–$10^{11}$ g cm^{-2} were essayed as well. The results are shown in Fig. 6.7, where dots indicate unstable states. Instability is predicted within a wide range of parameters for both neutron stars and black holes with s surface.

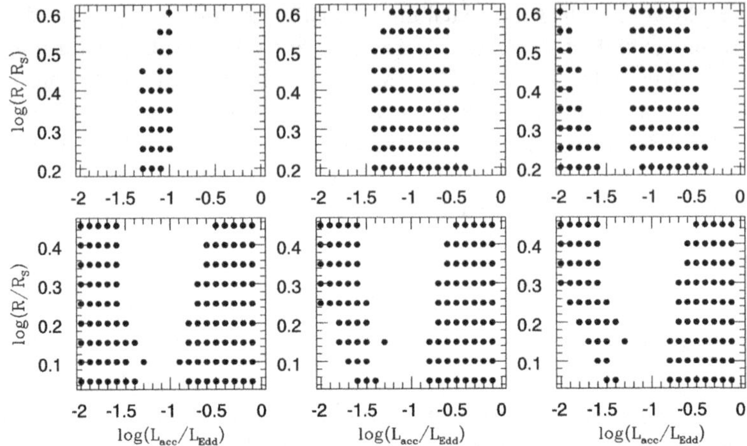

Fig. 6.7 Regions of unstable nuclear burning (*dots*) as a function of radius (in units of the Schwarzschild radius) and accretion luminosity (in Eddington units), for a $1.4M_\odot$ neutron star (*top*) and a $10M_\odot$ black hole candidate with a surface (*bottom*). Each *panel* corresponds to a different value of the temperature $T_{\rm in}$ at the base of the layer. *Top*, from left to right: $10^{8.5}$ K, 10^8 K, $10^{7.5}$ K. *Bottom*, from left to right: $10^{7.5}$ K, 10^7 K, $10^{6.5}$ K. From Narayan and Heyl (2002). Reproduced by permission of the AAS

Two points are worth remarking. First, neutron stars with $L_{\rm accr} \gtrsim 0.3 L_{\rm Edd}$ should not have bursts according to the model. This is in agreement with previous theoretical results and with the observational fact that the neutrons stars that do not burst are those that accrete at rates near or above the Eddington limit.[11] Second, black hole candidates in X-ray transients should produce Type I X-ray bursts as efficiently as neutron stars if they had a surface. During the passage from quiescence to the peak luminosity of and outburst[12] and back, the accretion rate sweeps a broad range of values from $\lesssim 10^{-3} \dot{M}_{\rm Edd}$ to $\sim \dot{M}_{\rm Edd}$. The path of the system in a plot like those in Fig. 6.7 should traverse the region of instability almost certainly. Furthermore, the time between Type I bursts predicted by the model is of about one day whereas the outbursts last typically for some months. The probability that black hole candidates have X-ray bursts but they have eluded detection appears to be practically ruled out.

The existence of the event horizon in black hole candidates, however, is not at all proved. The results of Narayan and Heyl (2002) are based on a series of assumptions that might not be valid. It might happen, for instance, that for some reason the instability propagates on the surface and triggers a burst in neutron stars but not in black hole candidates. Also an unusual chemical composition of the accretion flow could perhaps suppress the bursts, but this is unlikely to be the case

---

[11] These are the so-called "Z" sources. See, however, Kuulkers et al. (1997) and Lin et al. (2009).

[12] Here we refer to the outbursts described in Sect. 6.2.1, triggered by an increase in the accretion rate.

since both neutron stars and black holes accrete from "normal" companions. The magnetic field is another factor to be considered. Pulsars with very large fields generally do not burst; it is thought the cause is that the field channels the accretion flow increasing the local effective accretion rate. No pulsations have ever been observed in black hole candidates, so this scenario may be quite safely discarded. The effect of rotation (not included in the model), on the other hand, is largely unknown.

Other exotic possibilities may be imagined to explain the absence of Type I bursts in black hole candidates. Narayan (2003) analyzes a few of them, more alternatives are discussed in Abramowicz et al. (2002). Basically, they postulate that what we identify as black hole candidates are in fact other type of objects with a surface. Such objects could be made of some strange form of matter and have a surface and no horizon. Abramowicz et al. (2002) argue that gravastars, for example, would look much like black holes for a distant observer.[13] But since the bursts originate in "normal" matter accumulated in the surface it is not expected that the inner composition of the object plays any role, even if the strange matter extends up to near the surface. Narayan (2003) also consider the case of a compact object made of dark matter, or a material with similar properties, that only interacts extremely weakly with baryonic matter except gravitationally. The accretion flow would concentrate at the core, but even an object like this might eventually produce bursts. Finally, there is the possibility of having a naked singularity. Once again, the evidence in favor of the existence of the event horizon is inviting but not conclusive.

6.3 Evidence for a Supermassive Black Hole at the Galactic Center

The central few parsecs of the Milky Way are a complex region. There is a dense and luminous cluster of stars, neutral, ionized, and hot gas, molecular clouds, a young supernova remnant, and a supermassive black hole candidate. The surrounding gas and dust makes the galactic center invisible at optical frequencies, but otherwise it has been detected in the radio, millimeter and submillimeter, infrared, X-ray, and gamma-ray bands.

A strong radio source was detected in the 1950s towards the galactic center. Since it was located in the constellation Sagittarius, it was called Sagittarius A (Sgr A). Within this region Balick and Brown (1974) discovered a compact radio source, Sagittarius A* (Sgr A*). The position of Sgr A* matches that of the dynamical center of the Milky Way and usually the same name is given to the black hole candidate at the galactic center.

Early evidence of the presence of a compact, "point-like" very massive body at the galactic center was obtained from the analysis of the motion of neon emission lines (Lacy et al. 1980). Since then, the case for a supermassive black hole at the

---

[13]Gravastars are discussed in Sect. 7.7.

center of the Milky Way has grown overwhelming. We review some of the latest astonishing high-quality observations that lead to this conclusion.

6.3.1 Stellar Dynamics

The innermost young B-type stars in the dense cluster at the galactic center are called "S-stars". These stars describe highly eccentric elliptical orbits with Sgr A* at one focus in randomly distributed orbital planes. Some come so close to Sgr A* ($\lesssim 0.02$ pc) that their velocities may be as large as $\sim 10^3$–10^4 km s^{-1} (e.g. Genzel et al. 2010). The short orbital periods (on human timescales) of S-stars have allowed to trace partially and even completely the individual paths of more than 20 stars through high-resolution near infrared observations, thus providing an estimate of the mass concentrated at the galactic center.

The most accurate results so far have been obtained from the follow-up of the orbit of the star S0-2 (also called S2, see Fig. 6.8). With a period of 15.9 yr, S0-2 has been detected along the whole orbit (Ghez et al. 2008; Gillessen et al. 2009a, 2009b).[14] The latests fits of these data yield a mass $M = (4.40 \pm 0.27 \pm 0.5) \times 10^6 M_\odot$ for the black hole at the galactic center (Gillessen et al. 2009a). Because of how it is calculated, the value of the mass depends on the distance R_0 to the galactic center. The first uncertainty assigned to M is statistical at fixed R_0, and the second accounts for the systematic uncertainty in the determination of R_0. The distance to Sgr A* also follows from the fits to the orbits; Gillessen et al. (2009a) obtain $R_0 = (8.34 \pm 0.27 \pm 0.5)$ kpc. The orbit of S0-2 is compatible with a point mass; any extended mass component within the orbit is at most ~ 5 % of the point mass (Gillessen et al. 2009a). The position of the point mass and that of the radio source Sgr A* agree to ± 2 μarcsec.[15,16]

With the presently available resolution the orbits of S-stars are perfectly fit applying a Keplerian model to describe their motion. Very precise observations are being planned to detect relativistic perturbations. A good opportunity will arise in 2018 when S0-2 comes nearest to the galactic center black hole. Besides, future near infrared interferometers such as GRAVITY (Eisenhauer et al. 2011) are expected to be able to resolve stars with orbital periods of the order of just one year. For such short periods the relativistic effects on the orbit should be noticeable; relativistic precession in particular will be easily measurable.

---

[14]The discovery of star S0-102 with a period of only 11.5 yr has been recently reported by Meyer et al. (2012). S0-102 has also been tracked along the full orbit.

[15]At 8 kpc, 0.1 arcsec corresponds to $\sim 1.2 \times 10^{16}$ cm.

[16]The position of Sgr A* in the near infrared is much more difficult to determine. On the one hand, establishing an absolute reference frame in the near infrared at the galactic center is inherently difficult because of the lack of extragalactic reference sources; on the other hand Sgr A* is a very dim source at these wavelengths. The near infrared position matches the focus of the orbit os S0-2 within 0.01 arcsec. Various near infrared flares also agree with the position of Sgr A* at radio frequencies.

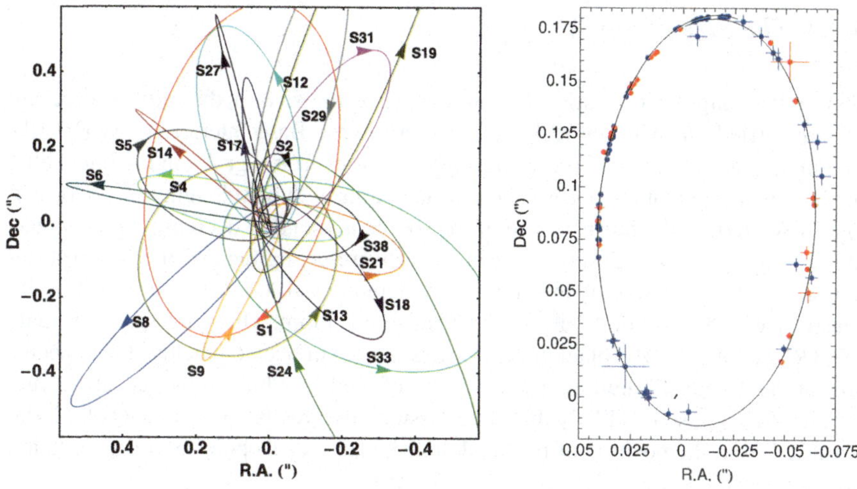

Fig. 6.8 *Left*: orbits of twenty S-stars; from Gillessen et al. (2009a). Reproduced by permission of the AAS. *Right*: the orbit of S0-2 reconstructed with data from Gillessen et al. (2009b) and Ghez et al. (2008); from Gillessen et al. (2009b). Reproduced by permission of the AAS

6.3.2 *The Motion of Sgr A**

Reid and Brunthaler (2004, see also Reid 2009) have measured the apparent motion in the sky of the radio source Sgr A* with respect to the background quasar J1745-283. Once the apparent motion is corrected to account for the motion of the Sun, the result is compatible with Sgr A* being virtually at rest at the dynamical center of the Milky Way as seen from the Sun-Earth system. Its intrinsic residual velocity perpendicular to the galactic plane is (-0.4 ± 0.9) km s^{-1}, whereas the component of the velocity in the galactic plane is (-7.2 ± 8.5) km s^{-1}. The latter is much more uncertain because of the also large uncertainties in the orbital velocity of the Sun in the galactic plane.[17]

These results may be compared to theoretical expectations and used to constrain the mass of Sgr A*. The motion of a massive object in the gravitational potential of a cluster of stars can be modeled as Brownian motion (Chatterjee et al. 2002a, 2002b; Dorband et al. 2003; Merritt et al. 2007). The system tends to an equilibrium state in which there is equipartition of kinetic energy among its components. Analytical calculations and simulations predict that each component of the velocity of an object of mass $4 \times 10^6 M_\odot$ should be ~ 0.2 km s^{-1}. The measured velocity of the radio source Sgr A* then requires that its mass is at least $4 \times 10^5 M_\odot$. It is also almost discarded that the source Sgr A* is not associated to this large mass for in that case we should detect its motion around it—except perhaps if it is in an extremely close orbit, but this again would require the presence of a very compact massive object.

---

[17]The components of the velocity of the Sun are $\sim$220–255 km s^{-1} and (7.16 ± 0.38) km s^{-1} in and perpendicular to the galactic plane, respectively.

6.3.3 The Size of Sgr A*

The highest angular resolution measurements of the size of the radio source Sgr A* are carried out with the technique of Very Large Base Interferometry (VLBI). The images are affected by the scattering of photons from Sgr A* in the interstellar medium in a way that the observed radius increases with wavelength as λ^2. Correcting for scattering, the intrinsic size of the source is found to decrease with decreasing λ. Only below $\lambda \sim 2$ mm scattering effects are negligible and the observations may reveal the real size of the source. Doeleman et al. (2008) determined an intrinsic size of Sgr A* of $37 \pm^{16}_{10}$ μarcsec at $\lambda = 1.3$ mm. This corresponds to only $\sim 3.8 R_{\mathrm{Schw}}$ for a $4.4 M_\odot$ black hole. As we shall see in Sect. 6.6 below, the expected size of the "shadow" (null emission region) of the black hole at the galactic center is ~ 4.5–$5 R_{\mathrm{Schw}}$, so it is likely that the emission observed by Doeleman et al. (2008) is shifted from the position of the black hole because of Doppler boosting or that it originates in a jet (e.g. Falcke and Markoff 2000).

Combining the size at 1.3 mm with the minimum mass of $4 \times 10^5 M_\odot$ inferred from the proper motion of Sgr A*, the lower limit for the mass density is $8 \times 10^{22} M_\odot$ pc^{-3}. The "density" of a Schwarzschild black hole of $4.3 \times 10^6 M_\odot$ is

$$\rho_{\mathrm{BH}} = \frac{M}{(4/3)\pi R^3_{\mathrm{Schw}}} \approx 1.5 \times 10^{25} M_\odot \; \mathrm{pc}^{-3}. \tag{6.4}$$

The derived lower limit for the density of Sgr A* is only about 100 times smaller than the critical value. Although not a proof that Sgr A* harbors a black hole, this result by itself allows to rule out many alternative models.

6.3.4 Alternatives to the Supermassive Black Hole Model

What could Sgr A* be if not a supermassive black hole?

The first natural option is that the compact concentration of mass at the galactic center is a cluster of very faint ("dark") stars,[18] brown dwarfs, white dwarfs, or very low-mass black holes for example. This possibility is almost completely ruled out by the extremely large mass density estimated for Sgr A*. The typical density of a globular cluster, for example, is $\sim 10^5 M_\odot$ pc^{-3}—large, but still many orders of magnitude below the mass density of Sgr A*. Furthermore, stars in dense clusters undergo collapse, collisions, and evaporation (continuous escape of stars due to weak gravitational scattering) that limit their lifetime (e.g. Maoz 1998). As a consequence, a putative cluster of stars of $\sim 1 M_\odot$ compatible with the current observations of the Sgr A* region would live less than 10^6 yr before evaporating (Reid 2009). This is even less than the lifetime of some stars in the galactic center.

---

[18]Sgr A* itself is very faint in the infrared, so an agglomerate of luminous stars is directly eliminated.

In order to avoid evaporation for $\sim 10^9$ yr the mass of the components should be $\sim 10^{-3} M_\odot$, what implies the implausible possibility of a cluster formed by $\sim 10^{10}$ objects similar to Jupiter (Reid 2009).

A more exotic alternative was proposed by Viollier et al. (1993). They introduced the idea of a "fermion ball", a massive object made of fermions (for example massive neutrinos) supported by degeneracy pressure. The maximum mass of such body depends on the mass of the constituent fermions, but fermion balls with $M \sim 10^6$–$10^9 M_\odot$ are allowed by the theory. A fermion ball made of neutrinos within the permitted mass range ($\sim$10–25 keV/c^2) and with the total mass of the black hole candidate in the galactic center, however, would have a radius $\sim$15 light days, much larger than the pericenter (point of closest approach) distance of S0-2 and other S-stars (Genzel et al. 2010). Fermion ball models as an explanation for the phenomena at the galactic center are nowadays almost excluded.

A model in the same spirit is that of Torres et al. (2000) who considered a "boson star" as an alternative to a black hole. Boson stars (see Liebling and Palenzuela 2012 for a recent review) are objects made of non-baryonic bosons that interact between them only gravitationally; they have no horizon, no singularity, and no solid surface. The radius of a boson star is only slightly larger than the radius of the event horizon of a black hole of the same mass. Supermassive boson stars are allowed by the model depending on the type of boson and their interactions. As an alternative to a black hole, a drawback of this scenario is that boson stars could collapse to a black hole should baryonic matter be added onto them by accretion (although Torres et al. 2000 discuss some ways to circumvent it). Furthermore, even though boson stars themselves do not have a solid surface, the accreted matter should cluster at its center eventually providing a surface where the accretion flow would come to rest. The liberation of the kinetic energy of the accretion flow should lead to thermal infrared emission—just as discussed for X-ray binaries in Sect. 6.2.2.1—but this is not observed. Actually, Broderick and Narayan (2006) and Broderick et al. (2009a) have argued for the presence of an event horizon in Sgr A* on these grounds.

6.4 Evidence for Extragalactic Supermassive Black Holes

It is nowadays thought that supermassive black holes lie at the center of every galaxy, whether active or not. Early speculations arose with the discovery of quasars. The huge power emitted from these sources appeared hard to be explained if not by accretion onto the deep potential well of a massive object. We have seen that an accretion flow may radiate away as much as 50 % of its gravitational energy—a really efficacious process compared to nuclear fusion that provides only $\sim$0.8 % of efficiency. Furthermore, the mass and density required to explain the power of quasars through nuclear reactions are so large that such a system would not remain stable during a physically significant time span. Yet another key characteristic of some active galactic nuclei, the ejection of collimated bipolar jets, suggests the presence of a spinning compact object operating as an "engine" that powers the outflows. A supermassive Kerr black hole naturally fits in this picture.

Evidence for the existence of supermassive black holes in the nuclei of other galaxies is searched largely in the same manner than in the Milky Way: the mass and the size of the gravitating center are measured (usually by observing the motion of objects in its environments) and alternatives are eliminated based on the result. The large distances involved call for the highest instrumental angular resolutions available if one desires to inspect the inner circumnuclear regions of the galaxies. The constraints imposed on mass and size are thus, in general, less stringent than for the black hole at the center of our galaxy, although under favorable conditions some very strong extragalactic black hole candidates have been identified. Additionally, the X-ray emission from the zones of the accretion flow under strong gravitational field might carry precious information about the nature of the compact object—even allowing to determine its spin.

6.4.1 Stellar and Gas Dynamics

The gravitational attraction of a black hole of mass M in the center of a galaxy effectively dominates the dynamics of matter around it inside a sphere of influence of radius roughly given by

$$R_{\text{infl}} = \frac{GM}{\sigma^2} \approx 4 \times \left(\frac{M}{10^7 M_\odot} \right) \left(\frac{100 \text{ km s}^{-1}}{\sigma} \right)^2 \text{ pc.} \qquad (6.5)$$

The variable σ is the *velocity dispersion*; it is the statistical dispersion (standard deviation of the velocity distribution) of line-of-sight velocities about the mean value for the objects in the region of interest. Inside this sphere stars and gas behave approximately as "test particles" orbiting the black hole.

The velocities of single stars in the nuclei of nearby galaxies are measured from the profiles of emission or absorption lines from the stellar atmospheres. Inferring the mass of the black hole from these data in not straightforward because the motion of stars is, in reality, influenced as well by the gravity of the rest of the galaxy. The value of M follows from fitting the data to models of orbits under the gravitational potential of a point mass plus a distribution of stars, dark matter, etc. Nevertheless, many robust mass determinations have been obtained by this method specially since the commissioning of the *Hubble Space Telescope*. The most massive black holes known so far have been recently discovered based on the motion of orbiting stars. They occupy the centers of the galaxies NGC 3842 and NGC 4889 with masses $\sim 9.7 \times 10^9 M_\odot$ and $\sim 2.1 \times 10^{10} M_\odot$ (McConnell et al. 2011), respectively, and NGC 1277 with a mass of $\sim 1.7 \times 10^{10} M_\odot$ (van den Bosch et al. 2012).

Whereas the motion of stars is exclusively determined by gravitation, gas is subject also to pressure gradients, radiation pressure, and viscous forces. The motion of gas, however, can probe shorter distances to the black hole candidate than stars. An additional advantage is that gas is usually distributed in a disk; this partially eliminates uncertainties related to the de-projection of the measured velocities.

The firmest candidate to host a supermassive black hole in its nucleus, along-side the Milky Way, is the spiral galaxy NGC 4258. Radio VLBI observations by Miyoshi et al. (1995) revealed the presence of clouds containing water masers in the accretion disk (viewed nearly edge-on from Earth) around the center of this galaxy. Masers are the analog of lasers in the microwave band—they emit by popu-lation inversion although astrophysical masers are largely incoherent. In the case of water vapor, the maser emission corresponds to the rotational transition 6_{16}–5_{23} at 22.23508 GHz ($\lambda = 1.35$ cm).

The motion of the clouds in the accretion disk of NGC 4258 perfectly fits Kep-lerian circular orbits, the innermost with a radius of only 0.13 pc and a tangential velocity of $\sim 10^3$ km s^{-1}. The mass of the gravitating center was determined to be $3.6 \times 10^7 M_\odot$. Such a large mass contained in a region that compact amounts to a density of $\sim 10^9 M_\odot$ pc^{-3}. This almost certainly eliminates the possibility that there is a dense star cluster at the dynamical center of NGC 4258. Although more ex-otic alternatives cannot be completely discarded, the huge mass density is strong evidence of the presence of a black hole.

The masses of a few other supermassive black hole candidates have been es-timated studying the motion of water masers, see for instance Greenhill et al. (1996) for NGC 1068 ($\sim 1.5 \times 10^7 M_\odot$), Kuo et al. (2011, six active galactic nu-clei with mass 0.75–6.5 $\times 10^7 M_\odot$), and recently Yamauchi et al. (2012) for IC 2560 ($\sim 3.5 \times 10^6 M_\odot$). Despite its success, the applicability of this technique is limited: on the one hand because the detection of maser emission is rare, and on the other hand because to be profitable it demands rather special observing conditions, such that the accretion disk is viewed almost edge-on. Besides, the velocity profile of masers sometimes depart from Keplerian suggesting that the gas moves under the gravitational influence of something else than a point mass. Values of mass cal-culated from the observational data in such cases are model-dependent and less reliable.

Another compelling case for a supermassive black hole is that of the nearby radio galaxy M87. The nucleus of M87 is surrounded by a disk of ionized gas. Macchetto et al. (1997) measured the rotation curve of the inner disk, at only ~ 5 pc from the dynamical center of the galaxy, from the emission line of Oxygen at $\lambda = 3727$ Å. Their best-fit model predicts that the gas forms a thin Keplerian disk around a $\sim 3.2 \times 10^9 M_\odot$ mass compressed within a region of 3.5 pc of radius. The implied mass density is $\sim 2 \times 10^7 M_\odot$ pc^{-3}, two orders of magnitude larger than that of dense stellar clusters.

To close our discussion, we briefly mention another kinematic technique to find the mass of black hole candidates in galaxies, called *reverberation mapping* (e.g. Peterson 1993). It is based on the analysis of the emission from the clouds in the broad line region. The velocity v_{BLR} of the clouds is measured from the Doppler broadening of the lines. The temporal variability of the continuum emission from the galactic core (the radiation that illuminates the clouds) is followed after a time lag by correspondent variability in the line emission. The retardation is equal to the travel time of light from the core to the clouds, so the distance of the clouds to the center of the galaxy is approximately $R_{\mathrm{BLR}} \sim c\Delta t$. An estimate of the mass of the black hole then follows from the Virial Theorem,[19] $M \sim R_{\mathrm{BLR}} v_{\mathrm{BLR}}^2 / G$. An advantage of the reverberation mapping method is that, being independent of the instrumental spatial resolution (the broad line region is never resolved with current telescopes), it may be applied to very distant galaxies (although it could take many years to collect the necessary data!).

6.4.2 Scaling Relations

When the available angular resolution is not enough to trace orbits of single stars (because the galaxy is very distant, for instance) the motion of groups of stars may be used to estimate the mass of black hole candidates in galactic nuclei. In some types of galaxies (ellipticals or with an spheroid) a correlation of the form

$$\log\left(\frac{M}{M_\odot}\right) = \alpha + \beta \log\left(\frac{\sigma}{200 \ \mathrm{km \, s^{-1}}}\right) \tag{6.6}$$

between the mass of the black hole candidate and the velocity dispersion σ in the bulge was discovered by Ferrarese and Merritt (2000) and Gebhardt et al. (2000). The $M - \sigma$ correlation is thought to reflect the history of joint evolution of the supermassive black holes and its surroundings. The values of α and β have varied with time as more systems were added to the sample; according to the latest fit by McConnell and Ma (2013), $\alpha = (8.33 \pm 0.05)$ and $\beta = (5.57 \pm 0.33)$; see Fig. 6.9.

The M–σ relation is not usually used to calculate black hole masses from scratch but just as a proxy; it is useful to normalize or as a consistency check for the values obtained by other methods. Notice, however, that the M–σ correlation underpredicts the mass of some the most massive (including the newly discovered) and less massive black hole candidates. The impact of these "outliers" on the interpretation of the M–σ is yet not clear.

The mass of supermassive black holes is also known to correlate with other properties of the host galaxy. The most frequently considered are the correlation between the black hole mass and the luminosity (M–L_{bulge} relation) and the mass (M–M_{bulge} relation) of the galactic bulge.

6.4.3 Fluorescense X-Ray Lines

The dynamics of stars and gas disclose the action of the black holes's gravitational attraction at distances of hundreds or thousands of gravitational radii. That far from

---

[19]The Virial Theorem states that in a system of particles that interact through a potential of the form $\propto 1/r$, the total time-averaged potential energy $\langle U \rangle$ and kinetic energy $\langle T \rangle$ satisfy $2\langle T \rangle + \langle U \rangle = 0$ in equilibrium state.

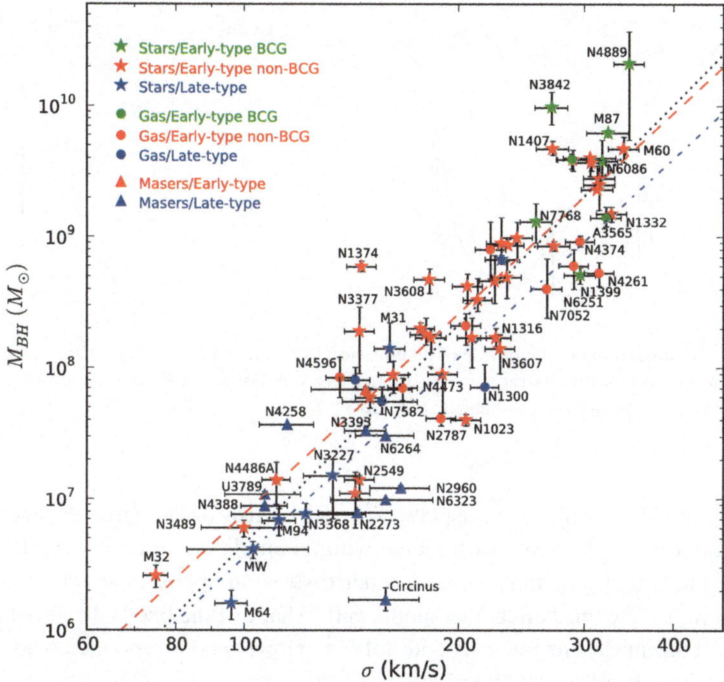

Fig. 6.9 Mass of the black hole candidate as a function of velocity dispersion for a sample of 72 galaxies. The masses were estimated from stellar orbits, gas dynamics, and maser emission in the galaxies marked with *stars*, *circles*, and *triangles*, respectively. "BCG" stands for "bright cluster galaxy". The *dotted black line* is the fit to the whole sample, the *red dotted line* only to the early-type galaxies, and the *blue dashed-dotted line* only to the late-type galaxies. From McConnell and Ma (2013). Reproduced by permission of the AAS

the black hole, as we have seen, the motion of these "test particles" (at least the part attributed to the presence of the black hole) is almost perfectly Keplerian. Accretion disks, however, may extend up to the last stable orbit that is only a few gravitational radii in size. Any radiation emitted from this region must be necessarily affected by the strong gravitational field of the black hole.

We have discussed in Chap. 4 how the accretion flow in the immediate vicinities of a compact object "inflates" to form a corona of very hot plasma that emits non-thermal radiation via Comptonization of photons from the accretion disk. A fraction of the scattered photons may impinge on the disk and trigger the emission of lines from ionized atoms. The strongest of the observed lines is the iron $K\alpha$ fluorescence line at $\sim$6.4 keV (X-rays) corresponding to the electron transition $2p \rightarrow 1s$.

Far from being sharp, the profile of the line is deeply distorted by a number of effects as seen in Fig. 6.10. To begin with the disk is rotating, so the emission from the approaching side with respect to the observer appears blueshifted and that from the receding side redshifted. This type of Doppler broadening is present even in Newtonian gravity. Depending on the inclination of the disk with respect to the

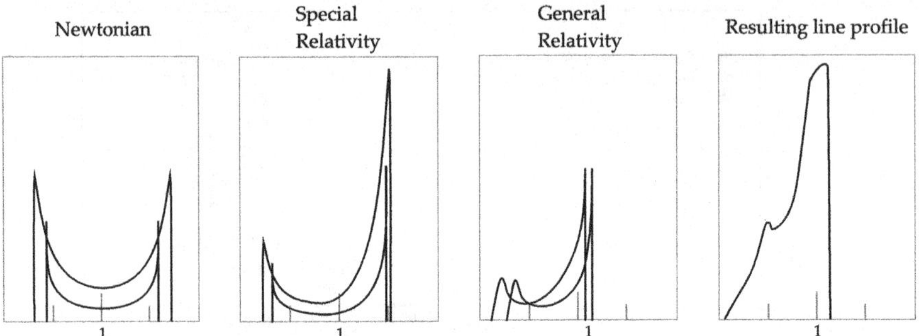

Fig. 6.10 Modifications in the profile of an emission line induced by gravitational and kinematic effects. The emissivity (in arbitrary units) is plotted as a function of the ratio of the observed to the emitted frequency. Adapted from Fabian et al. (2000)

line of sight, the profile may display two peaks. Also on a degree depending on the inclination, the intensity of the blue wing is amplified by relativistic beaming because the velocity of matter in the inner disk is an appreciable fraction of the speed of light. The third effect that modifies the shape of the line is the gravitational redshift: it diminishes its intensity and shifts it to lower photon energies. The overall result is a broad and skewed line profile.

The first clear detection of a distorted iron line from an extragalactic source[20] was made by Tanaka et al. (1995) in the galaxy MCG-6-30-15, one of the brightest AGN in X rays. Later observations showed that the line was so broadened that it must originate in material closer than $3R_{\mathrm{Schw}}$ to the black hole. This implies that either the black hole is rapidly spinning or that the line is emitted by material already plunging onto the black hole. Accepting the hypothesis of a Kerr black hole, its spin parameter must be $a_* > 0.9$. There is always of course the possibility that our modeling of the accretion flow is wrong and the accretion disk does not extend to the last stable orbit. However, lately Fabian et al. (2009, 2012) detected reverberation associated to the iron line emission in the AGN 1H0707-495. The variability pattern of the line emission lags $\sim$30 s behind the power-law continuum due to the difference in the lengths of the paths traversed by the photons that directly leave the corona and those reflected in the disk. For the mass of the black hole candidate in 1H0707-495 ($2 \times 10^6 M_\odot$) this time delay corresponds to $\sim$2R_{grav}, implying that the part of the corona that irradiates the disk is very close to the black hole and above the inner accretion disk. Clearly, X-ray spectroscopy is a powerful tool to explore very close to black holes and it is encouraging and suggestive that the line profiles largely agree with the predictions of General Relativity.

---

[20] Iron lines are also detected in galactic X-ray binaries.

6.5 Evidence for Intermediate-Mass Black Holes

6.5.1 Ultra-luminous X-Ray Sources

As we showed in Sect. 4.2.2, there is a limit to the maximum isotropic luminosity
of a source accreting spherically. This is the Eddington luminosity,

$$L_{\text{Edd}} \approx 1.3 \times 10^{38} \left(\frac{M}{M_\odot} \right) \text{erg s}^{-1}, \tag{6.7}$$

where M is the mass of the accretor. The radiation pressure on the infalling matter
would halt accretion if the luminosity were larger than this limit. The Eddington
luminosity of the most massive stellar-mass black holes ($\sim 20 M_\odot$) is some times
10^{39} erg s^{-1}.

Ultra-luminous X-ray sources (ULXs; see e.g. Fabbiano 2004; Mushotzky 2004;
Soria et al. 2005, and Feng and Soria 2011 for reviews) are usually defined as those
with X-ray luminosities 10^{39} erg s$^{-1} \le L_X \lesssim 10^{41}$ erg s^{-1}. Since they are brighter
than accreting stellar-mass black holes but less luminous than active galactic nuclei,
sometimes they are also called *intermediate-luminosity X-ray objects*. The nature of
ULXs is unknown and subject to speculation and debate. A number of alternatives
have been proposed, and a few and exciting appear plausible.

All ULXs detected so far are outside the Milky Way. They are generally located off the centers of the galaxies; this rules out the possibility that they are low-
luminosity AGN. For the same reason they cannot be underluminous accreting supermassive black holes: unless very far away, such an object would fall to the center
of its host galaxy due to dynamical friction (e.g. Binney and Tremaine 1987) in less
than a Hubble time.

The idea that ULXs are accreting black hole binaries naturally suggests itself and
some observational results lend support to the hypothesis. One is the variability (on
timescales of years to hours and periodic in a few cases, e.g. Liu et al. 2002; Han
et al. 2012), and the other is that some ULXs show spectral components (power-law
and thermal) and transitions between soft and hard X-ray states similar (although
far from identical) to the observed in stellar-mass black hole X-ray binaries (e.g.
Feng and Kaaret 2009; Soria 2011). Now, if we assume that ULXs are powered
by accretion and that they radiate X-rays isotropically below or at the Eddington
luminosity, the black hole must be of intermediate mass.[21]

Isotropic emission from the accretion flow onto an intermediate-mass black hole
is, nonetheless, not the only way to explain ULXs. Begelman (2002), for example,
has proposed that ULXs are stellar-mass XRBs that radiate isotropically above their

---

[21]The existence of IMBHs in binaries, though, poses theoretical difficulties since no known binary
evolutionary path leads to such system. The black hole could have been born isolated and later
captured a companion (e.g. Hopman et al. 2004), but in this case all ULXs today should be found
in globular or dense star clusters and most are not. Because of their large mass it is also unlikely
that IMHBs have been born in clusters but later ejected (e.g. Miller and Hamilton 2002).

Eddington limit via a radiation pressure dominated accretion disk. King et al. (2001) conjectured that ULXs are XRBs with high-mass donor stars going through a particular transient phase of their evolution, characterized by a super-Eddington mass transfer rate on thermal timescales. This is likely to occur in all high-mass XRBs after the wind accretion phase ends. The large luminosities of ULXs could be explained if the emission from the accretion disk in such system were not isotropic but modestly beamed. In line with this idea, Körding et al. (2002) suggested that ULXs might be stellar-mass black hole XRBs that emit highly beamed radiation in relativistic jets. Indeed, radio emission compatible with synchrotron radiation from a jet has been detected in at least two ULXs (Webb et al. 2012; Kaaret et al. 2003). It cannot be excluded, of course, that the jets are actually launched from an accreting intermediate-mass black hole.

Which is, then, the evidence for intermediate-mass black holes we have gathered from ultra-luminous X-ray sources? Here are the most relevant facts.

- **Dynamical constraints.** Measuring the mass of the compact objects would be the strongest argument in favor of the existence of intermediate-mass black holes in ultra-luminous X-ray sources. Up to date there are only ambiguous estimates, mostly because applying the method of the mass function to ULXs is complicated. Although there are many ULXs with optical counterparts, they are too distant to measure their radial velocities from absorption lines. The radial velocity may be measured from the profiles of emission lines as well, but in ULXs the spectrum is expected to be contaminated with emission lines from the X-ray irradiated outer parts of the accretion disk. Fortunately, these lines emitted in the disk are useful since they provide information about the radial velocity semi-amplitude V_c of the compact object. Then, an equivalent form of the mass function

$$f(M_*) = \frac{V_c^3 \, P_{\mathrm{orb}}}{2\pi \, G} = \frac{M_*^3 \sin^3 i}{(M_* + M_c)^2} \tag{6.8}$$

  may be applied to set limits on the mass of the black hole and the donor star if the orbital parameters are known. Recently, Liu et al. (2012) performed fits to the optical light curve of the ULX NGC 1313 X-2—the only with measured orbital period and an estimated value for V_c. Allowing for quite large variations in V_c around the measured value ~ 100 km s^{-1}, they obtain acceptable fits for black hole masses between a few tens and more than a thousand solar masses. This is too uncertain to draw any conclusion. Other attempts on the same basis have yielded similar results, see for instance Roberts et al. (2011). The first dynamical determination of the mass of the black hole in an ULX still awaits.

- **Temperature of the innermost accretion disk.** In general, the X-ray spectrum of ULXs is dominated by single a power-law component. An excess at low energies observed in some sources has been interpreted as the thermal emission from an accretion disk. Adopting the standard model of a thin disk that radiates as a blackbody, from direct manipulation of the equations it follows that (e.g. Makishima et al. 2000)

$$kT_{\rm in} \approx 1.2\alpha^{1/2}\left(\frac{\kappa}{1.7}\right)\left(\frac{\xi}{0.41}\right)^{1/2}\left(\frac{L_{\rm d}}{L_{\rm Edd}}\right)^{1/4}\left(\frac{M}{10M_{\odot}}\right)^{-1/4} {\rm keV}. \qquad (6.9)$$

Here $T_{\rm in}$ is the maximum temperature of the disk, $R_{\rm in} = \alpha 3 R_{\rm Schw}$ its inner radius, $L_{\rm d}$ its bolometric luminosity, ξ is a factor related to the boundary condition assumed at $R_{\rm in}$,[22] and κ (called the hardening factor) is the ratio of the color (thermal) temperature to the effective temperature of the disk. The value of $T_{\rm in}$ may be obtained fitting the observational data, in which case Eq. (6.9) provides an estimate of the mass of the accretor. Notice that standard accretion disks around intermediate-mass black holes should be colder than disks around stellar-mass black holes ($kT \sim 1$ keV).

The method has been applied with uneven success. There is not a solid understanding of the origin of the X-ray emission in the different spectral states of ULXs, so the results are model-dependent and often more than one model can fit the data. In some ULX the addition of a thermal component with $kT_{\rm in} \sim 0.17$–0.3 keV to a power-law significantly improves the goodness of the fits (e.g. Miller et al. 2003, 2004). If the disk extends up to the last stable orbit of a Schwarzschild black hole, an intermediate-mass black hole with $M \sim 10^2$–$10^4 M_{\odot}$ is implied. The spectrum of other ULXs, however, may be successfully fit only with a thin disk model. In seven sources modeled in this way by Makishima et al. (2000), for example, the best-fit inner temperature turned out to be $kT_{\rm in} \sim 1.0$–1.8 keV. This is too high for a disk that continues up to the last stable orbit of an intermediate-mass black hole—it is in fact of the order of the temperatures observed during the high state of stellar-mass black hole binaries. Makishima et al. (2000) suggested that the results could be reconciled if the black hole was rotating, in which case the last stable orbit would have a smaller radius and a larger temperature. Nevertheless, the discovery that the disk luminosity and the inner disk temperature do not follow the relation $L_{\rm d} \propto T_{\rm in}^4$ characteristic of a thermal spectrum (e.g. Mizuno et al. 2001; Feng and Kaaret 2007b), weakens the interpretation of the soft excess as simply the emission from a standard accretion disk.

Also the supposition that the inner disk radius coincides with the innermost stable orbit of the black hole should be considered with care. The thermal component in most ULXs carries only a small fraction of the power, the emission being largely dominated by the non thermal power-law. It is likely that, as in stellar-mass black holes binaries, the immediate surroundings of the compact object are filled with an ADAF-like corona. The inner radius of the thin disk could be much larger than assumed, invalidating the results of the fits. There are a couple of very luminous ULXs, however, that settle to a thermal-dominated state with only a power-law tail in the X-ray spectrum (e.g. Feng and Kaaret 2010). Davis

---

[22]When the torque is taken to be zero at $R_{\rm in}$, the maximum temperature is not achieved at the disk inner edge but at a slightly larger radius.

et al. (2011) have studied one of them, the source HLX-1. They fit the thermal component applying a relativistic disk model that allows for spinning black holes. Accounting for uncertainties in the values of the various parameters, they obtain $3 \times 10^3 M_\odot \lesssim M \lesssim 10^5 M_\odot$.

- **No beaming.** Evidence that ULXs radiate isotropically (without beaming) would support the argument in favor of an accreting intermediate-mass black hole. Isotropic emission in some ULXs is indeed inferred from the properties of the surrounding medium. Several ULXs are associated with expanding bubbles or emission-line optical nebulae that appear to be powered by the isotropic (or at most modestly beamed) X-ray and UV emission of the accretion flows (e.g. Pakull and Mirioni 2002; Kaaret et al. 2004; Kaaret and Corbel 2009; Cseh et al. 2012). In one source, the ULX IC 342 X-1, a synchrotron radio nebula is detected as well (Cseh et al. 2012). Both in the optical and radio bands the shape of the nebula is elongated, suggesting that it is being inflated by a jet outflow. A similar mechanism is known to be in action in the microquasars SS433 (and at smaller scale in Cygnus X-1) in our galaxy and S26 in the galaxy NGC 7793.

- **Quasi-periodic oscillations.** *Quasi-periodic oscillations* (QPOs; e.g. van der Klis 2005, 2006) are finite-width peaks in the power density spectrum (i.e. the squared magnitude of the Fourier transform of the X-ray flux as a function of time) of a source. Quasi-periodic oscillations with frequencies from Hz to kHz are observed in black hole and neutron star X-ray binaries. Many models have been developed to explain QPOs; they are usually associated to some of the characteristic frequencies (or beats between them) of the system, such as the orbital frequency of the accretion disk near the inner radius and the spin of the compact object, or to oscillations in the disk. Quasi-periodic oscillations peaked at frequencies from ~ 0.1–100 mHz have been identified in some ULXs (e.g. Strohmayer and Mushotzky 2003; Liu et al. 2005; Feng and Kaaret 2007a; Rao et al. 2010). The detection of QPOs and the long variability timescales (compared to XRBs) favor the hypothesis of disk accretion (no beaming) onto an intermediate-mass black hole.

There is general agreement that the existence of intermediate-mass black holes in ULXs is far from being established. It is likely that ULXs are, in fact, a population composed of various types of sources. Among them, the strongest intermediate-mass black hole candidates are found in the brightest ULXs. These sources have X-ray luminosities ($L_X \sim 10^{41}$ erg s^{-1}, e.g. Sutton et al. 2012) too many orders of magnitude above the Eddington limit as to be stellar-mass black hole X-ray binaries, even allowing for beaming.

6.6 What Comes Next

We have reviewed part of the enormous, almost overwhelming, corpus of data that suggests that the compact objects that lie at the centers of galaxies, in some X-ray binaries, and perhaps in a fraction of ultra-luminous X-ray sources, are black holes.

Most of this evidence, though, is indirect, does not probe the strong field regime, and may even sometimes be accounted for alleging the presence of other types of (exotic) objects. The main goal for the near future is to prove that the objects we identify as black hole have indeed the properties predicted by the relativistic theory of gravitation, the key feature being the event horizon. It is hard to say if the final, definite proof will ever be find; after all black holes should be *black*. In the worst of the cases, however, the phenomena we shall describe in the next sections will serve as spectacular tests of General Relativity.

6.6.1 Imaging the Shadow of a Black Hole

Photons emitted by a source of large angular size placed behind a black hole will be deflected by the strong gravitational field. Those with the smallest impact parameters, however, will not be able to pass by and will be trapped by the black hole. An astonished observer contemplating that part of the sky would see a black spot of radius somewhat larger than that of the event horizon around the position of the black hole. This dark zone is called the "shadow" of the black hole.

It was Bardeen in the 1970's who considered the observational appearance of a Schwarzschild black hole in front of a planar radiating source (Bardeen 1973, 1974). He showed that the diameter of the shadow is only $\sqrt{27} \sim 5.2$ Schwarzschild radii. If the black hole is spinning the shadow departs from circular because prograde photons can reach closer to the black hole than retrograde photons. The contour of the shadow also depends on the inclination of the observer's line of sight with respect to the rotation axis of the black hole.

Astrophysical black holes are as a rule surrounded by a rich radiating environment.[23] Falcke et al. (2000, but see already Luminet 1979) presented realistic images of the appearance of Sgr A* when illuminated by the radiation of an accretion flow. Using a general relativistic ray-tracing code, they calculated the shadows of Kerr and Schwarzschild black holes for two configuration of optically thin plasma: quasi-spherical in free fall and purely rotating spherical shells with the equatorial Keplerian frequency. To asses the effect of instrumental limitations, they also simulated the images as would be obtained with a putative VLBI array at $\lambda = 0.6$ mm and $\lambda = 1.3$. The results are shown in Fig. 6.11.

The dip in the intensity is clear in both cases, although much deeper for the Kerr black hole. The effective radius of the shadow is $\sim 10 R_{\mathrm{grav}}$ (~ 30 μarcsec), largely independent of the spin. Notice the change in the simulated VLBI images with the observing wavelength. At 1.3 mm the effect is almost washed out, but it is clearly evidenced already at 0.6 mm. Recently, other authors have simulated the appearance of Sgr A* in mm and sub-mm wavelengths with more sophisticated models of the accretion flow, see for example Broderick and Loeb (2006), Broderick et al. (2009b,

---

[23]Paczyński (1986) considered the possibility of detecting the shadow of black holes produced by the cosmic microwave background. This might be a way to spot isolated black holes.

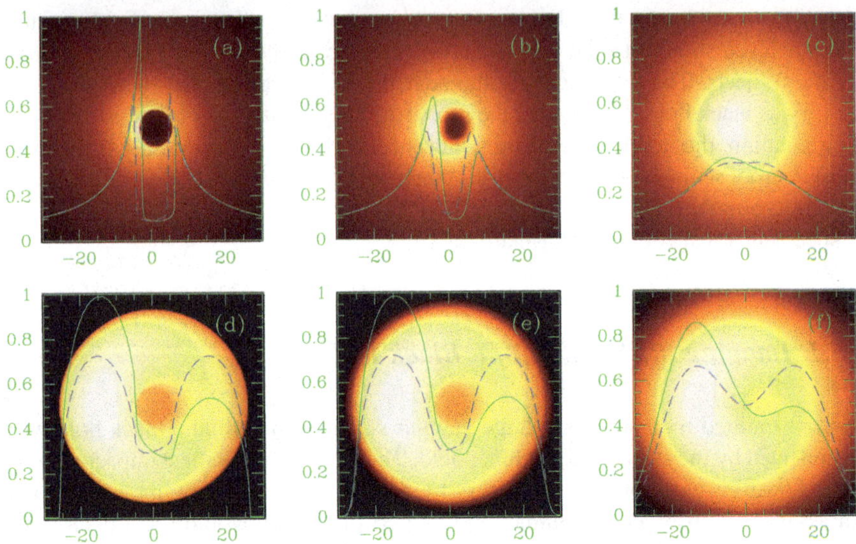

Fig. 6.11 Images of a Kerr black hole with $a_* = 0.998$ (*upper row*) and a Schwarzschild black hole (*bottom row*). The radiating plasma is in free fall and rotating in spherical shells, respectively. Panels **b** and **e**, and **c** and **f** are simulated VLBI images at 0.6 mm and 1.3 mm, respectively. The *curves* show the intensity as a function of distance to the black hole normalized to its gravitational radius. The mass and distance to the black hole are those of Sgr A* and the viewing angle is 45° with respect to the rotation axis when it corresponds. From Falcke et al. (2000). Reproduced by permission of the AAS

2011), Moscibrodzka et al. (2009), Dexter and Fragile (2011), Shcherbakov et al. (2012). The goal is to study how the shadow depends on the characteristic parameters of the black hole (mass and spin) and the accretion flow (e.g. tilt of the accretion disk, presence of hot spots). Also Broderick and Narayan (2006) have investigated how the shadow would look like if Sgr A* were not a black hole but a compact object with a solid surface accreting in a RIAF regime.

The best prospects of imaging the shadow are offered by supermassive black holes, in particular in our galaxy and in M87 (see Fig. 6.12). The angular resolution required to observe these gravitational effects is of the order of the μarcsec. This is within the capabilities of imminently available terrestrial VLBI arrays and space-borne telescopes.

The Event Horizon Telescope[24] is a project to combine existing and planned mm and sub-mm telescopes into a worldwide VLBI array. Among its prior science goals is to image the shadows of the black holes in Sgr A* and M87 and the accretion flow in their surroundings. The best angular resolution achieved up to date with the current configuration of the Event Horizon Telescope is 60 μarcsec, but this is expected to improve substantially by adding more telescopes to enlarge the baseline and by

---

[24]www.eventhorizontelescope.com.

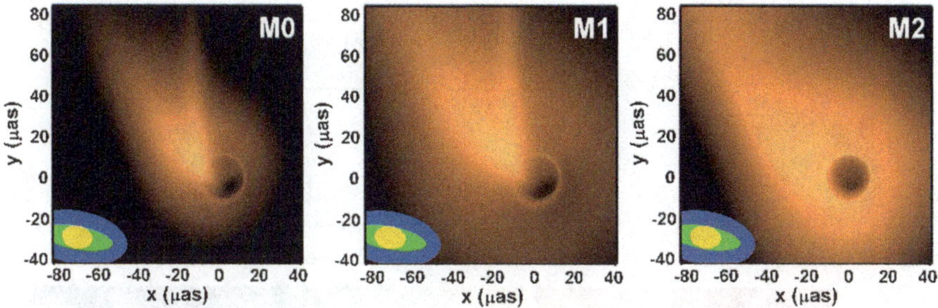

Fig. 6.12 Simulated VLBI image at 0.87 mm of a force-free jet launched from the supermassive black hole in the radio galaxy M87. Models M0 and M1 correspond to a Kerr black hole with $a_* = 0.998$ and model M2 to a Schwarzschild black hole. The *ellipses in the left corner* are the estimated beam size of different arrays of antennas. From Broderick and Loeb (2009). Reproduced by permission of the AAS

moving to higher frequencies. One of the extensions of the array, with baseline joining France and the South Pole, will provide an angular resolution of 15 µarcsec at 345 GHz ($\sim$8.5 mm). This is just $1.5\,R_{\mathrm{Schw}}$ radii of Sgr A*!

Another interferometer that will achieve sufficiently small angular resolutions to observe close to supermassive black holes is the space-based RadioAstron,[25] an international mission led by Russia. This interferometer is composed by a 10 m radiotelescope on board the Spacecraft Spectr-R (launched in July 2011) working together with radiotelescopes on Earth. It is expected to reach an angular resolutions of some µarcsec at cm wavelengths.

6.6.2 Prospects of Detecting Gravitational Waves

We have already described gravitational waves as solutions of the linearized Einstein's equations in Sect. 1.11. Is there any hope of detecting them? It seems we are indeed approaching to it.

There are plenty of sources of gravitational waves of all wavelengths in the Universe. Figure 6.13 shows a summary of them together with the appropriate detection techniques in each frequency range. We shall come to the detectors later; it suffices to say now that the search is one for *extremely* weak signals and demands incredible instrumental precision. The key to the detection of gravitational waves is the evolution of the waveform with time, a property that uniquely discloses its origin.

Figure 6.14 is a cartoon of the phases of the merger of two black holes and the gravitational wave signal expected from the process. While the black holes are well separated the wave is approximately sinusoidal, of constant amplitude and at the orbital frequency. But as they rapidly inspiral towards each other the frequency and the amplitude of the wave rise creating a "chirp" signal. The two black holes

---

[25]http://www.asc.rssi.ru/radioastron.

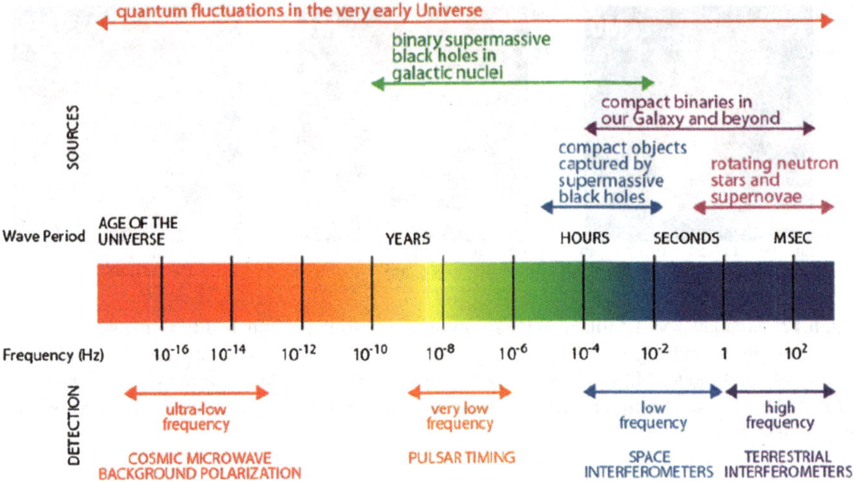

Fig. 6.13 Sources of gravitational waves and detection techniques appropriate to each frequency range. Reprinted with permission from Centrella (2011). Copyright 2011, American Institute of Physics

Fig. 6.14 Spectrum of gravitational waves emitted during the merger of two black holes. Reprinted with permission from Baumgarte (2006). Copyright 2006, American Institute of Physics

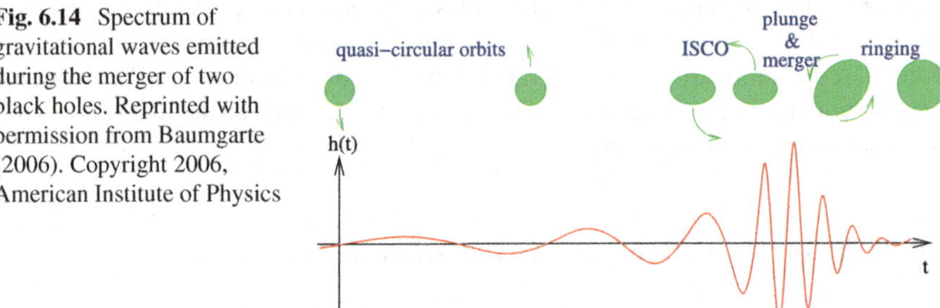

finally merger to form a single "excited" one that dissipates energy ("rings down") by emitting gravitational waves until it settles to the steady state. The calculation of the gravitational wave spectrum from each stage is carried out analytically under some approximations or directly through numerical simulations. The analysis of the waveform reveals the dynamics of the merger and the characteristic parameters of the system (mass, spin).

Gravitational waves interact weakly with matter so they can be detected from very distant sources. This advantage is also a main drawback, because the perturbation that the wave induces in the detector is also exceedingly small. Current gravitational wave detectors operate as interferometers much as the famous Michelson's device. When a gravitational wave hits the interferometer it modifies the length of the arms and therefore the path of the laser beams. When the beams recombine they are out of phase and produce an interference pattern. The change in the length L of the arms is of the order of $\sim hL/2$ where h is the amplitude of the wave. This amplitude is extremely small, and even though the arms of the interferometers are some km long, the disturbances are a hundredth of the diameter of the proton! Eliminating all other sources of perturbations (e.g. seismic activity) is indeed a phenomenal challenge.

Despite the apparently insurmountable difficulties, there are several ground-based gravitational wave interferometers currently operative, among them the Laser Interferometer Gravitational-Wave Observatory (LIGO) in the USA,[26] Virgo within the European Gravitational Observatory in Italy,[27] and GEO 600 in Germany.[28] It is expected that a couple of positive detections are announced soon. Both LIGO and Virgo have achieved their design sensitivity goals and upgrades (Advanced LIGO and Advanced Virgo) are scheduled to start working in the next years. These upgrades will improve the sensitivity by a factor of about 10. An important feature of the detection of gravitational waves is that a single detector cannot locate the source. LIGO itself is in fact two detectors separated by 3000 km. Joint observations with LIGO and Virgo have a localization accuracy of $\sim$100 deg^2 in the sky; a future array of interferometers may reduce it to $\sim$1 deg^2 (Centrella 2011).

Ground-based interferometers can detect high frequency gravitational waves in the range $\sim$1–10^4 Hz. They are suitable to investigate supernova collapses and mergers of stellar-mass compact objects (two black holes, a black hole and a neutron star, or two neutron stars). To detect lower frequency signals (1–10^{-5} Hz) from mergers of massive black holes or the capture of a stellar-mass black hole by a massive one, the detectors must be taken to space. The Laser Interferometer Space Antenna (LISA, recently renamed New Gravitational wave Observatory, NGO) is a project currently under study by the European Space Agency. LISA is planned to be an interferometer with arms $\sim$10^6 km long formed by three spacecraft in orbit around the Sun. Similar to LISA is the DECI-Hertz Interferometer Gravitational wave Observatory (DECIGO), a Japanese space-borne interferometer working between 0.1 and 10 Hz. If the projects prosper, LISA and DECIGO are to be launched perhaps in the 2020's.

To detect even lower frequency (10^{-7}–10^{-9} Hz) gravitational waves the proposal is to use observations of millisecond pulsars. This technique is called Pulsar Timing Array (PTA). Millisecond pulsars have rotational periods ($\sim$1–10 ms) that are very stable over time. The passage of a gravitational wave between the Earth and the pulsar will cause a variation in the period that might be measured after long-term observations. Presently, about 40 ms pulsars are being monitored by a few research groups. The most likely sources to be detected in the frequency range probed by PTA are (very) supermassive black holes.

The number of detections we can expect in a given time span is not yet certain. Next generation ground-based interferometers might detect of the order of tens of binaries of compact objects per year. Depending on the their final design capability,

---

[26]http://www.ligo.caltech.edu/.

[27]http://www.ego-gw.it/.

[28]www.geo600.org/.

a space-borne interferometer like LISA might detect 1–30 supermassive black hole mergers and ~50 events of stellar-mass black hole capture by supermassive black holes per year, and many more galactic compact binaries (Centrella 2011).

References

A.A. Abdo, M. Ackermann, M. Ajello et al., Science **326**, 1512 (2009a)

A.A. Abdo, M. Ackermann, M. Ajello et al., Astrophys. J. **701**, L123 (2009b)

A.A. Abdo, M. Ackermann, M. Ajello et al., Astrophys. J. **706**, L56 (2009c)

J. Abraham et al. (Pierre Auger Collaboration), Science **318**, 938 (2007)

M.A. Abramowicz, W. Kluźniak, J.-P. Lasota, Astron. Astrophys. **396**, L31 (2002)

V.A. Acciari, M. Beilicke, G. Blaylock et al., Astrophys. J. **679**, 1427 (2008)

F.A. Aharonian, Mon. Not. R. Astron. Soc. **332**, 215 (2002)

F.A. Aharonian, A.G. Akhperjanian, A.R. Bazer-Bachi et al., Astron. Astrophys. **460**, 743 (2006)

J. Albert, E. Aliu, H. Anderhub et al., Science **312**, 1771 (2006)

J. Albert, E. Aliu, H. Anderhub et al., Astrophys. J. **665**, L51 (2007)

R. Antonucci, Annu. Rev. Astron. Astrophys. **31**, 473 (1993)

J.M. Bardeen, in *Gravitational Radiation and Gravitational Collapse*, ed. by C. De Witt (1973), p. 132

J.M. Bardeen, in *Black Holes*, ed. by C. De Witt, B.S. De Witt (1974), p. 215

B. Balick, R.L. Brown, Astrophys. J. **194**, 265 (1974)

T.W. Baumgarte, AIP Conf. Proc. **861**, 161 (2006)

M. Begelman, Astrophys. J. **568**, L97 (2002)

T.M. Belloni, S.E. Motta, T. Muñoz-Darias, Bull. Astron. Soc. India **39**, 409 (2011)

J. Binney, S. Tremaine, *Galactic Dynamics* (Princeton University Press, Princeton, 1987)

V. Bosch-Ramon, G.E. Romero, J.M. Paredes, Astron. Astrophys. **447**, 263 (2006)

M. Böttcher, Astrophys. Space Sci. **309**, 95 (2007)

A.E. Broderick, A. Loeb, Mon. Not. R. Astron. Soc. **367**, 905 (2006)

A.E. Broderick, A. Loeb, Astrophys. J. **697**, 1164 (2009)

A.E. Broderick, R. Narayan, Astrophys. J. **638**, L21 (2006)

A.E. Broderick, A. Loeb, R. Narayan, Astrophys. J. **701**, 1357 (2009a)

A.E. Broderick, V.L. Fish, S.S. Doeleman, A. Loeb, Astrophys. J. **697**, 45 (2009b)

A.E. Broderick, V.L. Fish, S.S. Doeleman, A. Loeb, Astrophys. J. **735**, 110 (2011)

J. Casares, in *Highlights of Spanish Astrophysics V*, ed. by J.M. Diego, L.J. Goicoechea, J.I. González-Serrano, J. Gorgas. Astrophys. Space Sci. Proc. (2010), p. 3

P. Chatterjee, L. Hernsquist, A. Loeb, Phys. Rev. Lett. **88**, 121103 (2002a)

P. Chatterjee, L. Hernsquist, A. Loeb, Astrophys. J. **572**, 371 (2002b)

D. Cseh, S. Corbel, P. Kaaret, C. Lang, F. Grisé, Z. Paragi, A. Tzioumis, V. Tudose, H. Feng, Astrophys. J. **749**, 17 (2012)

J. Centrella, AIP Conf. Proc. **1381**, 98 (2011)

S. Corbel, M.A. Nowak, R.P. Fender, A.K. Tzioumis, S. Markoff, Astron. Astrophys. **400**, 1007 (2003)

G.B. Cook, S.L. Shapiro, S.A. Teukolsky, Astrophys. J. **424**, 823 (1994)

A. Cumming, Nucl. Phys. B, Proc. Suppl. **132**, 435 (2004)

S.W. Davis, R. Narayan, Y. Zhu, D. Barret, S.A. Farrell, O. Godet, M. Servillat, N.A. Webb, Astrophys. J. **734**, 111 (2011)

D. de Martino, T. Belloni, M. Falanga, A. Papitto, S. Motta, A. Pellizzoni, Y. Evangelista, G. Piano, N. Masetti, J.-M. Bonnet-Bidaud, M. Mouchet, K. Mukai, A. Possenti, Astron. Astrophys. **550**, A89 (2013)

J. Dexter, P.C. Fragile, Astrophys. J. **730**, 36 (2011)

S. Doeleman, J. Weintroub, A.E.E. Rogers, R. Plambeck, R. Freund, R.P.J. Tilanus, P. Friberg, L.M. Ziurys, J.M. Moran, B. Corey, K.H. Young, D.L. Smythe, M. Titus, D.P. Marrone,

R.J. Cappallo, D.C.J. Bock, G.C. Bower, R. Chamberlin, G.R. Davis, T.P. Krichbaum, J. Lamb, H. Maness, A.E. Niell, A. Roy, P. Strittmatter, D. Werthimer, A.R. Whitney, D. Woody, Nature **455**, 78 (2008)

E.N. Dorband, M. Hemsendorf, D. Merritt, J. Comput. Phys. **185**, 484 (2003)

G. Dubus, Astron. Astrophys. **456**, 801 (2006)

D. Eichler, M. Livo, T. Piran, D.N. Schramm, Nature **340**, 126 (1989)

F. Eisenhauer et al., Messenger **143**, 16 (2011)

A.C. Fabian, K. Iwasawa, C.S. Reynolds, A.J. Young, Publ. Astron. Soc. Pac. **112**, 1145 (2000)

A.C. Fabian, A. Zoghbi, R.R. Ross, P. Uttley, L.C. Gallo, W.N. Brandt, A.J. Blustin, T. Boller, M.D. Caballero-Garcia, J. Larsson, J.M. Miller, G. Miniutti, G. Ponti, R.C. Reis, C.S. Reynolds, Y. Tanaka, A.J. Young, Nature **459**, 540 (2009)

A.C. Fabian, A. Zoghbi, D. Wilkins, T. Dwelly, P. Uttley, N. Schartel, G. Miniutti, L. Gallo, D. Grupe, S. Komossa, M. Santos-Lleó, Mon. Not. R. Astron. Soc. **419**, 116 (2012)

G. Fabbiano, Rev. Mex. Astron. Astrophys. Ser. Conf. **20**, 46 (2004)

M. Falanga, D. de Martino, J.M. Bonnet-Bidaud, T. Belloni, M. Mouchet, N. Masetti, K. Mukai, G. Matt, in *Proceedings of the 8th INTEGRAL Workshop "The Restless Gamma-Ray Universe"* (2010), p. 54

H. Falcke, S. Markoff, Astron. Astrophys. **362**, 113 (2000)

H. Falcke, F. Melia, E. Agol, Astrophys. J. **528**, L13 (2000)

H. Falcke, E. Körding, S. Markoff, Astron. Astrophys. **414**, 895 (2004)

R. Fender, E. Körding, T. Belloni, P. Uttley, I. McHardy, T. Tzioumis, PoS(MQW6) 011 (2007)

H. Feng, P. Kaaret, Astrophys. J. **668**, 941 (2007a)

H. Feng, P. Kaaret, Astrophys. J. **660**, L113 (2007b)

H. Feng, P. Kaaret, Astrophys. J. **696**, 1712 (2009)

H. Feng, P. Kaaret, Astrophys. J. **712**, L169 (2010)

H. Feng, R. Soria, New Astron. Rev. **55**, 166 (2011)

L. Ferrarese, D. Merritt, Astrophys. J. **539**, L9 (2000)

J.L. Friedman, J.R. Ipser, Astrophys. J. **314**, 594 (1987)

M.R. García, J.E. McClintock, R. Narayan, P. Callanan, D. Barret, S.S. Murray, Astrophys. J. **553**, L47 (2001)

K. Gebhardt, R. Bender, G. Bower, A. Dressler, S.M. Faber, A.V. Filippenko, R. Green, C. Grillmair, L.C. Ho, J. Kormendy, T.R. Lauer, J. Magorrian, J. Pinkney, D. Richstone, S. Tremaine, Astrophys. J. **539**, L13 (2000)

R. Genzel, F. Eisenhauer, S. Gillessen, Rev. Mod. Phys. **82**, 3121 (2010)

S. Gillessen, F. Eisenhauer, S. Trippe, T. Alexander, R. Genzel, F. Martins, T. Ott, Astrophys. J. **692**, 1075 (2009a)

S. Gillessen, F. Eisenhauer, T.K. Fritz, H. Bartko, K. Dodds-Eden, O. Pfuhl, T. Ott, R. Genzel, Astrophys. J. **707**, L114 (2009b)

A.M. Ghez, S. Salim, N.N. Weinberg, J.R. Lu, T. Do, J.K. Dunn, K. Matthews, M. Morris, S. Yelda, E.E. Becklin, T. Kremenek, M. Milosavljevic, J. Naiman, Astrophys. J. **689**, 1044 (2008)

L.J. Greenhill, C.R. Gwinn, R. Antonucci, R. Barvainis, Astrophys. J. **472**, L21 (1996)

X. Han, T. An, J.-Y. Wang, J.-M. Lin, M.-J. Xie, H.-G. Xu, X.-Y. Hong, S. Frey, Res. Astron. Astrophys. **12**, 1 (2012)

S. Heinz, R.A. Sunyaev, Mon. Not. R. Astron. Soc. **343**, L59 (2003)

C. Hopman, S.F. Portegies Zwart, T. Alexander, Astrophys. J. Lett. **604**, L101 (2004)

P. Kaaret, S. Corbel, Astrophys. J. **697**, 950 (2009)

P. Kaaret, S. Corbel, A.H. Prestwich, A. Zezas, Science **299**, 365 (2003)

P. Kaaret, M.J. Ward, A. Zezas, Mon. Not. R. Astron. Soc. **351**, L38 (2004)

V. Kalogera, G. Baym, Astrophys. J. **470**, L61 (1996)

M.M. Kaufman-Bernadó, G.E. Romero, I.F. Mirabel, Astron. Astrophys. **385**, L10 (2002)

A.R. King, M.B. Davies, M.J. Ward, G. Fabbiano, M. Elvis, Astrophys. J. **552**, L109 (2001)

E. Körding, H. Falcke, S. Markoff, Astron. Astrophys. **382**, L13 (2002)

E. Kuulkers, M. van der Klis, T. Oosterbroek, J. van Paradijs, W.H.G. Lewin, Mon. Not. R. Astron. Soc. **287**, 495 (1997)

C.Y. Kuo, J.A. Braatz, J.J. Condon, C.M.V. Impellizzeri, K.Y. Lo, I. Zaw, M. Schenker, C. Henkel, M.J. Reid, J.E. Greene, Astrophys. J. **727**, 20 (2011)

J.H. Lacy, C.H. Townes, T.R. Geballe, D.J. Hollenbach, Astrophys. J. **241**, 132 (1980)

W.H.G. Lewin, J. van Paradijs, R.E. Taam, Space Sci. Rev. **62**, 223 (1993)

D. Lin, D. Altamirano, J. Homan, R.A. Remillard, R. Wijnands, T. Belloni, Astrophys. J. **699**, 60 (2009)

S.L. Liebling, C. Palenzuela, Living Rev. Relativ. **15**, 6 (2012)

J.-F. Liu, J.N. Bregman, J. Irwin, P. Seitzer, Astrophys. J. **581**, L93 (2002)

J.-F. Liu, J.N. Bregman, E. Lloyd-Davies, J. Irwin, C. Espaillat, P. Seitzer, Astrophys. J. **621**, L17 (2005)

J.-F. Liu, J. Orosz, J.N. Bregman, Astrophys. J. **745**, 22 (2012)

Q.Z. Liu, J. van Paradijs, E.P.J. van den Heuvel, Astron. Astrophys. **455**, 1165 (2006)

Q.Z. Liu, J. van Paradijs, E.P.J. van den Heuvel, Astron. Astrophys. **469**, 807 (2007)

J.-P. Luminet, Astron. Astrophys. **75**, 228 (1979)

A.I. MacFadyen, S.E. Woosley, Astrophys. J. **524**, 262 (1999)

F. Macchetto, A. Marconi, D.J. Axon, A. Capetti, W. Sparks, P. Crane, Astrophys. J. **489**, 579 (1997)

K. Makishima, A. Kubota, T. Mizuno, T. Ohnishi, M. Tashiro, Y. Aruga, K. Asai, T. Dotani, K. Mitsuda, Y. Ueda, S. Uno, K. Yamaoka, K. Ebisawa, Y. Kohmura, K. Okada, Astrophys. J. **535**, 632 (2000)

E. Maoz, Astrophys. J. **494**, L181 (1998)

J. McClintock, R. Remillard, in *Compact Stellar X-Ray Sources*, ed. by W.H.G. Lewin, M. van der Klis (Cambridge University Press, Cambridge, 2006)

M.L. McConnell, J.M. Ryan, W. Collmar, V. Schoenfelder, H. Steinle, A.W. Strong, H. Bloemen, W. Hermsen, L. Kuiper, K. Bennett, B.F. Phlips, J.C. Ling, Astrophys. J. **543**, 928 (2000)

N.J. McConnell, C.-P. Ma, Astrophys. J. **764**, 184 (2013)

N.J. McConnell, C.-P. Ma, K. Gebhardt, S.A. Wright, J.D. Murphy, T.R. Lauer, J.R. Graham, D.O. Richstone, Nature **480**, 215 (2011)

A. Merloni, S. Heinz, T. Di Matteo, Mon. Not. R. Astron. Soc. **345**, 1057 (2003)

D. Merritt, P. Berczik, F. Laun, Astron. J. **133**, 553 (2007)

L. Meyer, A.M. Ghez, R. Schoedel, S. Yelda, A. Boehle, J.R. Lu, T. Do, M.R. Morris, E.E. Becklin, K. Matthews, Science **338**, 84 (2012)

S. Migliari, R.P. Fender, Mon. Not. R. Astron. Soc. **366**, 79 (2006)

S. Migliari, J.C.A. Miller-Jones, D.M. Russell, Mon. Not. R. Astron. Soc. **415**, 2407 (2011)

J.M. Miller, G. Fabbiano, M.C. Miller, A.C. Fabian, Astrophys. J. **585**, L37 (2003)

J.M. Miller, A.C. Fabian, M.C. Miller, Astrophys. J. **607**, 931 (2004)

M.C. Miller, D.P. Hamilton, Mon. Not. R. Astron. Soc. **330**, 232 (2002)

I.F. Mirabel, L.F. Rodríguez, Nature **392**, 673 (1998)

I.F. Mirabel, L.F. Rodríguez, B. Cordier, J. Paul, F. Lebrun, Nature **358**, 215 (1992)

M. Miyoshi, J. Moran, J. Herrnstein, L. Greenhill, N. Nakai, P. Diamond, M. Inoue, Nature **373**, 127 (1995)

T. Mizuno, A. Kubota, K. Makishima, Astrophys. J. **554**, 1282 (2001)

M. Moscibrodzka, C.F. Gammie, J.C. Dolence, H. Shiokawa, P.K. Leung, Astrophys. J. **706**, 497 (2009)

A. Mücke, R.J. Protheroe, R. Engel, J.P. Rachen, T. Stanev, Astropart. Phys. **18**, 593 (2003)

R. Mushotzky, Prog. Theor. Phys. Suppl. **155**, 27 (2004)

E. Nakar, Phys. Rep. **442**, 166 (2007)

R. Narayan, Astron. Geophys. **44**, 6.22 (2003)

R. Narayan, J.S. Heyl, Astrophys. J. **574**, L139 (2002)

R. Narayan, J.S. Heyl, Astrophys. J. **599**, 419 (2003)

R. Narayan, M.R. García, J.E. McClintock, Astrophys. J. **478**, L79 (1997)

R. Narayan, M.R. García, J.E. McClintock, in *Proceedings of the Ninth Marcel Grossmann Meeting*, ed. by V.G. Gurzadyan, R.T. Jantzen, R. Ruffini (World Scientific, Singapore, 2002), p. 405

P.L. Nolan et al. (Fermi-LAT Collaboration), Astrophys. J. Suppl. Ser. **199**, 31 (2012)

J.A. Orosz, J.E. McClintock, J.P. Aufdenberg, R.A. Remillard, M.J. Reid, R. Narayan, L. Gou, Astrophys. J. **742**, 84 (2011)

F. Özel, D. Psaltis, R. Narayan, J.E. McClintock, Astrophys. J. **725**, 1918 (2010)

B. Paczyński, Nature **321**, 419 (1986)

B. Paczyński, Astrophys. J. **494**, L45 (1998)

M.W. Pakull, L. Mirioni, in *Proceedings of the Symposium "New Visions of the X-Ray Universe in the XMM-Newton and Chandra Era"* (2002)

J.M. Paredes, J. Marti, M. Ribo, M. Massi, Science **288**, 2340 (2000)

B.M. Peterson, Publ. Astron. Soc. Pac. **105**, 247 (1993)

M.J. Reid, Int. J. Mod. Phys. D **18**, 889 (2009)

M.J. Reid, A. Brunthaler, Astrophys. J. **616**, 872 (2004)

R.A. Remillard, J.E. McClintock, Annu. Rev. Astron. Astrophys. **44**, 49 (2006)

F. Rao, H. Feng, P. Kaaret, Astrophys. J. **722**, 620 (2010)

M.T. Reynolds, J.M. Miller, Astrophys. J. Lett. **734**, L17 (2011)

M.M. Reynoso, M.C. Medina, G.E. Romero, Astron. Astrophys. **531**, A30 (2011)

C.E. Rhoades, R. Ruffini, Phys. Rev. Lett. **32**, 324 (1974)

T.P. Roberts, J.C. Gladstone, A.D. Goulding, A.M. Swinbank, M.J. Ward, M.R. Goad, A.J. Levan, Astron. Nachr. **332**, 398 (2011)

G.E. Romero, M. Orellana, Astron. Astrophys. **439**, 237 (2005)

G.E. Romero, G.S. Vila, Astron. Astrophys. **485**, 623 (2008)

G.E. Romero, A.T. Okazaki, M. Orellana, S.P. Owocki, Astron. Astrophys. **474**, 15 (2007)

G.E. Romero, F.L. Vieyro, G.S. Vila, Astron. Astrophys. **519**, A109 (2010)

S. Sabatini, M. Tavani, E. Striani et al., Astrophys. J. **712**, L10 (2010)

B.J. Sams, A. Eckart, R.A. Sunyaev, Nature **382**, 47 (1996)

R.V. Shcherbakov, R.F. Penna, J.C. McKinney, Astrophys. J. **755**, 133 (2012)

R. Soria, Astron. Nachr. **332**, 330 (2011)

R. Soria, M. Cropper, C. Motch, Chin. J. Astron. Astrophys. **5**, 153 (2005)

R. Soria, M.W. Pakull, J.W. Broderick, S. Corbel, C. Motch, Mon. Not. R. Astron. Soc. **409**, 541 (2010)

T.E. Strohmayer, R.F. Mushotzky, Astrophys. J. **586**, L61 (2003)

A.D. Sutton, T.P. Roberts, D.J. Walton, J.C. Gladstone, A.E. Scott, Mon. Not. R. Astron. Soc. **432**, 1154 (2012)

Y. Tanaka, K. Nandra, A.C. Fabian, H. Inoue, C. Otani, T. Dotani, K. Hayashida, K. Iwasawa, T. Kii, H. Kunieda, F. Makino, M. Matsuoka, Nature **375**, 659 (1995)

M. Tavani, A. Bulgarelli, G. Piano et al., Nature **462**, 620 (2009)

D.F. Torres, S. Capozziello, G. Lambiase, Phys. Rev. D **62**, 104012 (2000)

C.M. Urry, P. Padovani, Publ. Astron. Soc. Pac. **107**, 803 (1995)

R.C.E. van den Bosch, K. Gebhardt, K. Gültekin, G. van de Ven, A. van der Wel, J.L. Walsh, Nature **491**, 729 (2012)

M. van der Klis, Astron. Nachr. **326**, 798 (2005)

M. van der Klis, in *Compact Stellar X-Ray Sources*, ed. by W.H.G. Lewin, M. van der Klis (Cambridge University Press, Cambridge, 2006)

F.L. Vieyro, G.E. Romero, Astron. Astrophys. **542**, A7 (2012)

G.S. Vila, G.E. Romero, Mon. Not. R. Astron. Soc. **403**, 1457 (2010)

G.S. Vila, G.E. Romero, N.A. Casco, Astron. Astrophys. **538**, A97 (2012)

R.D. Viollier, D. Trautmann, G.B. Tupper, Phys. Lett. B **306**, 79 (1993)

N. Webb, D. Cseh, E. Lenc, O. Godet, D. Barret, S. Corbel, S. Farrell, R. Fender, N. Gehrels, I. Heywood, Science **337**, 554 (2012)

A. Yamauchi, N. Nakai, Y. Ishihara, P. Diamond, N. Sato, Publ. Astron. Soc. Jpn. **64**, 103 (2012)

Y.-J. Yang, A.K.H. Kong, D.M. Russell, F. Lewis, R. Wijnands, Mon. Not. R. Astron. Soc. **427**, 2876 (2012)

Chapter 7
Wormholes and Exotic Objects

7.1 Historical Remarks

The first paper on 'wormhole' solutions of Einstein's field equations was published in 1935 by Einstein himself and Nathan Rosen (Einstein and Rosen 1935). Einstein and Rosen used the word "bridge" to describe their solution. They were looking for solutions able to represent physical particles in pure geometrical terms of space-time. These solutions should be singularity-free, in order to avoid the divergence problem of particles in classical field theory.

The so-called "Einstein-Rosen bridge" was built through a coordinate change in the Schwarzschild solution (2.10): $u^2 = r - 2M$ (units of $G = c = 1$). This renders the metric in the form

$$ds^2 = \frac{u^2}{u^2 + 2M}dt^2 - 4(u^2 + 2M)du^2 + (u^2 + 2M)d\Omega^2. \qquad (7.1)$$

Here the new variable range is $(-\infty, +\infty)$. The region including the curvature singularity, $r \in [0, 2M)$, is excluded and the range of the variable covers twice the asymptotically flat region $r \in [2M, +\infty)$. The whole solution is like two Schwarzschild black holes cut and pasted at the event horizon. The region near $u = 0$ is the "bridge" that connects the two asymptotically flat regions (at $u = -\infty$ and $u = +\infty$). The Einstein-Rosen bridge, however, is not traversable: any particle moving along u from $-\infty$ to $+\infty$ will experience infinite tidal forces (see, for instance, Visser 1996).

The expression "wormhole", used to describe topologically non-trivial space-time regions was introduced by Misner and Wheeler (1957). Their objective in that work was to explain *all* classical physics through geometric concepts. Specifically, they used the source-free Maxwell equations coupled with General Relativity formulated on a multiple-connected manifold to build classical models of elementary particles. This was a new attempt to follow Einstein's ideas of expressing all physical phenomena in a geometric framework. The result was the so-called "geometrodynamics" (Wheeler 1962). The wormholes conceived by Misner and Wheeler,

G.E. Romero, G.S. Vila, *Introduction to Black Hole Astrophysics*,
Lecture Notes in Physics 876, DOI 10.1007/978-3-642-39596-3_7,
© Springer-Verlag Berlin Heidelberg 2014

however, were doomed to collapse into black holes making geometrodynamics inviable. The so-called "topological censorship theorems" (e.g. Visser 1996) imply that any space-time where the average null energy condition (ANEC) is satisfied cannot contain traversable wormholes.

The ANEC condition can be stated as follows:

$$\text{ANEC holds on } \Gamma \iff \int_\Gamma T_{\mu\nu}k^\mu k^\nu d\lambda \geq 0, \tag{7.2}$$

where Γ is an null curve, k^μ is the corresponding tangent vector, and λ is a generalized affine parametrization of Γ. If the curve is time-like, instead of null, the condition is called the average weak energy condition (AWEC).

General macroscopic wormhole solutions of Einstein's field equations that explicitly violate AWEC were first found by Morris and Thorne (1988). In that paper, the authors narrate the genesis of these solutions, related to the writing of Carl Sagan's novel *Contact* (Sagan 1985). We shall discuss these solutions in what follows.

7.2 Wormhole Metric

As we have mentioned, a wormhole is a region of space-time with non-trivial topology. It has two mouths connected by a throat (see Figs. 7.1 and 7.2). The mouths are not hidden by event horizons, as in the case of black holes, and, in addition, there is no singularity to avoid the passage of particles, or travelers, from one side to the other. Contrary to black holes, wormholes are holes in space-time, i.e. their existence implies a multiple-connected space-time.

There are many types of wormhole solutions of Einstein's field equations (see Visser 1996). Let us consider the static spherically symmetric line element,

$$ds^2 = e^{2\Phi(l)}c^2dt^2 - dl^2 - r(l)^2 d\Omega^2$$

where l is a proper radial distance that covers the entire range $(-\infty, \infty)$. In order to have a wormhole which is traversable in principle, we need to demand that:

1. $\Phi(l)$ be finite everywhere, to be consistent with the absence of event horizons.
2. For the spatial geometry to tend to an appropriate asymptotically flat limit, it must happen that

$$\lim_{r \to \infty} r(l)/l = 1$$

and

$$\lim_{r \to \infty} \Phi(l) = \Phi_0 < \infty.$$

The radius of the wormhole is defined by $r_0 = \min\{r(l)\}$, where we can set $l = 0$.

Fig. 7.1 Embedding
diagrams of wormholes.
From Misner et al. (1973)

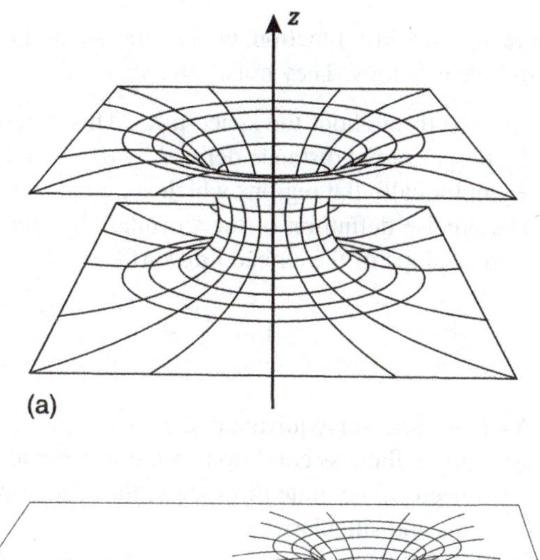

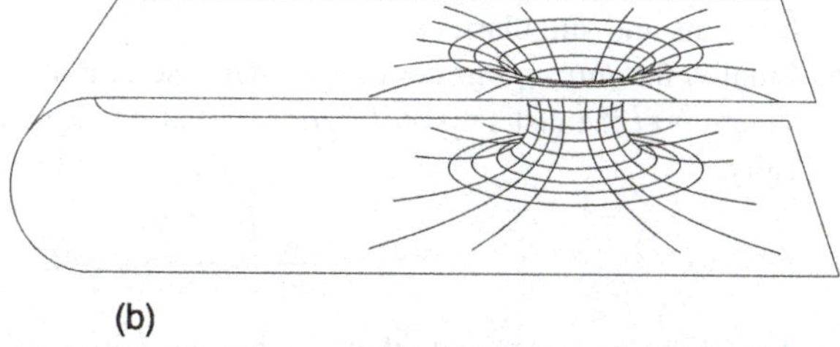

Fig. 7.2 Embedding of a
Lorentzian wormhole

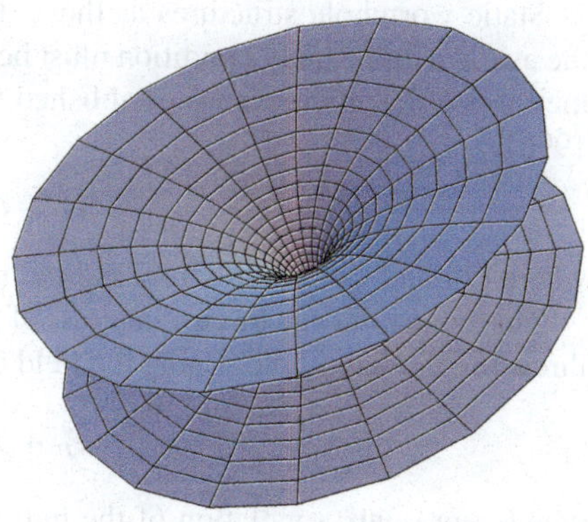

To formulate wormholes which can be traversable in practice, we should introduce additional engineering constraints. Notice that for simplicity we have considered both asymptotic regions as interchangeable. This is the best choice of coordinates for the study of wormhole geometries because calculations result considerably
simplified. In general, two patches are needed to cover the whole range of l, but this
is not noticed if both asymptotic regions are assumed to be similar. The static line
element is:

$$ds^2 = e^{2\Phi(r)}c^2 dt^2 - e^{2\Lambda(r)}dr^2 - r^2 d\Omega^2, \tag{7.3}$$

where the redshift function Φ and the shape-like function $e^{2\Lambda}$ characterize the wormhole topology. They must satisfy:

1. $e^{2\Lambda} \geq 0$ throughout the space-time. This is required to ensure the finiteness of the proper radial distance defined by $dl = \pm e^{\Lambda}\, dr$. The $\pm$ signs refer to the two asymptotically flat regions which are connected by the wormhole throat.
2. The precise definition of the wormhole's throat (minimum radius, r_{th}) entails a vertical slope of the embedding surface:

$$\lim_{r \to r_{\mathrm{th}}^+} \frac{dz}{dr} = \lim_{r \to r_{\mathrm{th}}^+} \pm\sqrt{e^{2\Lambda} - 1} = \infty. \qquad (7.4)$$

3. As $l \to \pm\infty$ (or equivalently, $r \to \infty$), $e^{2\Lambda} \to 1$ and $e^{2\Phi} \to 1$. This is the asymptotic flatness condition on the wormhole space-time.
4. $\Phi(r)$ needs to be finite throughout the space-time to ensure the absence of event horizons and singularities.
5. Finally, the *flaring out* condition, that asserts that the inverse of the embedding function $r(z)$ must satisfy $d^2r/dz^2 > 0$ at or near the throat. Stated mathematically,

$$-\frac{\Lambda' e^{-2\Lambda}}{(1 - e^{-2\Lambda})^2} > 0. \qquad (7.5)$$

This is equivalent to state that $r(l)$ has a minimum.

Static wormhole structures as those described by the above metric require that the average null energy condition must be violated in the wormhole throat. From the metric coefficients it can be established that (e.g. Morris and Thorne 1988; Visser 1996):

$$G_{tt} + G_{rr} < 0, \qquad (7.6)$$

where G_{tt} and G_{rr} are the time and radial components of the Einstein tensor.

This constraint can be cast in terms of the energy-momentum tensor of the matter threading the wormhole. Using the field equations, it reads:

$$T_{tt} + T_{rr} < 0, \qquad (7.7)$$

which represents a violation of the null energy condition. This implies also a violation of the weak energy condition (see Visser 1996 for details). Plainly stated, it means that the matter threading the wormhole must exert gravitational repulsion in order to stay stable against collapse.

Although there are known violations of the energy conditions (e.g. the Casimir effect), it is far from clear at present whether large macroscopic amounts of "exotic matter" exist in nature. If natural wormholes exist in the universe (e.g. if the original topology after the Big Bang was multiply connected), then there should be observable electromagnetic signatures of such objects (e.g. Torres et al. 1998b). Currently, the observational data allow to establish an upper bound on the total amount of exotic matter under the form of wormholes of $\sim 10^{-36}$ g cm^{-3}. The production of this kind of matter in the laboratory is completely out of the current technical possibilities, at least in significant macroscopic quantities.

A simple choice of $\Phi(r)$ and $\Lambda(r)$ is (e.g. Morris and Thorne 1988; Hong and Kim 2006):

$$\Phi(r) = \frac{1}{2}\ln\left(1 - \frac{b(r)}{r}\right), \tag{7.8}$$

$$e^{2\Lambda(r)} = \left(1 - \frac{b(r)}{r}\right)^{-1}, \tag{7.9}$$

where

$$b(r) = b(r_0) = \text{const} = B > 0. \tag{7.10}$$

The wormhole shape function has a minimum at $r = r_0$, where the exotic matter is concentrated.

Another possibility is the so-called "absurdly benign" wormhole (Morris and Thorne 1988):

$$b(r) = b_0\left[\frac{1 - (r - b_0)}{a_0}\right]^2, \qquad \Phi(r) = 0, \quad \text{for } b_0 \leq r \leq b_0 + a_0, \tag{7.11}$$

$$b = \Phi = 0, \quad \text{for } r \geq b_0 + a_0. \tag{7.12}$$

There is a vast literature on wormhole solutions. The reader is referred to Lobo (2008) for further readings.

7.3 Detectability

The idea that wormholes can act as gravitational lenses and induce a microlensing signature on a background source was first suggested by Kim and Cho (1994). Cramer et al. (1995) carried out more detailed analysis of negative mass wormholes and considered the effects they can produce on background point sources, at non-cosmological distances. The generalization to a cosmological scenario was carried out by Torres et al. (1998a), although lensing of point sources was still used. The first and only bound on the possible existence of negative masses, imposed using astrophysical databases, was given by Torres et al. (1998b). These authors showed that the effective gravitational repulsion of light rays from background gamma-ray emitting AGNs creates two bursts, which are individually asymmetric under time reversal. Then, Anchordoqui et al. (1999) searched in existent gamma-ray bursts databases for signatures of wormhole microlensing. Although they detected some interesting candidates, no conclusive results were obtained. Peculiarly asymmetric gamma-ray bursts (Romero et al. 1999), although highly uncommon, might be probably explained by more conventional hypothesis, like precessing jets (see, for instance, Reynoso et al. 2008).

In the following subsections we shall discuss the physics of gravitational microlensing of background sources by natural wormholes. We follow the treatment given by Safonova et al. (2002).

7.3.1 Lensing by a Point Negative Mass

We shall consider lensing by a point negative mass lens, so we can adopt all the assumptions concurrent with the treatment of the Schwarzschild lens:

- *Geometrical optics approximation*—the scale over which the gravitational field changes is much larger than the wavelength of the light being deflected.
- *Small-angle approximation*—the total deflection angle is small. The typical bending angles involved in gravitational lensing of cosmological interest are less than $<1'$; therefore we can describe the lens optics in the paraxial approximation.
- *Geometrically-thin lens approximation*—the maximum deviation of the ray is small compared to the length scale on which the gravitational field changes. Although the scattering takes place continuously over the trajectory of the photon, the appreciable bending occurs only within a distance of the order of the impact parameter.

We define two planes, the source and the lens plane. These planes, described by Cartesian coordinate systems (ξ_1, ξ_2) and (η_1, η_2), respectively, pass through the source and deflecting mass and are perpendicular to the optical axis (the straight line extended from the source plane through the deflecting mass to the observer). Since the components of the image position and the source positions are much smaller than the distances to the lens and source planes, we can write the coordinates in terms of the observed angles. Therefore, the image coordinates can be written as (θ_1, θ_2) and those of the source as (β_1, β_2).

7.3.2 Effective Refractive Index of the Gravitational Field of a Negative Mass and the Deflection Angle

The "Newtonian" potential of a negative point mass lens is given by

$$\Phi(\xi, z) = \frac{G|M|}{(b^2 + z^2)^{1/2}}, \tag{7.13}$$

where b is the impact parameter of the unperturbed light ray and z is the distance along the unperturbed light ray from the point of closest approach. Here the potential is positive defined and approaching zero at infinity. In view of the assumptions stated above, we can describe light propagation close to the lens in a locally Minkowskian space-time perturbed by the *positive* gravitational potential of the lens to first post-Newtonian order. In this weak field limit, we describe the metric of a negative mass body in orthonormal coordinates $x^0 = ct$, $\mathbf{x} = (x^i)$ by

$$ds^2 \approx \left(1 + \frac{2\Phi}{c^2}\right)c^2 dt^2 - \left(1 - \frac{2\Phi}{c^2}\right)dl^2, \tag{7.14}$$

where $dl = |\mathbf{x}|$ denotes the Euclidean arc length. The effect of the space-time curvature on the propagation of light can be expressed in terms of an effective index of refraction n_{eff}, given by

$$n_{\mathrm{eff}} = 1 - \frac{2}{c^2}\Phi. \tag{7.15}$$

Thus, the effective speed of light in the field of a negative mass is

$$v_{\mathrm{eff}} = c/n_{\mathrm{eff}} \approx c + \frac{2}{c}\Phi. \tag{7.16}$$

Because of the increase in the effective speed of light in the gravitational field of a negative mass, light rays would arrive faster than those following a similar path in vacuum. This leads to a very interesting effect when compared with the propagation of a light signal in the gravitational field of a positive mass. In that case, light rays are delayed relative to propagation in vacuum—the well known *Shapiro time delay*. In the case of a negative mass lensing, this effect is replaced by a new one, which is called by Safonova et al. (2002) *time gain*.

Defining the deflection angle as the difference of the initial and final ray direction

$$\boldsymbol{\alpha} \equiv \hat{\mathbf{e}}_{\mathrm{in}} - \hat{\mathbf{e}}_{\mathrm{out}}, \tag{7.17}$$

where $\hat{\mathbf{e}} \equiv d\mathbf{x}/dl$ is the unit tangent vector of a ray $\mathbf{x}(l)$, we obtain the deflection angle as the integral along the light path of the gradient of the gravitational potential

$$\boldsymbol{\alpha} = \frac{2}{c^2}\int \boldsymbol{\nabla}_\perp \Phi\, dl, \tag{7.18}$$

where $\boldsymbol{\nabla}_\perp \Phi$ denotes the projection of $\boldsymbol{\nabla}\Phi$ onto the plane orthogonal to the direction $\hat{\mathbf{e}}$ of the ray. We find

$$\boldsymbol{\nabla}_\perp \Phi(b, z) = -\frac{G|M|\mathbf{b}}{(b^2 + z^2)^{3/2}}. \tag{7.19}$$

Then, the deflection angle is

$$\boldsymbol{\alpha} = -\frac{4G|M|\mathbf{b}}{c^2 b^2}. \tag{7.20}$$

In the case of the negative mass lensing, the term "deflection" has its rightful meaning—the light is deflected *away* from the mass, unlike in the positive mass lensing, where it is bent *towards* the mass.

Fig. 7.3 Lensing geometry of a negative mass. O is the observer, S is the source, W is the negative mass lens, I_1 is one of the images. β is the angle between the source and the lens-position of the source, θ is the angle between the source and the image-position of the image, and α is the deflection angle. b is the impact parameter and D_1, D_s and D_{ls} are angular diameter distances. Other quantities are auxiliary

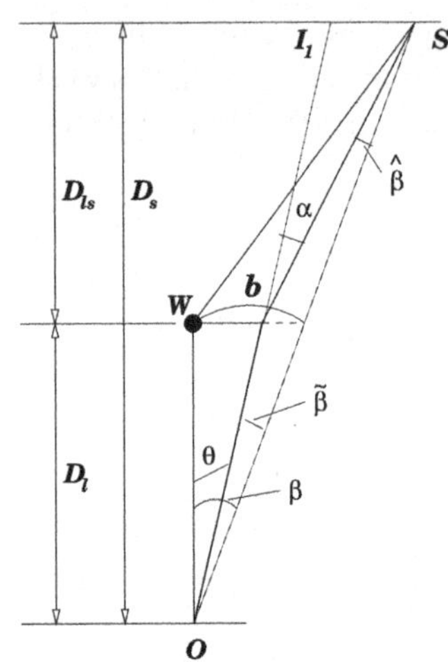

7.3.3 Lensing Geometry and Lens Equation

In Fig. 7.3 we show the lensing geometry for a point-like negative mass. From this figure and the definition of the deflection angle, we can obtain the relation between the positions of the source and the image:

$$(\beta - \theta)D_s = -\alpha D_{ls} \tag{7.21}$$

or

$$\beta = \theta - \frac{D_{ls}}{D_s}\alpha. \tag{7.22}$$

With the deflection angle, we can write the lens equation as

$$\beta = \theta + \frac{4G|M|}{c^2\xi}\frac{D_{ls}}{D_s} = \theta + \frac{4G|M|}{c^2}\frac{D_{ls}}{D_s D_l}\frac{1}{\theta}. \tag{7.23}$$

7.3.4 Einstein Radius and the Formation of Images

A natural angular scale in this problem is given by the quantity

$$\theta_E^2 = \frac{4G|M|}{c^2}\frac{D_{ls}}{D_s D_l}, \tag{7.24}$$

which is called the Einstein angle. In the case of a positive point mass lens, this corresponds to the angle at which the Einstein ring is formed, happening when source, lens and observer are perfectly aligned. This does not happen if the mass of the lens is negative. There are other differences as well. A typical angular separation of images is of order $2\theta_E$ for a positive mass lens. Sources which are closer than about θ_E to the optical axis are significantly magnified, whereas sources which are located well outside the Einstein ring are magnified very little. All this is different with a negative mass lens, but nonetheless, the Einstein angle remains an useful scale for the description of the various regimes in the present case.

The Einstein angle corresponds to the Einstein radius in the linear scale (in the lens plane):

$$R_E = \theta_E D_l = \sqrt{\frac{4GM}{c^2}\frac{D_{ls}D_l}{D_s}}. \tag{7.25}$$

In terms of Einstein angle the lens equation takes the form

$$\beta = \theta + \frac{\theta_E^2}{\theta}, \tag{7.26}$$

which can be solved to obtain two solutions for the image position θ:

$$\theta_{1,2} = \frac{1}{2}\left(\beta \pm \sqrt{\beta^2 - 4\theta_E^2}\right). \tag{7.27}$$

Unlike in the lensing by positive masses, there are now three distinct regimes here and, thus, we can classify the lensing phenomenon as follows:

I: $\beta < 2\theta_E$ There is no real solution for the lens equation. It means that there are no images when the source is inside twice the Einstein angle.

II: $\beta > 2\theta_E$ There are two solutions, corresponding to two images both on the same side of the lens and between the source and the lens. One is always inside the Einstein angle, the other is always outside it.

III: $\beta = 2\theta_E$ This is a degenerate case, $\theta_{1,2} = \theta_E$; two images merge at the Einstein angular radius, forming the *radial* arc.

We also obtain two important scales, one is the Einstein angle (θ_E)—the angular radius of the radial critical curve, the other is twice the Einstein angle ($2\theta_E$)—the angular radius of the caustic. Thus, we have two images, one is always inside the θ_E, one is always outside; and as a source approaches the caustic ($2\theta_E$) from the positive side, two images coming closer and closer together, and nearer and nearer the critical curve, thereby brightening. When the source crosses the caustic, the two images merge on the critical curve (θ_E) and disappear.

7.3.5 Magnifications

Light deflection not only changes the direction but also the cross section of a bundle of rays. For an infinitesimally small source, the ratio between the solid angles gives the flux amplification due to lensing,

$$|\mu| = \frac{d\omega_i}{d\omega_s}.$$ (7.28)

For an infinitesimal source at angular position $\boldsymbol{\beta}$ and image at angular position $\boldsymbol{\theta}$, the relation between the two solid angles is determined by the area distortion, given in turn by the determinant of the Jacobian matrix $\mathcal{A}$ of the lens mapping $\boldsymbol{\theta} \mapsto \boldsymbol{\beta}$,

$$\mathcal{A} \equiv \frac{\partial \boldsymbol{\beta}}{\partial \boldsymbol{\theta}}.$$ (7.29)

For a point mass lens magnification is given by

$$\mu^{-1} = \left| \frac{\beta}{\theta} \frac{d\beta}{d\theta} \right|.$$ (7.30)

The image is thus magnified or de-magnified by a factor $|\mu|$. If a source is mapped into several images, the total amplification is given by the sum of the individual image magnifications. From the lens equation we find

$$\frac{\beta}{\theta} = \frac{\theta^2 + \theta_{\mathrm{E}}^2}{\theta^2}, \qquad \frac{d\beta}{d\theta} = \frac{\theta^2 - \theta_{\mathrm{E}}^2}{\theta^2}.$$ (7.31)

Thus,

$$\mu_{1,2}^{-1} = \left| 1 - \frac{\theta_{\mathrm{E}}^4}{\theta_{1,2}^4} \right|.$$ (7.32)

The total magnification is

$$\mu_{\mathrm{tot}} = |\mu_1| + |\mu_2| = \frac{u^2 - 2}{u\sqrt{u^2 - 4}}$$ (7.33)

where $u = \beta/\theta_{\mathrm{E}}$.

7.3.6 Microlensing

When the angular separation between the images $d\theta$

$$d\theta = \sqrt{\beta^2 - 4\theta_{\mathrm{E}}^2}$$ (7.34)

is of the order of milliarcsecs, we cannot resolve the two images with existing telescopes and we can only observe the lensing effect through their combined light intensity. This effect is called *microlensing*. Both the lens and the source are moving with respect to each other (as well as the observer). Thus, images change their position and brightness. Of particular interest are sudden changes in luminosity, which occur when a compact source crosses a critical curve. For the positive mass lensing the situation is quite simple (for a review on the positive mass microlensing and its applications, see Schneider et al. 1992).

For a negative mass lens the situation is different. We define a dimensionless minimum impact parameter B_0, expressed in terms of the Einstein radius, as the shortest distance between the path line of the source and the lens. For three different values of B_0 we have three different lensing configurations (see Safonova et al. 2002).

We define the time scale of the microlensing event as the time it takes the source to move across the Einstein radius, projected onto the source plane, $\xi_0 = \theta_E D_s$,

$$t_v = \frac{\xi_0}{V}. \tag{7.35}$$

The angle β changes with time as

$$\beta(t) = \sqrt{\left(\frac{Vt}{D_s}\right)^2 + \beta_0^2}. \tag{7.36}$$

Here the moment $t = 0$ corresponds to the smallest angular distance β_0 between the lens and the source. Normalizing to θ_E,

$$u(t) = \sqrt{\left(\frac{Vt}{\theta_E D_s}\right)^2 + \left(\frac{\beta_0}{\theta_E}\right)^2}, \tag{7.37}$$

where u is a dimensionless impact parameter. Including the time scale t_v and defining

$$B_0 = \frac{\beta_0}{\theta_E}, \tag{7.38}$$

we obtain

$$u(t) = \sqrt{B_0^2 + \left(\frac{t}{t_v}\right)^2}. \tag{7.39}$$

Finally, the total amplification as a function of time is given by

$$A(t) = \frac{u(t)^2 - 2}{u(t)\sqrt{u(t)^2 - 4}}. \tag{7.40}$$

In Fig. 7.4 we show the light curves for the point source for four source trajectories with different minimum impact parameters B_0. As can be seen from the light curves, when the distance from the point mass to the source trajectory is larger than $2\theta_E$, the light curve is identical to that of a positive mass lens light curve. However, when the distance is less than $2\theta_E$ (or in other terms, $B_0 \leq 2.0$), the light curve shows significant differences. Such events are characterized by the *asymmetrical* light curves, which occur when a compact source crosses a critical curve. A very

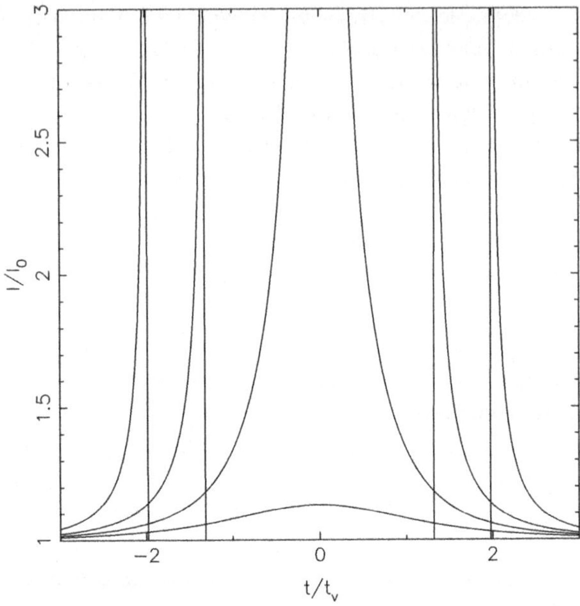

Fig. 7.4 Light curves for the negative mass lensing of a point source. From the center of the graph towards the corners the curves correspond to $B_0 = 2.5$, 2.0, 1.5, 0.0. The time scale here is ξ_0 divided by the effective transverse velocity of the source. From Safonova et al. (2002)

interesting, eclipse-like, phenomenon occurs here; a zero intensity region (disappearance of images) with an angular radius θ_0

$$\theta_0 = \sqrt{4\theta_E^2 - \beta_0^2},\tag{7.41}$$

or in terms of normalized unit θ_E,

$$d = \sqrt{4 - B_0^2}.\tag{7.42}$$

7.3.7 Extended Source

Many astrophysical sources appear as extended, and although their size may be small compared to the relevant length scales of a lensing event, this extension has an impact on the light curves.

We define the dimensionless source radius, $\tilde{R}$, as

$$\tilde{R} = \frac{\rho}{\theta_E} = \frac{R}{\xi_0},\tag{7.43}$$

where ρ and R are the angular and the linear physical size of the source, respectively, and ξ_0 is the length unit in the source plane.

It is convenient to write the lens equation in the scaled scalar form

$$y = x + \frac{1}{x}, \tag{7.44}$$

where we normalized the coordinates to the Einstein angle:

$$x = \frac{\theta}{\theta_E}, \qquad y = \frac{\beta}{\theta_E}. \tag{7.45}$$

The lens equation can be solved analytically for any source position. The amplification factor, and thus the total amplification, can be readily calculated for point sources. However, as we are interested in extended sources, this amplification has to be integrated over the source, Eq. (7.46), and furthermore, as we want to build the light curves, the total amplification for an extended source has to be calculated for many source positions. The amplification $\mathcal{A}$ of an extended source with surface brightness profile $I(\mathbf{y})$ is given by

$$\mathcal{A} = \frac{\int d^2 y\, I(\mathbf{y}) \mathcal{A}_0(\mathbf{y})}{\int d^2 y\, I(\mathbf{y})}, \tag{7.46}$$

where $\mathcal{A}_0(\mathbf{y})$ is the amplification of a point source at position $\mathbf{y}$.

In Fig. 7.5 we show the images of an extended source with a Gaussian brightness distribution for an effective dimensionless source radius $\tilde{R}_S = 3.0$ (frames a to e), together with the corresponding light curve. Here the source path passes through the lens ($B_0 = 0$), which lies exactly in the center of each frame. The source's extent in the lens plane is greater than the Einstein radius of the lens. Annotated wedges provide color scale for the images. We notice there that there is an eclipse-like phenomenon, occurring most notably when most of the source is near or exactly behind the lens. This is consistent with the light curve (frame f), where there is a de-magnification.

In Fig. 7.6 we compare light curves for three different radially symmetric source profiles (uniform, Gaussian and exponential) for two dimensionless source radii $\tilde{R} = \tilde{R}_S^{\text{gauss}} = \tilde{R}_S^{\text{expon}} = 0.1$ and $\tilde{R} = \tilde{R}_S^{\text{gauss}} = \tilde{R}_S^{\text{expon}} = 1.0$. As a reference curve we show the light curve of the point source. All curves are made for the impact parameter $B_0 = 0$. We can see the larger noise in the uniform source curve, since the source with uniform brightness has extremely sharp edge, whereas Gaussian and exponential sources are extremely smooth. Though we considered the sources with the same effective radius, we can see from the plot that for a small source size, the maximum magnification is reached by the source with exponential profile (upper panel), which is explained by the fact that this profile has a more narrow central peak than the Gaussian.

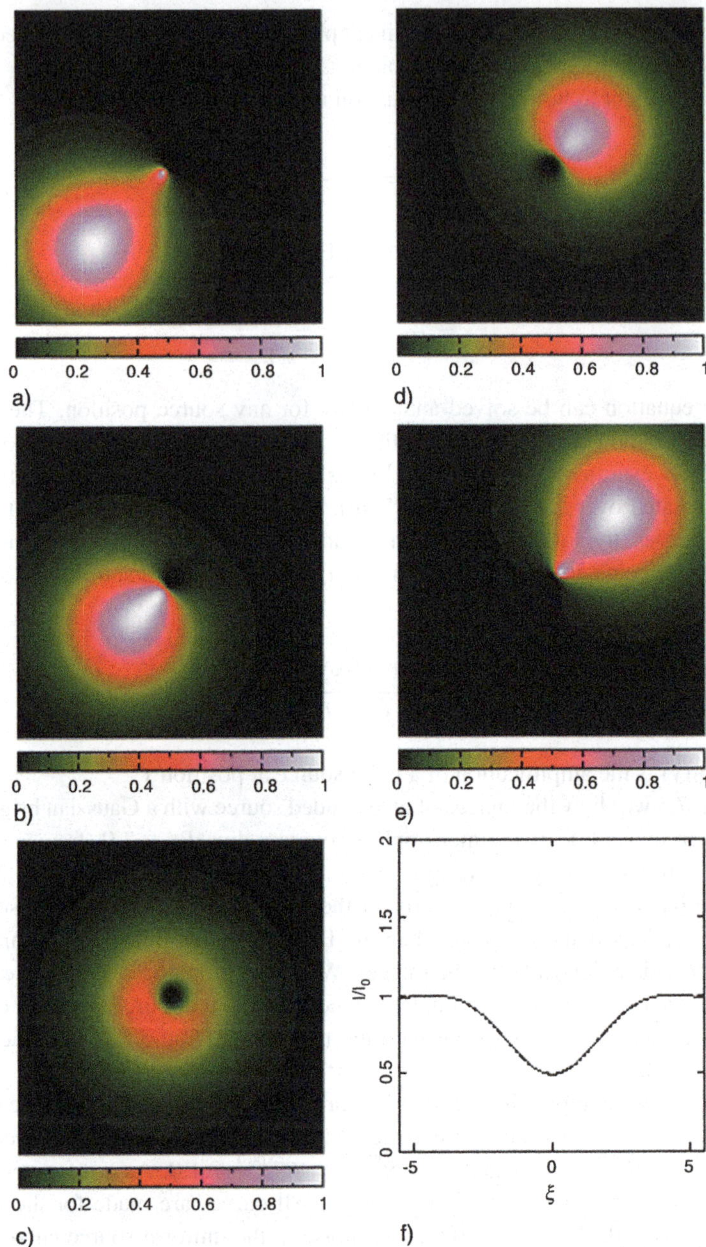

Fig. 7.5 Image configurations (frames **a** to **e**) and a corresponding light curve (frame **f**) for a Gaussian source with effective radius $\tilde{R}_S = 3.0$, in units of the Einstein angle. The source is moving from the lower left corner (frame **a**) to the right upper corner (frame **e**), passing through the lens ($B_0 = 0$). The lens is in the center of each frame. Size of each frame is 5×5, in the normalized units. Wedges to each frame provide the color scale for the images. Note the eclipse-like phenomenon, consistent with the incomplete demagnification showed in the light curve (frame **f**). From Safonova et al. (2002)

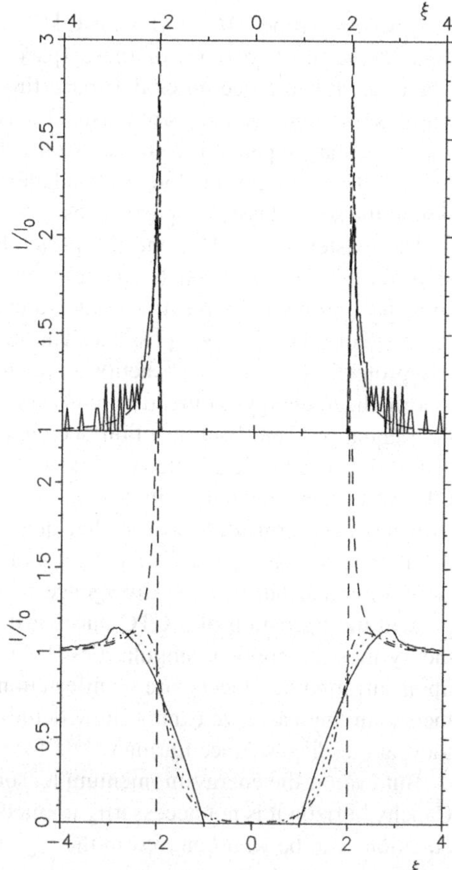

Fig. 7.6 Light curves for the point mass source (*dashed line*), source with constant surface brightness (*solid*), source with Gaussian brightness distribution (*dash-dotted*) and exponential brightness distribution (*dotted*) for two different effective dimensionless source radii, 0.1 (*upper panel*) and 1.0 (*bottom panel*). From Safonova et al. (2002)

7.4 Closed Time-Like Curves and Time Travel

Closed time-like curves (CTCs) are worldlines of any physical system in a temporally orientable space-time which, moving always in the future direction, ends arriving back at some point of its own past. Although solutions of Einstein's field equations where CTCs exist are known at least since Gödel's (1949) original work on rotating universes, it has been only since the last decade of the 20th century that physicists have shown a strong and sustained interest on this topic. The renewed attraction of CTCs and their physical implications stem from the discovery, at the end of the 1980's, of traversable wormhole space-times. A wormhole can be immediately transformed into a time machine inducing a time-shift between the two mouths. This can be made through relativistic motion of the mouths (a Special Relativity effect) or by exposing one of them to an intense gravitational field (see Morris et al. 1988 for further details; for the paradoxes of time travel, see Romero and Torres 2001).

Any space-time (M, g_{ab}) with CTCs is called a *chronology-violating* space-time. There are two types of these space-times: those where CTCs exist everywhere (as, for instance, in Gödel space-time), and those where CTCs are confined within some regions and there exists at least one region free of them. The regions with CTCs are separated from the "well-behaved" space-time by Cauchy horizons (wormhole space-times belong to this latter type). Here we shall restrict the discussion to the second type of space-times.

The existence of CTCs and the possibility of backward time travel have been objected by several scientists championed by Hawking (1992), who proposed the so-called chronology protection conjecture: the laws of physics are such that the appearance of CTCs is never possible. The suggested mechanism to enforce chronology protection is the back-reaction of vacuum polarization fluctuations: when the renormalized energy-momentum tensor is fed back to the semi-classical Einstein's field equations, the back-reaction accumulates energy in such a way that it may distort the space-time geometry so strongly as to form a singularity, destroying the CTC at the very moment of its formation.

It has been argued, however, that quantum gravitational effects would cut the divergence off saving the CTCs (Kim and Thorne 1991). By other hand, Li et al. (1993) pointed out that the divergence of the energy-momentum tensor does not prevent the formation of a CTC but is just a symptom that a full quantum gravity theory must be applied: singularities, far from being physical entities that can act upon surrounding objects, are manifestations of the breakdown of the gravitational theory. In any case, we cannot draw definitive conclusions with the semi-classical tools at our disposal (see Earman 1995 for additional discussion).

But even if the energy-momentum tensor of vacuum polarization diverges at the Cauchy horizon it is not necessarily implied that CTCs must be destroyed, since the equations can be well-behaved in the region inside the horizon (Li et al. 1993). In particular, wormhole space-times could be stabilized against vacuum fluctuations introducing reflecting boundaries between the wormhole mouths (Li 1994) or using several wormholes to create CTCs (Thorne 1992; Visser 1997).

Even Hawking has finally recognized that back-reaction does not necessarily enforce chronology protection (Cassidy and Hawking 1998). Although the quest for finding an effective mechanism to avoid CTCs continues, it is probable that the definitive solution to the problem should wait until a complete theory of quantum gravity can be formulated. In the meantime, the profound physical consequences of time travel in General Relativity should be explored in order to push this theory to its ultimate limits, to the region where the very foundations of the theory must be revisited.

A different kind of objection to CTC formation is that they allow illogical situations like the "grandfather" paradox[1] which would be expressing that the corresponding solutions of the field equations are "non-physical". This is a commonplace and has been conveniently refuted by Earman (1995), among others (see also Nahin

---

[1]The grandfather paradox: a time traveler goes to his past and kills his young grandfather then avoiding his own birth and, consequently, the time travel in which he killed his grandfather.

1999 and references therein). Grandfather-like paradoxes do not imply illogical situations. In particular, they do not mean that local determinism does not operate in chronology-violating space-times because it is always possible to choose a neighborhood of any point of the manifold such that the equations that represent the laws of physics have appropriate solutions. Past cannot be changed (the space-time manifold is unique) but it can be causally affected from the future, according to General Relativity. The grandfather paradox, as pointed out by Earman 1995, is just a manifestation of the fact that consistency constraints must exist between the local and the global order of affairs in space-time. This leads directly to the so-called Principle of Self-Consistency (PSC).

7.4.1 The Principle of Self-Consistency

In space-times with CTCs, past and future are no longer globally distinct. Events on CTCs should causally influence each other along a time-loop in a self-adjusted, consistent way in order to occur in the real universe. This has been stated by Friedman et al. (1990) as a general principle of physics:

Principle of Self-Consistency: the only solutions to the laws of physics that occur locally in the real universe are those which are globally self-consistent.

When applied to the grandfather paradox, the PSC says that the grandfather cannot be killed (a local action) because in the far future this would generate an inconsistency with the global world line of the time traveler. Just consistent histories can develop in the universe. An alternative way to formulate the PSC is to state that (Earman 1995):

The laws of physics are such that any local solution of their equations that represents a feature of the real universe must be extensible to a global solution.

The principle is not tautological or merely prescriptive, since it is clear that local observations can provide information of the global structure of the world: it is stated that there is a global-to-local order in the universe in such a way that certain local actions are ruled out by the global properties of the space-time manifold.

If the PSC is neither a tautological statement nor a methodological rule, what is then its epistemological status? It has been suggested that it could be a basic law of physics—in the same sense that Einstein's field equations are laws of physics—(Earman 1995). This would imply that there is some "new physics" behind the PSC. On the other hand, Carlini et al. (1995) have proposed, on the basis of some simple examples, that the PSC could be a consequence of the Principle of Minimal Action. In this case, no new physics would be involved. Contrary to these opinions, that see in the PSC a law statement, or at least a consequence of law statements, we suggest that this principle actually is a *metanomological statement*, like the Principle of General Covariance among others (see Bunge 1961 for a detailed discussion on metanomological statements). This means that the reference class of the PSC is not formed by physical systems, but by laws of physical systems. The usual laws are restrictions to the state space of physical systems. Metanomological statements are laws of laws, i.e. restrictions on the global network of laws that thread the universe. The requirement of consistency constraints would then be pointing out the existence of deeper level super-laws, which enforce the harmony between local and global affairs in space-time. Just in this sense it is fair to say that "new physics" is implied.

7.4.2 Causal Loops: Self-Existent Objects

Although the PSC eliminates grandfather-like paradoxes from chronology-violating space-times, other highly perplexing situations remain. The most obscure of these situations is the possibility of an ontology with self-existent objects (Romero and Torres 2001). Let us illustrate with an example what we understand by such an object:

> *Suppose that, in a space-time where CTCs exist, a time traveler takes a ride on a time machine carrying a book with her. She goes back to the past, forgets the book in—what will be—her laboratory, and returns to the future. The book remains then hidden until the time traveler finds it just before starting her time trip, carrying the book with her.*

It is not hard to see that the primordial origin of the book remains a mystery. Where does the book come from? This puzzle has been previously mentioned in philosophical literature by Nerlich (1981) and MacBeath (1982). Physicists, instead, have not paid much attention to it, despite the interesting fact that the described situation is apparently not excluded by the PSC: the local and global structures of the loop are perfectly harmonious and there are not causality violations. There is just a book never created, never printed, but, somehow, existing in space-time. It has been suggested (Nerlich 1981) that if CTCs exist, then we are committed to accept an ontology of self-existent objects: they are just out there, trapped in space-time. There is no sense in asking where they come from. Even energy is conserved if we admit that the system is not only the present time-slice of the manifold but rather the two slices connected by the time loop: the energy removed from the present time ($M_{\mathrm{book}}c^2$) is deposited in the past.

The acceptance of such a bizarre ontology, however, proceeds from an incorrect application of the PSC. This principle is always discussed within the context of General Relativity, although actually it should encompass *all* physical laws. A fully correct formulation of the PSC should read *laws of nature* where *laws of physics* appears in the formulation given above. What should be demanded is total consistency and not only consistency in the solutions of Einstein's field equations. In particular, when thermodynamics is included in the analysis, the loop of a self-existent objects becomes inconsistent because of the entropic degradation, that makes the final and initial states of the object not to match (Romero and Torres 2001). Even more strange paradoxes, related to human self-reproduction like the amazing Jocasta paradox (Harrison 1979), can be shown to be non-consistent when the laws of genetics are taken into account (see Nahin 1999 and references therein). These consideration, nonetheless, seem not to apply when non-interacting elementary particles are involved, since the concept of entropy can be formulated only for statistical systems.

7.4.3 *Information Loops and the PSC*

Consider the local light cone of a time traveler. There are three, and only three, possible final destinations for a backwards time trip. The arrival point could be (a) within the past light cone, (b) on the edge of it, or (c) elsewhere out of the cone. In case (a) the time traveler can transmit information at a velocity $v \leq c$ and affect its own past. Information transmitted in case (b) that propagates at the speed of light, instead, will arrive at the very moment when the time travel started. In case (c), the information flux can only reach the future of the time traveler. However, even in the latter case, the past might be affected by an information flux from the future if several time machines are available (for instance as in the situation known as a Roman ring, where there are two wormholes in relative motion). The conclusion, then, is that whatever the final destiny of the time-traveler is, in principle, she could always affect her own past. Otherwise stated, chronology-violating space-times generally admit information loops: they cannot be excluded on the sole basis of the PSC.

Although the PSC does not preclude that within chronology-violating regions information steaming from the future can affect the past, it at least imposes constraints on the way this can be done. In fact, any physical process causally triggered by the backwards information flux must be consistent with the past history of the universe. This means that if a time traveler goes back to her past and tries, for instance, to communicate the contents of the theory of Special Relativity to the scientific community before 1905, she will fail *because* at her departure it was historically clear that the first paper on Special Relativity was published by Albert Einstein in June 1905. The details of her failure will depend on the details of her travel and attempt, in the same way that the details of the failure of a perpetual motion machine depends on the approach used by the imprudent inventor. All we can say *a priori* is that the laws of physics are such that these attempts *cannot* succeed and information cannot propagate arbitrarily in space-time. It is precisely because of the PSC that we know that our past is not significantly affected from the future: we know that until now, knowledge has been generated by evolutionary processes, i.e. there is small room in history for information loops. This does not necessarily mean that the same is valid for the entire space-time.

7.5 White Holes

The analytical extension of the Schwarzschild solution in Kruskal-Szekeres coordinates shows a region that is singular and covered by an horizon but from where geodesic lines emerge. Such an extension can be interpreted as a time-reversal image of a black hole: the matter from an expanding cloud began to expand *from* the horizon. Such space-time regions are known as *white holes* (Novikov 1964; Ne'eman 1965). White holes cannot result from the collapse of physical objects in the real Universe, but they could be imagined to be intrinsic features of space-time. A white hole acts as a source that ejects matter from its event horizon. The sign of the acceleration is invariant under time reversal, so both black and white holes attract matter.

Fig. 7.7 Embedding diagram
representing a white hole

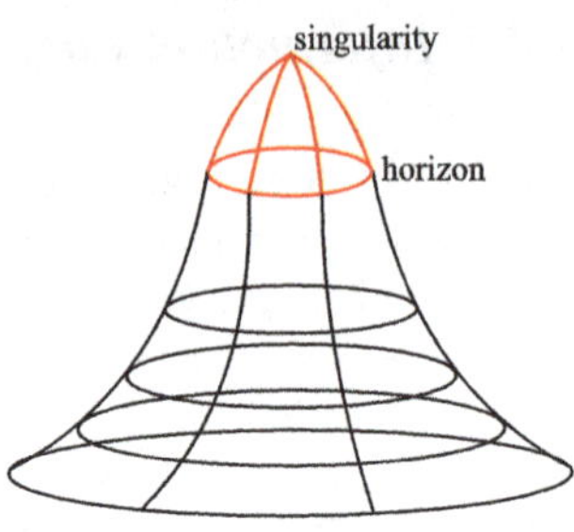

The only potential difference between them is in the behavior at the horizon. White
hole horizons recede from any incoming matter at the local speed of light, in such
a way that the infalling matter never crosses the horizon. The infalling matter is
then scattered and re-emitted at the death of the white hole, receding to infinity after
having come close to the final singular point where the white hole is destroyed. The
total proper time until an infalling object encounters the singular endpoint is the
same as the proper time to be swallowed by a black hole, so the white hole picture
does not say what happens to the infalling matter. In Fig. 7.7 we show an embedding
diagram of a white hole.

The existence of white holes is doubtful since they are unstable. The instability
of white holes results from both classical processes caused by the interaction of the
surrounding matter (Frolov 1974) and from processes of quantum particle creation
in the gravitational field of the holes (Zel'dovich et al. 1974). The accretion of matter
into white holes causes the instability and converts them into black holes. The reader
is referred to Frolov and Novikov (1998) for a detailed discussion. In addition, we
notice that the entropy of a black hole is related to the horizon area in Planck units,
and this is the maximum entropy which a given region can contain. When an object
flies out of a white hole, the area of the horizon always decreases by more than the
maximum possible entropy that can be squeezed into the object, which is a time-
reversed statement of the Bekenstein bound.[2] White holes, then, appear to violate
the second law of thermodynamics.

7.6 Topological Black Holes

In four-dimensional General Relativity there are several theorems restricting the
topology of the event horizon of a black hole (Heusler 1996). In the stationary case,
black holes must have a spherical horizon. Toroidal horizons are allowed only for
very short times, while the recently formed black hole (e.g. by a neutron star merger)
evolves towards the final steady configuration of trivial topology. However, in higher
dimensional relativity or in some gravitational theories other than General Relativity, black holes with non-trivial topologies might exist.

---

[2]The Bekenstein bound says that the maximum possible entropy of a black hole is $S = A/4$, where
A is the two-dimensional area of the black hole's event horizon in units of the Planck area.

The Kerr solution in four-dimensional Einstein's theory, which describes asymptotically flat rotating black holes, was generalized to higher-dimensions ($D \geq 5$) by Myers and Perry (1986). The essential difference from the four-dimensional Kerr solution is that there are several independent rotation planes, whose number N depends on the space-time dimensions as $N = [(D - 1)/2]$, in higher dimensions. Therefore, this solution is specified by the mass parameter μ and N spin parameters a_i ($i = 1, \ldots, N$). The forms of the metric in odd and even dimensions are slightly different.

When the space-time dimension, D, is odd, the metric takes the form (Tomizawa and Ishihara 2011):

$$ds^2 = dt^2 - \left(r^2 + a_i^2\right)\left(d\mu_i^2 + \mu_i^2 d\varphi_i^2\right) - \frac{\mu r^2}{\Pi F}\left(dt - a_i \mu_i^2 d\varphi_i\right)^2$$

$$- \frac{\Pi F}{\Pi - \mu r^2} dr^2, \tag{7.47}$$

where the functions F and Π are defined as

$$F(r, \mu_i) = 1 - \frac{a_i^2 \mu_i^2}{r^2 + a_i^2}, \qquad \Pi = \prod_{i=1}^{N}\left(r^2 + a_i^2\right) \tag{7.48}$$

and $i = 1, \ldots, N$ and μ_i have to satisfy the constraint $\sum_i \mu_i^2 = 1$.

For D even, the metric is:

$$ds^2 = -dt^2 + \left(r^2 + a_i^2\right)\left(d\mu_i^2 + \mu_i^2 d\varphi_i^2\right) + \frac{\mu r^2}{\Pi F}\left(dt - a_i \mu_i^2 d\varphi_i\right)^2$$

$$+ \frac{\Pi F}{\Pi - \mu r} dr^2, \tag{7.49}$$

where μ_i satisfies $\sum_i \mu_i^2 + \alpha^2 = 1$ with the constant satisfying $-1 \leq \alpha \leq 1$. The ADM mass[3] and angular momenta for the i-th rotational plane (x_{2i-1}, x_{2i}) are given by

$$M = \frac{(D - 2)\Omega_{D-2}}{16\pi G}\mu, \qquad J_i = \frac{\Omega_{D-2}}{16\pi G}\mu a_i. \tag{7.50}$$

The horizons exist at the place where g^{rr} vanishes, as in the four-dimensional case. For even D with $D \geq 6$, the location of the horizons is determined by the roots of the equation

$$\Pi - \mu r = 0. \tag{7.51}$$

---

[3]Roughly speaking, the ADM mass is the total mass-energy contained in a asymptotically flat space-like region. The ADM mass is a conserved quantity. For a formal definition see Visser (1996, p. 111).

The left-hand side is a polynomial of order $2N$, so that this equation for arbitrary dimensions would have no general analytic solution.

Since curvature singularities appear at $r = 0$, the existence of horizons requires that Eq. (7.51) should have at least one solution for positive r. For positive r the function $\Pi - \mu r$ has only a single local minimum, the value of the mass parameter μ assumed to be positive.

For odd D with $D \geq 5$, the horizons exist at the values of r of the equation

$$\Pi - \mu r^2 = 0. \tag{7.52}$$

The left-hand side is a polynomial of $(D-1)/2$ order in $l = 2r^2$, so that in general for $D = 5, 7, 9$ only, the above equation would have analytic solutions. In particular, for $D = 5$, an analytic solution can be found as

$$r_{\pm}^2 = \frac{\mu - a_1^2 - a_2^2 \pm \sqrt{(\mu - a_1^2 - a_2^2)^2 - 4a_1^2 a_2^2}}{2}. \tag{7.53}$$

Therefore the presence (absence) of the horizon requires the parameters should lie in the range

$$\mu > 0, \quad |a_1| + |a_2| \begin{cases} > \sqrt{\mu} & \text{no horizon,} \\ = \sqrt{\mu}, \ a_1 a_2 = 0 & \text{no horizon,} \\ = \sqrt{\mu}, \ a_1 a_2 \neq 0 & \text{one degenerate horizon,} \\ < \sqrt{\mu} & \text{two horizons.} \end{cases} \tag{7.54}$$

For arbitrary D larger than 5, when all of spin parameters a_i are non-vanishing, the function $\Pi(l) - 2\mu l$ has only a single local minimum at $l = l_*$, which is determined from $\partial_l \Pi(l_*) - 2\mu = 0$.

The vacuum Einstein's equations in 5-dimensions admit a solution describing a stationary asymptotically flat spacetime regular on and outside an event horizon of topology $S^1 \times S^2$ (Emparan and Reall 2002). This solution describes a rotating "black ring". It is an example of a stationary asymptotically flat vacuum solution with an event horizon of non-spherical topology. There is a range of values for the mass and angular momentum for which there exist two black ring solutions as well as a black hole solution. Therefore the uniqueness theorems valid in four dimensions do not have simple 5-dimensional generalizations. It has been suggested that increasing the spin of a 5-dimensional black hole beyond a critical value results in a transition to a black ring, which can have an arbitrarily large angular momentum for a given mass. To keep a balance against its self-gravitational attractive force by centrifugal force, the black ring must be rotating along the S^1 direction.

The metric of the black ring rotating along the S^1 direction can be written in terms of several convenient coordinate systems. In the C-metric coordinates, the metric of the Emparan-Reall solution is given by

$$ds^2 = \frac{F(y)}{F(x)}\left(dt + CR\frac{1+y}{F(y)}d\psi\right)^2$$
$$-\frac{R^2 F(x)}{(x-y)^2}\left[\frac{G(y)}{F(y)}d\psi^2 + \frac{G(x)}{F(x)}d\varphi^2 + \frac{dx^2}{G(x)} + \frac{dy^2}{G(y)}\right], \qquad (7.55)$$

where the functions F and G are defined by

$$F(\xi) = 1 + \lambda\xi, \qquad G(\xi) = \left(1 - \xi^2\right)(1 + \nu\xi), \qquad (7.56)$$

and the constant C is

$$C = \sqrt{\lambda(\lambda - \nu)\frac{1+\lambda}{1-\lambda}}. \qquad (7.57)$$

The (x, y) coordinates span the range

$$-1 \le x \le 1, \qquad -\infty < y \le -1, \qquad (7.58)$$

and the parameters lie in the range

$$0 < \nu \le \lambda < 1. \qquad (7.59)$$

This black ring space-time admits three mutually commuting Killing vectors, stationary Killing vector field ∂_t, and two independent axial Killing vectors ∂_ψ, ∂_φ with closed integral curves. It turns out that there exists no closed time-like curves in the domain of outer communication. Three parameters, R, ν and λ, are not independent, which comes from the requirement for the absence of the conical singularities. To avoid conical singularities at the ψ-axis ($y = -1$) and the outer φ-axis ($x = -1$), the coordinates ψ and φ must have a periodicity of

$$\Delta\varphi = \Delta\psi = 2\pi\frac{\sqrt{1-\lambda}}{1-\nu}. \qquad (7.60)$$

This condition also assures that the space-time is asymptotically flat. However, even if this condition is satisfied, in general, the space-time still possesses a disk-shaped conical defect at the inner axis ($x = 1$) of the black ring. Therefore, the absence of conical singularities at the inner axis ($x = 1$) should be required. This imposes the following constraint to the angular coordinate φ:

$$\Delta\varphi = 2\pi\frac{\sqrt{1+\lambda}}{1+\nu}. \qquad (7.61)$$

Hence, combining Eqs. (7.60) and (7.61), it is found that regularity requires that the parameters must satisfy

$$\lambda = \frac{2\nu}{1+\nu^2}. \qquad (7.62)$$

Fig. 7.8 Illustration of
embedded 5-D black rings

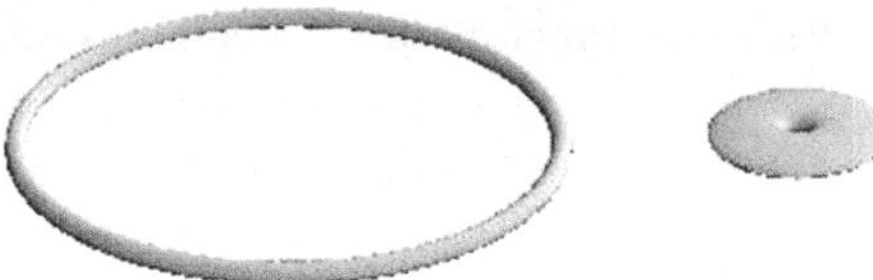

This can be interpreted as an *equilibrium (balance) condition* for a black ring: the radius of the ring is dynamically fixed by the balance between the centrifugal and tensional forces.

Unlike rotating black holes, a rotating black ring space-time has only a single horizon at y satisfying $G(y) = 0$, namely $y = -1/\nu$. This solution has two non-vanishing charges, the mass and angular momentum:

$$M = \frac{3\pi R^2}{4G} \frac{\lambda}{1 - \nu}, \qquad J_\psi := J_1 = \frac{\pi R^3}{2G} \frac{\sqrt{\lambda(\lambda - \nu)(1 + \lambda)}}{(1 - \nu)^2}. \tag{7.63}$$

The SLS is at $y = -1/\lambda$ where $F(y) = 0$. Figure 7.8 shows an illustration of embedded black rings.

7.7 Gravastars

A different type of hypothetical objects are Gravitational Vacuum Stars or "gravastars". Gravastars were first proposed by Mazur and Motolla (2001) as an alternative to black holes. They are mathematically constructed as compact objects with an interior de Sitter condensate phase and an exterior Schwarzschild geometry of arbitrary total mass M. These are separated by a phase boundary with a small but finite thickness of fluid with equation of state $p = +\rho c^2$, replacing both the Schwarzschild and de Sitter classical horizons. The interior region has an equation of state $p = -\rho c^2$, in such a way that the energy conditions are violated and the singularity theorems do not apply. The solution, then, has no singularities and no event horizons. The system is sustained against collapse by the negative pressure of the vacuum. The assumption required for this solution to exist is that gravity undergoes a vacuum rearrangement phase transition in the vicinity of $r = r_{\text{Schw}}$. Recent theoretical work, however, has shown that gravastars as well as other alternative black hole models are not stable when they rotate (Cardoso et al. 2008). This might be interpreted as a "no go theorem" for them.

References

L. Anchordoqui, G.E. Romero, D.F. Torres, I. Andruchow, Mod. Phys. Lett. A **14**, 791 (1999)
M. Bunge, Am. J. Phys. **29**, 518 (1961)
V. Cardoso, P. Pani, M. Cadoni, M. Cavaglià, Phys. Rev. D **77**, id. 124044 (2008)

A. Carlini, V.P. Frolov, M.B. Mensky, I.D. Novikov, H.H. Soleng, Int. J. Mod. Phys. D **4**, 557 (1995)

M.J. Cassidy, S.W. Hawking, Phys. Rev. D **57**, 2372 (1998)

J. Cramer, R. Forward, M. Morris, M. Visser, G. Benford, G. Landis, Phys. Rev. D **51**, 3117 (1995)

J. Earman, *Bangs, Crunches, Whimpers, and Shrieks: Singularities and Acausalities in Relativistic Space-Times* (Oxford University Press, New York, 1995)

A. Einstein, N. Rosen, Phys. Rev. **48**, 73 (1935)

R. Emparan, H.S. Reall, Phys. Rev. Lett. **88**, 101101 (2002)

J.L. Friedman, M.S. Morris, I.D. Novikov, F. Echeverria, G. Klinkhammer, K.S. Thorne, U. Yurtsever, Phys. Rev. D **42**, 1915 (1990)

V.P. Frolov, Zh. Eksp. Teor. Fiz. **66**, 813 (1974)

V.P. Frolov, I.D. Novikov, *Black Hole Physics* (Kluwer Academic, Dordrecht, 1998)

K. Gödel, Rev. Mod. Phys. **21**, 447 (1949)

J. Harrison, Analysis **39**, 65 (1979)

S.W. Hawking, Phys. Rev. D **46**, 603 (1992)

M. Heusler, *Black Hole Uniqueness Theorems* (Cambridge University Press, Cambridge, 1996)

S.-T. Hong, S.-W. Kim, Mod. Phys. Lett. A **21**, 789 (2006)

S.-W. Kim, Y.M. Cho, in *Evolution of the Universe and Its Observational Quest* (Universal Academy Press and Yamada Science Foundation, Tokyo, 1994), p. 353

S.-W. Kim, K.S. Thorne, Phys. Rev. D **43**, 3929 (1991)

L.-X. Li, Phys. Rev. D **50**, R6037 (1994)

L.-X. Li, J.-M. Xu, L. Liu, Phys. Rev. D **48**, 4735 (1993)

F.S.N. Lobo, in *Classical and Quantum Gravity Research*, ed. by M.N. Christiansen, T.K. Rasmussen (Nova Science, New York, 2008)

P.O. Mazur, E. Motolla, arXiv:gr-qc/0109035v5 (2001)

M. MacBeath, Synthese **51**, 397 (1982)

C.W. Misner, K.S. Thorne, J.A. Wheeler, *Gravitation* (Freeman, New York, 1973)

C.W. Misner, J.A. Wheeler, Ann. Phys. **2**, 525 (1957)

M.S. Morris, K.S. Thorne, Am. J. Phys. **56**, 395 (1988)

M.S. Morris, K.S. Thorne, U. Yurtsever, Phys. Rev. Lett. **61**, 1446 (1988)

R.C. Myers, M.J. Perry, Ann. Phys. **172**, 304 (1986)

P.J. Nahin, *Time Machines: Time Travel in Physics, Metaphysics and Science Fiction* (Springer/AIP Press, New York, 1999)

Y. Ne'eman, Astrophys. J. **141**, 1303 (1965)

G. Nerlich, Pac. Philos. Q. **62**, 227 (1981)

I.D. Novikov, Astron. Ž. **43**, 911 (1964)

M.M. Reynoso, G.E. Romero, H.R. Christiansen, Mon. Not. R. Astron. Soc. **387**, 1745 (2008)

G.E. Romero, D.F. Torres, I. Andruchow, L. Anchordoqui, B. Link, Mon. Not. R. Astron. Soc. **308**, 799 (1999)

G.E. Romero, D.F. Torres, Mod. Phys. Lett. A **16**, 1213 (2001)

M. Safonova, D.F. Torres, G.E. Romero, Phys. Rev. D **65**, 023001 (2002)

C. Sagan, *Contact* (Simon & Schuster, Inc., New York, 1985)

P. Schneider, J. Ehlers, E.E. Falco, *Gravitational Lenses* (Springer, Berlin, 1992)

K.S. Thorne, in *General Relativity and Gravitation*, ed. by J.L. Gleiser et al. (Institute of Physics, Bristol, 1992)

S. Tomizawa, H. Ishihara, Prog. Theor. Phys. Suppl. **189**, 7 (2011)

D.F. Torres, G.E. Romero, L.A. Anchordoqui, Phys. Rev. D **58**, 123001 (1998a)

D.F. Torres, G.E. Romero, L.A. Anchordoqui, Mod. Phys. Lett. A **13**, 1575 (1998b)

M. Visser, *Lorentzian Wormholes* (AIP Press, New York, 1996)

M. Visser, Phys. Rev. D **55**, 5212 (1997)

J.A. Wheeler, *Geometrodynamics* (Academic Press, New York, 1962)

Y.A. Zel'dovich, I.D. Novikov, A.A. Starobinsky, Zh. Èksp. Teor. Fiz. **66**, 1897 (1974)

Chapter 8
Black Holes and Cosmology

8.1 Overview of Current Cosmology

Modern cosmology results from the application of General Relativity to the universe as a whole. The so-called Standard Big Bang model is the current dynamical model that purports to describe the evolution and formation of structure of the universe. This model is supported by three main pieces of evidence: the increasing redshift of distant galaxies, the existence and properties of the cosmic microwave background (CMB) radiation, and the primordial nucleosynthesis. The first is interpreted in the model as due to the expansion of space-time. The second as the result of photon escape after the last scattering with charged particles occurred when the temperature of the universe dropped, again due to the expansion, to a level that allowed the combination of ions and electrons forming atoms. And the third as the effect of nuclear fusion reactions that occurred during the early phase of expansion; these reactions can explain the observed abundances of Hydrogen, Deuterium and Helium with great precision. The model says nothing about how the expansion started. The recent observation of a large number of high redshift supernovae (of Type Ia) led to the assumption that, if the universe is approximately homogeneous and isotropic (a bold hypothesis), then the current expansion should be accelerated (Reiss et al. 1998; Perlmutter et al. 1999). The Big Bang model based on Einstein's equations with a cosmological positive term plus the adoption of cold (i.e. non-relativistic) dark matter to explain some dynamic features of the rotation of galaxies and gravitational lensing is called the ΛCDM Big Bang model. Such a model is not free of problems (e.g. the value of the cosmological constant predicted by quantum field theories disagrees with what is observed by 120 order of magnitude) and several alternatives have been proposed (e.g. Amendola and Tsujikawa 2010).

The assumption of an expanding homogeneous and isotropic space-time leads to the Friedmann-Lemaitre-Robertson-Walker (FLRW) metric (see, e.g., Weinberg 1972):

$$ds^2 = c^2 dt^2 - a^2(t)\left[\frac{dr^2}{1 - kr^2} + r^2(d\theta)^2 + \sin^2\theta d\varphi^2\right]. \tag{8.1}$$

G.E. Romero, G.S. Vila, *Introduction to Black Hole Astrophysics*,
Lecture Notes in Physics 876, DOI 10.1007/978-3-642-39596-3_8,
© Springer-Verlag Berlin Heidelberg 2014

Here, t is the co-moving (with the cosmic fluid) time, $a(t)$ is a scale factor, and k is a normalized curvature index whose values are 0, $+1$, or -1 for flat, positively, and negatively curved spatial sections of the universe. If the cosmic fluid is an ideal one with the standard energy-momentum tensor $T_{\mu\nu} = (P + \rho)u_\mu u_\nu - P g_{\mu\nu}$, then Einstein's equations can be written as:

$$3\left(\dot{a}^2 + c^2 k\right) = \frac{8\pi G \rho a^2}{c^2}, \tag{8.2}$$

$$2a\ddot{a} + \dot{a}^2 + c^2 k = -\frac{8\pi G P a^2}{c^2}, \tag{8.3}$$

where the dot indicates a derivative with respect to t. An usual, alternative form of these equations, called Friedmann's equations, is:

$$\left(\frac{\dot{a}}{a}\right)^2 + \frac{c^2 k}{a^2} = \frac{8\pi G \rho}{3c^2}, \tag{8.4}$$

$$\frac{\ddot{a}}{a} + \frac{1}{2}\left(\frac{\dot{a}}{a}\right)^2 + \frac{c^2 k}{2a^2} = -\frac{8\pi G P}{2c^2}. \tag{8.5}$$

We can eliminate $\dot{a}/a$ and get an equation for the acceleration of the universe:

$$\frac{\ddot{a}}{a} = -\frac{4\pi G}{3c^2}(\rho + 3P). \tag{8.6}$$

From this equation we see that if the universe is dominated by non-relativistic matter ($\rho \gg P$), the universe decelerates:

$$\frac{\ddot{a}}{a} = -\frac{4\pi G}{3c^2}\rho < 0. \tag{8.7}$$

In the case of a relativistic gas ($P = \rho/3$):

$$\frac{\ddot{a}}{a} = -\frac{8\pi G}{3c^2}\rho < 0, \tag{8.8}$$

and the deceleration of the expansion is twice the previous value. For a universe dominated by vacuum energy ($P = -\rho$, see Sect. 1.5) we obtain:

$$\frac{\ddot{a}}{a} = \frac{16\pi G}{3c^2}\rho > 0, \tag{8.9}$$

i.e., the universe accelerates its expansion.

We can introduce a few useful parameters in the equations above. The first one is the Hubble parameter H_0:

$$H_0 \equiv \left[\frac{\dot{a}}{a}\right]_{t_0}, \tag{8.10}$$

where t_0 is the local time. Then, with $a(t_0) = a_0$, we define:

$$\hat{a}(t) \equiv \frac{a(t)}{a_0},\tag{8.11}$$

which is usually called the "reduced scale factor". The present-day normalized densities of cold matter, relativistic matter, and vacuum (or "dark") energy[1] are given by:

$$\Omega_i \equiv \frac{\rho_i(a_0)}{3c^2 H_0^2 / 8\pi G},\tag{8.12}$$

where $i = $ M, R, and Λ, respectively, and the normalization is to the "critical density" $\rho_c = 3c^2 H_0^2 / 8\pi G$ obtained for $k = 0$ and $t = t_0$ in Eq. (8.4). The total energy density is:

$$\Omega \equiv \Omega_M + \Omega_R + \Omega_\Lambda = 1.\tag{8.13}$$

The cosmological case with $k = \Omega_M = \Omega_R = 0$ and $\Lambda > 0$ yields:

$$\left(\frac{\dot{a}}{a}\right)^2 = \frac{c^2 \Lambda}{3},\tag{8.14}$$

$$\frac{\ddot{a}}{a} + \frac{1}{2}\left(\frac{\dot{a}}{a}\right)^2 = \frac{c^2 \Lambda}{2}.\tag{8.15}$$

The solution of these equations is:

$$a(t) = \text{constant} \times \exp\left[\left(\frac{c^2}{3}\Lambda\right)^{1/2} t\right],\tag{8.16}$$

which describes an empty universe in exponential expansion with positive cosmological constant. Such a model is known as *de Sitter* universe.

From Eqs. (8.4) and (8.5) we obtain an equation for the local energy conservation:

$$\frac{\partial \rho}{\partial t} = 3(\rho + P)\left(\frac{\dot{a}}{a}\right).\tag{8.17}$$

In the case of de Sitter space-time, since $P = -\rho$, the evolution of the energy density with time is:

$$\rho_\Lambda(t) = \rho_\Lambda(t_0) = \text{cte.}\tag{8.18}$$

For universes dominated by cold matter ($\rho \gg |P|$) and relativistic matter ($P = \rho/3$), the solutions of Eq. (8.17) are:

$$\rho_M(a) = \rho_M(a_0)\hat{a}^{-3},\tag{8.19}$$

---

[1] The expression "dark energy" is an abuse of language, since strictly speaking "energy" is a property, not a kind of matter or field. A more correct expression would be "dark field energy density".

$$\rho_R(a) = \rho_R(a_0)\hat{a}^{-4}.\tag{8.20}$$

As expected, the energy density of expanding universes with matter decays with the expansion. Instead, the energy density of a de Sitter universe remains constant since its value is independent of the number density of particles.

The standard Bing Bang model presents at least three serious problems, called the *flatness problem*, the *horizon problem*, and the *monopole problem*. The first problem is to explain why the density parameter Ω is so close to 1. The second problem is related to the following puzzle: why the universe is so homogeneous and isotropic at large scale, since distant regions are completely isolated from each other? The third problem concerns the production of monopoles in the early universe, predicted by grand unified theories, and not observed today. In the early 1980s it was proposed that a brief and accelerated expansion phase by a factor $\sim 10^{50}$ in the very early universe might solve these problems (Starobinsky 1980, 1982, Guth 1981). Models based on this assumption are called *inflationary models* (Linde 1990).

Inflationary models are obtained introducing a scalar field φ with a Lagrangian of the type

$$L = \frac{1}{2}g^{\mu\nu}\nabla_\mu\varphi\nabla_\nu\varphi - V(\varphi).\tag{8.21}$$

The corresponding energy-momentum tensor for φ is

$$T_{\mu\nu} = \nabla_\mu\varphi\nabla_\nu\varphi - \frac{1}{2}g_{\mu\nu}\nabla_\alpha\varphi\nabla^\alpha\varphi - g_{\mu\nu}V(\varphi).\tag{8.22}$$

The associated energy density and pressure are

$$\rho_\varphi = T_{\mu\nu}u^\mu u^\nu = \frac{\dot{\varphi}^2}{2} + V(\varphi),\tag{8.23}$$

$$P_\varphi = \frac{T_{ii}}{g_{ii}} = \frac{\dot{\varphi}^2}{2} - V(\varphi).\tag{8.24}$$

During the inflation $\dot{\varphi} \approx 0$ and $\rho \approx V(\varphi)$, and the scalar field potential is taken to be positive. This field is usually known as the *inflaton*. Friedmann's equations for the inflaton era are:

$$\left(\frac{\dot{a}}{a}\right)^2 + \frac{c^2 k}{a^2} = \frac{8\pi G}{3c^2}\left[\frac{\dot{\varphi}^2}{2} + V(\varphi)\right],\tag{8.25}$$

$$\frac{\ddot{a}}{a} + \frac{1}{2}\left(\frac{\dot{a}}{a}\right)^2 + \frac{c^2 k}{2a^2} = -\frac{8\pi G}{2c^2}\left[\frac{\dot{\varphi}^2}{2} - V(\varphi)\right].\tag{8.26}$$

Eliminating $\dot{a}/a$, we get:

$$\frac{\ddot{a}}{a} = -\frac{8\pi G}{3c^2}\left[\dot{\varphi}^2 - V(\varphi)\right].\tag{8.27}$$

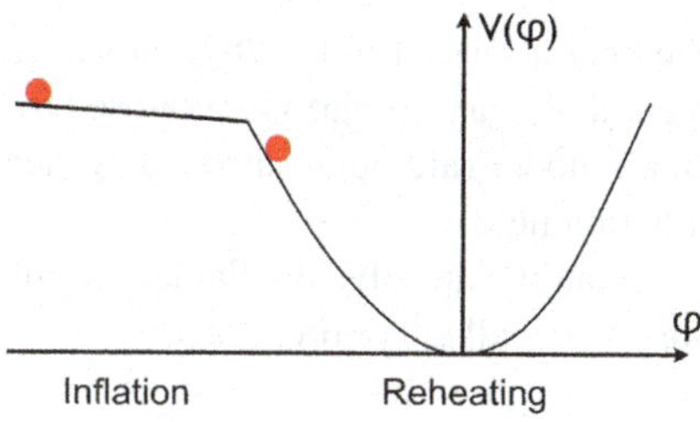

Fig. 8.1 A possible form for the potential of the scalar field responsible for the inflationary evolution of the scale factor of the universe

Adopting the *slow-roll approximation*:

$$\dot{\varphi}^2 \ll V(\varphi), \tag{8.28}$$

Eq. (8.27) reduces to

$$\frac{\ddot{a}}{a} \approx \frac{8\pi G}{3c^2} V(\varphi). \tag{8.29}$$

Hence, we see that the scale factor will accelerate, since $\ddot{a}(t) > 0$. The evolution of the scale factor under the inflaton can be obtained from Eq. (8.25), neglecting the $\dot{\varphi}$ and curvature terms:

$$\left(\frac{\dot{a}}{a}\right)^2 \approx \frac{8\pi G}{3c^2} V(\varphi), \tag{8.30}$$

from where we get

$$a(t) = a_0 \exp(\zeta\, t), \quad \zeta = \sqrt{\frac{8\pi G}{3c^2} V(\varphi)}, \tag{8.31}$$

i.e. the scale factor undergoes an exponential expansion.

The inflaton obeys a Klein-Gordon equation:

$$\ddot{\varphi} + 3H\dot{\varphi} + \frac{dV}{d\varphi} = 0, \tag{8.32}$$

where $H = \dot{a}/a$. A specific inflationary model is characterized by the form of the potential V. An illustrative form is

$$V(\varphi) = \lambda\left(\varphi^2 - \sigma^2\right)^2, \tag{8.33}$$

where λ and σ are constants. This potential is represented in Fig. 8.1. Since at $V(0)$ the potential is out of a minimum, the scalar field starts to increase. At first the rate of increment is slow, but it grows as the field goes to the minimum of the potential at $\varphi = \sigma$. There the field oscillates dissipating energy that goes to re-heating the universe, that was cooled down during the expansion period. The expansion sets up at $t \approx 10^{-34}$ s and is damped at around 10^{-32} s. By then, the scale factor has inflated by a factor of 10^{50}. This picture is oversimplified since the potential actually should change with the temperature as well, i.e. $V = V(\varphi,\, T)$ (e.g. Albrecht and

Steinhardt 1982; Linde 1982); however, it is enough to provide a feeling about how the inflationary scenario is supposed to solve the main problems of the standard Big Bang model (although introducing new ones such as the origin and nature of the inflation field).

Inflation can solve the flatness problem because the rapid growth of the universe suppresses all curvature. Since

$$|\Omega - 1| = \frac{|k|}{a^2 H^2} \tag{8.34}$$

inflation leads to $\Omega \cong 1$. Inflation, then, predicts a universe extremely close to spatial flatness. In addition, because of inflation, parts of the universe that are now causally disconnected were once in thermal equilibrium. This solves the horizon problem. Finally, the inflation also dilutes any relic particles to a point that they cannot be observed today. This solves the monopole problem.

Despite solving these problems, the inflationary scenario generates its own. In particular, the nature of the inflaton is not clear, and inflation seems to worsen the problem of the low entropy in the past required by the presence of irreversible process in the present.

8.2 The First Massive Black Holes

Supermassive black holes are found in host galaxies showing a striking correlation between the black hole mass and galactic properties such as bulge mass and the stellar velocity dispersion of the hot stellar component (Volonteri 2010 and references therein, see also Sect. 6.4.2). These correlations suggest a co-evolution of both the galaxies and massive black holes. However, supermassive black holes of $\sim 10^9 M_\odot$ have been found at redshifts larger than 6, i.e. when the universe was about 1 Gyr old. How these first very massive black hole might have been formed on such a short timescales? If we consider that a black hole accreting at the Eddington rate increases its mass as

$$M(t) \approx M_0 \exp\left(\frac{9t}{t_{Edd}}\right), \tag{8.35}$$

we find that black holes with masses in the range of 10^2–$10^5 M_\odot$ should have existed as early as 0.5 Gyr, i.e. well within the dark ages, before the re-ionization of the universe, when galaxies and the first stars were starting to form.

Three main scenarios for early massive black hole formation have been suggested (Volonteri 2010, 2012, and references therein): (1) the direct collapse of the first generation of stars (Population III stars), (2) black hole formation triggered by gas dynamical instabilities, and (3) black hole assembling by stellar-dynamical processes. We shall briefly review these scenarios.

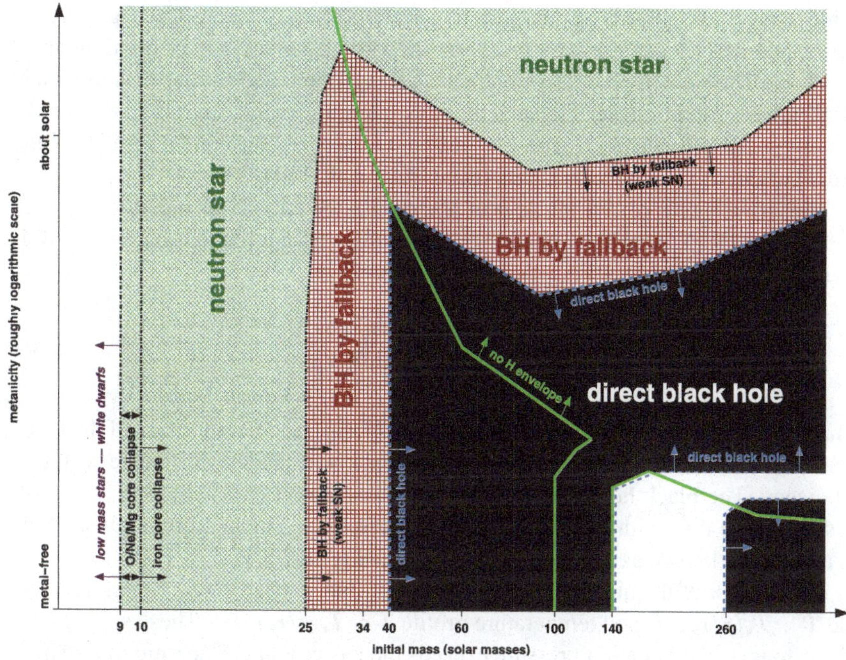

Fig. 8.2 Plot of metallicity vs. the initial mass of the star, showing the final stellar state. From Heger et al. (2003). Reproduced by permission of the AAS

8.2.1 Population III Stellar Collapse

The first generation of stars formed out of mini-halos of zero-metallicity gas, at $z \sim 20$–50. These halos resulted from the collapse of the highest peaks of the primordial density field. Low-metallicity stars of 25–$140 M_\odot$ are expected to collapse directly to black holes with a mass of around a half the mass of the star. This would produce black holes up to $70 M_\odot$. Between 140 and $260 M_\odot$ pair instability produces supernovae that completely disrupt the stars leaving no compact object behind (Bond et al. 1984). If the mass of the Population III stars is above $260 M_\odot$ direct black hole formation is again possible. Stars lighter than $25 M_\odot$, on the other hand, end as neutron stars or white dwarfs. All this is summarized in Fig. 8.2.

The initial mass function of Pop III stars, however, is not known, and it remains an open problem whether extremely massive stars were present at very early times, in order to provide seeds for the supermassive black holes observed at $z > 6$.

8.2.2 Black Holes from Gas-Dynamical Processes

Self-gravitating gas in dark matter haloes can lose angular momentum rapidly via runaway, global dynamical instabilities, the so-called "bars within bars" mechanism

(Shlosman et al. 1989). This leads to the rapid build-up of a dense, self-gravitating core supported by gas pressure and surrounded by a radiation pressure dominated envelope. The system gradually contracts and is compressed further by subsequent infall. These conditions lead to such high temperatures in the central region that the gas cools catastrophically by thermal neutrino emission, causing the formation and rapid growth of a central black hole (Begelman et al. 2006, 2008).

The growth rate of the black hole for typical conditions in metal-free haloes with $T_{\mathrm{vir}} \sim 10^4$ K, which are the most likely to be susceptible to runaway infall, is (Begelman et al. 2006)

$$\dot{M} \sim 3\alpha \frac{c^3}{\varepsilon G} \left(\frac{M}{M_\star} \right)^2 \left(\frac{v_\star}{c} \right)^4, \qquad (8.36)$$

where α is the viscosity parameter, $v_\star(t)^2 \sim GM_\star(t)/r_\star$, M is the mass of the black hole, $M_\star$ is the mass accumulated around the black hole, and $\varepsilon \sim 0.1$ is the accretion efficiency. The black hole will grow at super Eddington rate as long as $M_\star > M$. A system like this, with a radiation supported envelope and an accreting black hole as a source is known as a "quasi-star" (Begelman et al. 2008).

The interior of the quasi-star has a density profile $\rho_\star \sim \rho(r/r_\star)^{-2}$, pressure profile $P \sim P_\star(r/r_\star)^{-2}$, and temperature profile $T \sim T_\star(r/r_\star)^{-1/2}$. These scalings apply as long as the radiation pressure exceeds the gas pressure. The ratio of radiation pressure to gas pressure decreases with decreasing r,

$$\frac{P_{\mathrm{rad}}}{P_{\mathrm{gas}}} \sim \left(\frac{M_\star}{M_\odot} \right)^{1/2} \left(\frac{r}{r_\star} \right)^{1/2}, \qquad (8.37)$$

implying that $P_{\mathrm{gas}} \sim P_{\mathrm{rad}}$ at small enough radii.

The initial black hole should have a mass $< 20M_\odot$, but in principle could grow at a super-Eddington rate until it reaches $M \sim 10^4$–$10^6 M_\odot$. Rapid growth may be limited by feedback from the accretion process and disruption of the mass supply by star formation or halo mergers. Even if super-Eddington growth stops at $<10^3$–$10^4 M_\odot$, this process would give black holes ample time to attain quasar-size masses by a redshift of 6, and could also provide the seeds for the massive black holes seen in the local universe.

8.2.3 Black Holes by Stellar-Dynamical Processes

The gas fragmentation and formation of low-mass stars after the first generation of Pop III stars provide an alternative channel for the formation of the first massive black holes in the universe. According to this view (Devecchi and Volonteri 2009), the high redshift black hole seeds form as a result of multiple successive instabilities that occur in low-metallicity ($Z \sim 10^{-5} Z_\odot$) proto-galaxies. In relatively massive halos after the very first stars in the universe have completed their evolution, the second generation of stars form efficiently only in the very inner core of the proto-galaxy. There, very compact stellar clusters form. The typical star cluster masses are of the order of $10^5 M_\odot$ and the typical half mass radii of ~ 0.1 pc. A large fraction of these very dense clusters undergo core collapse before stars are able to complete stellar evolution. Runaway star-star collisions eventually lead to the formation of a very massive star, leaving behind a massive black hole remnant. Clusters unstable to runaway collisions are always the first, less massive ones that form. As the metallicity of the universe increases, the critical density for fragmentation decreases and stars start to form in the entire proto-galactic disk so that accretion of gas in the center is no longer efficient and the core collapse timescale increases. Typically a fraction ~ 0.05 of the proto-galaxies at $z \sim 10$–20 are expected to host black hole seeds, with masses in the range $M \sim 1000$–$2000 M_\odot$, allowing black hole growth by accretion during the quasar epoch, so at $z \sim 6$ supermassive ($\sim 10^9 M_\odot$) black holes can exist in galaxies.

8.3 Black Holes and Re-ionization

When the universe expanded, the primordial plasma cooled down until combination of ions and electrons was possible. The CMB radiation could then escape and the "dark ages" of the universe started $\sim 3.8 \times 10^5$ years after the Big Bang. This phase of the universe, now filled with neutral hydrogen, lasted for $\sim 10^9$ years, when the complex process of re-ionization of the intergalactic medium (IGM) was complete (see, for instance, Loeb 2010, for a review).

It is usually thought that most of the re-ionization was caused by the ultraviolet radiation from massive stars formed in the first generations of galaxies. It is uncertain, however, what fraction of ionizing ultraviolet photons could escape from primitive galaxies to produce and maintain the ionization far from galaxies in low-density regions of the IGM. Observations with the *Hubble Space Telescope* suggest that the rest-frame ultraviolet radiation from the most distant galaxies detected so far at the heart of the dark ages is not enough to heat and ionize the IGM over large volumes of space. Mirabel et al. (2011) have proposed that X-rays from accreting black holes in binary systems might have played a major role in the re-ionization of the universe, since the X-rays have a longer mean free path than ultraviolet photons.

Recent hydrodynamic simulations of the formation of the first generations of stars show that a substantial fraction of stars in primordial galaxies formed as binaries with typical masses of tens of solar masses (e.g. Krumholz et al. 2009). Models of single stars with very low metal content and initial masses of a few tens of solar masses show that they collapse directly with no energetic natal kicks, and end as black holes (e.g. Heger et al. 2003). It is then expected that the fraction of black holes to neutron stars and the fraction of black hole binaries to solitary black holes, should increase with redshift. That is, the rate of formation of bright BH-HMXBs was likely much larger in the early universe than at present.

The energy output from one of those BH-HMXBs during its whole lifetime can be more than 10^{54} erg (i.e. orders of magnitude larger than the energy from a typical core collapse supernova). In addition, an accreting black hole in a high-mass binary emits a total number of ionizing photons that is comparable to its progenitor star, but one X-ray photon emitted by an accreting black hole may cause the ionization of several tens of hydrogen atoms in a fully neutral medium.

The most important effect of BH-HMXBs in the early universe is the heating of the IGM to temperatures of $\sim 10^4$ K, which limits the recombination rate of Hydrogen and keeps the IGM ionized. An interesting prediction of Mirabel's et al. proposal is that the high temperatures achieved through X-ray ionization prevent the formation of faint galaxies at high redshifts. The total mass of dwarf galaxies should be $\geq 10^9 M_\odot$.

An additional effect of metallicity in the formation of BH-HMXBs (Mirabel 2010) is to boost the formation of BH-BH binaries as more likely sources of gravitational waves than NS-NS systems (Belczynski et al. 2010).

In this way, black holes seem to have played a major role in shaping the universe, as we observe it today.

8.4 The Future of Black Holes

According to Eq. (3.29), an isolated black hole with $M = 10 M_\odot$ would have a lifetime of more than 10^{66} yr. This is 56 orders of magnitude longer than the age of the universe.[2] However, if the mass of the black hole is small, then it could evaporate within the Hubble time. A primordial black hole, created by extremely energetic collisions short after the Big Bang, should have a mass of at least 10^{15} g in order to exist today. Less massive black holes must have already evaporated. What happens when a black hole losses its mass so it cannot sustain an event horizon anymore? As the black hole evaporates, its temperature raises. When it is cold, it radiates low energy photons. When the temperature increases, more and more energetic particles will be emitted. At some point gamma rays would be produced. If there is a population of primordial black holes, their radiation should contribute to the diffuse gamma-ray background. This background seems to be dominated by the contribution of unresolved active galactic nuclei and current observations indicate that if there were primordial black holes their mass density should be less than $10^{-8}\Omega$, where Ω is the cosmological density parameter (~ 1). After producing gamma rays, the mini black hole would produce leptons, quarks, and super-symmetric particles, if they exist. At the end, the black hole would have a quantum size and the final remnant will depend on the details of how gravity behaves at Planck scales. The final product might be a stable, microscopic object with a mass close to the Planck mass. Such particles might contribute to the dark matter present in our galaxy and in other galaxies and clusters. The cross-section of black hole relics is extremely

---

[2]We assume that the universe originated at the Big Bang, although, of course, this needs not to be necessarily the case.

small: 10^{-66} cm^2 (Frolov and Novikov 1998), hence they would be basically non-interacting particles.

A different possibility, advocated by Hawking (1974), is that, as a result of the evaporation nothing is left behind: all the energy is radiated. This creates a puzzle about the fate of the information stored in the black hole: is it radiated away during the black hole lifetime or does it simply disappear from the universe?

Actually, the very question is likely meaningless: information is not a property of physical systems. Information is a property of languages, and languages are human constructs. The physical property usually confused with information is entropy. The reason for the confusion is probably that a same mathematical formalism can be used to describe both properties. Of course, this does not mean that these quite different properties are identical. Black holes, as we have seen, have huge entropy. Is the entropy of the universe decreasing when a black hole evaporates? We think that the answer is the same one given by Bekenstein and already mentioned: *the total generalized entropy never decreases*. The entropy of the universe was increasing because of the black hole evaporation through a simple process of thermalization. The disappearance of the horizon is simply the end of such a process. Information is related to our capability of describing the process through a mathematical language, not to the process itself.

Independently of the problem of mini black hole relics, it is clear that the fate of stellar-mass and supermassive black holes is related to fate of the whole universe. In an ever expanding universe or in an accelerating universe as it seems to be our actual universe, the fate of the black holes will depend on the acceleration rate. The local physics of the black hole is related to the cosmic expansion through the cosmological scale factor $a(t)$. A Schwarzschild black hole embedded in a FLRW universe can be represented by a generalization of the McVittie metric (e.g. Gao et al. 2008):

$$ds^2 = \frac{[1 - \frac{2GM(t)}{a(t)c^2r}]^2}{[1 + \frac{2GM(t)}{a(t)c^2r}]^2} c^2 dt^2 - a(t)^2 \left[1 + \frac{2GM(t)}{a(t)c^2r}\right]^4 \left(dr^2 + r^2 d\Omega^2\right). \qquad (8.38)$$

Assuming that $M(t) = M_0 a(t)$, with M_0 a constant, the above metric can be used to study the evolution of the black hole as the universe expands. Adopting an equation of state for the cosmic fluid given by $P = \omega\rho c^2$, with ω constant, for $\omega < -1$ the universe accelerates its expansion in such a way that the scale factor diverges in a finite time. This time is known as the Big Rip. If $\omega = -1.5$, then the Big Rip will occur in 35 Gyr. The event horizon of the black hole and the cosmic apparent horizon will coincide for some time $t < t_{\mathrm{Rip}}$ and then the inner region of the black hole would be visible to observers in the universe. Unfortunately for curious observers, Schwarzschild black holes surely do not exist in nature, since all astrophysical bodies have some angular momentum and is reasonable then to expect that only Kerr black holes exist in the universe. Equation (8.38) does not describe a cosmological embedded Kerr black hole. Although no detailed calculations exist for such a case, we can speculate that the observer would be allowed to have a look at the second horizon of the Kerr black hole before being ripped apart along with the rest of the cosmos. A rather dark view for the Doomsday.

In case of $\omega > -1$ the expansion will continue during an infinite time. Black holes will become more and more isolated. As long as their temperature is higher than that of the CMB, they will accrete photons and increase their mass. When, because of the expansion, the CMB temperature falls below that of the black holes, they will start to evaporate. On the very long run, all black holes will disappear. If massive particles decay into photons on such long timescales, the final state of the universe will be that of a dilute photon gas. Cosmic time will cease to make any sense for such a state of the universe, since whatever exists will be on a null surface. Without time, there will be nothing else to happen.

References

A. Albrecht, P. Steinhardt, Phys. Rev. Lett. **48**, 1220 (1982)

L. Amendola, S. Tsujikawa, *Dark Energy* (Cambridge University Press, Cambridge, 2010)

M.C. Begelman, M. Volonteri, M.J. Rees, Mon. Not. R. Astron. Soc. **370**, 289 (2006)

M.C. Begelman, E.M. Rossi, P.J. Armitage, Mon. Not. R. Astron. Soc. **387**, 1649 (2008)

K. Belczynski, M. Dominik, T. Bulik, R. O'Shaughnessy, C. Fryer, D.E. Holz, Astrophys. J. **715**, L138 (2010)

J.R. Bond, W.D. Arnett, B.J. Carr, Astrophys. J. **280**, 825 (1984)

B. Devecchi, M. Volonteri, Astrophys. J. **694**, 302 (2009)

V.P. Frolov, I.D. Novikov, *Black Hole Physics* (Kluwer Academic, Dordretch, 1998)

C. Gao, X. Chen, V. Faraoni, Y.-G. Shen, Phys. Rev. D **78**, 024008 (2008)

A. Guth, Phys. Rev. D **23**, 347 (1981)

S.W. Hawking, Nature **248**, 30 (1974)

A. Heger, C.L. Fryer, S.E. Woosley, N. Langer, D.H. Hartmann, Astrophys. J. **591**, 288 (2003)

M.R. Krumholz, R.I. Klein, C.F. McKee, S.S.R. Offner, A.J. Cunningham, Science **323**, 754 (2009)

A.D. Linde, Phys. Lett. B **108**, 389 (1982)

A.D. Linde, *Particle Physics and Inflationary Cosmology* (Hardwood Academic, Chur, 1990)

A. Loeb, *How Did the First Stars and Galaxies Form?* (Princeton University Press, Princeton, 2010)

I.F. Mirabel, in *Jets at All Scales*. Proceedings of the IAU Symposium, vol. 275, ed. by G.E. Romero, R.A. Sunyaev, T. Belloni (Cambridge University Press, Cambridge, 2010), p. 2

I.F. Mirabel, M. Dijkstra, P. Laurent, A. Loeb, J.R. Pritchard, Astron. Astrophys. **528**, 149 (2011)

S. Perlmutter et al., Astrophys. J. **517**, 565 (1999)

A.G. Reiss et al., Astron. J. **116**, 1009 (1998)

I. Shlosman, J. Frank, M.C. Begelman, Nature **338**, 45 (1989)

A.A. Starobinsky, Phys. Lett. B **91**, 99 (1980)

A.A. Starobinsky, Phys. Lett. B **117**, 175 (1982)

M. Volonteri, Annu. Rev. Astron. Astrophys. **18**, 279 (2010)

M. Volonteri, Science **337**, 544 (2012)

S. Weinberg, *Gravitation and Cosmology: Principles and Applications of the General Theory of Relativity* (Wiley, New York, 1972)

Appendix A
Topology and Manifolds

In this appendix we provide some definitions that might be useful for readers interested in a more formal approach to General Relativity and black holes. For detailed treatments we refer to the books of Isham (2005), Frankel (2012), and Nash and Sen (2011).

A.1 Topology

Topology is the study of those properties of a geometric shape that are unchanged under continuous deformation. In more technical terms, topology deals with topological spaces. One of the main aspects of topology is that it allows to make qualitative predictions when quantitative ones are impossible or extremely difficult.

A.1.1 Topological Spaces

Let X be any set and $T = \{X_\alpha\}$ a collection, finite or infinite, of subsets of X. Then the ordered pair (X, T) forms a *topological space* iff:

1. $X \in T$.
2. $\emptyset \in T$.
3. Any finite or infinite sub-collection $\{X_1, X_2, \ldots, X_n\}$ of the X_α is such that $\bigcup_1^n X_i \in T$.
4. Any *finite* sub-collection $\{X_1, X_2, \ldots, X_n\}$ of the X_α is such that $\bigcap_1^n X_i \in T$.

The set X is called a topological space and the X_α are called *open sets*. The assignment of T to X is said to "give" a topology to X.

A function f mapping from the topological space X onto the topological space X^* is continuous if the inverse image of an open set in X^* is an open set in X.

If a set X has two topologies $T^1 = \{X_\alpha\}$ and $T^2 = \{X_\alpha^*\}$ such that $T^1 \supset T^2$, we say that T^1 is *stronger* than T^2.

G.E. Romero, G.S. Vila, *Introduction to Black Hole Astrophysics*,
Lecture Notes in Physics 876, DOI 10.1007/978-3-642-39596-3,

A.1.2 Neighborhoods

Given a topology T on X, then N is a *neighborhood* of a point $x \in X$ if $N \subset X$ and there is some $X_\alpha \subset N$ such that $x \in X_\alpha$. Notice that it is not necessary for N to be an open set. However, all open sets X_α which contain x are neighborhoods of x since they are contained in themselves. Thus, neighborhoods are more general than open sets.

A.1.3 Closed Sets

Let T be a topology on X. Then any $U \subset X$ is *closed* if the complement of U in X $(\bar{U} = X - U)$ is an open set. Since $\bar{\bar{U}} = U$ then a set is open when its complement is closed. The sets X and $\emptyset$ are open and closed regardless the topology T.

A.1.4 Closure of a Set

Given a set U, there will be in general many closed sets that contain U. Let F_α be the family of closed sets that contain U. The *closure* of U is $\tilde{U} = \bigcap_\alpha F_\alpha$. The closure is the smallest closed set that contains U. Notice that $\tilde{\tilde{U}} = \tilde{U}$.

A.1.5 Boundary and Interior

The *interior* U^0 of a set U is the union of all open sets O_α of U: $U_0 = \bigcup_\alpha O_\alpha$. The interior of U is the largest open set of U.

The *boundary* $b(U)$ of a set U is the complement of the interior of U in the closure of U: $b(U) = \tilde{U} - U^0$. Closed sets always contain their boundaries:

$$U \cap b(U) = \emptyset \iff U \text{ is open,}$$

$$b(U) \subset U \iff U \text{ is closed.}$$

Notice that the sets (a, b), $[a, b)$, $(a, b]$, and $[a, b]$ all have the same boundary: $b = a, b$.

A.1.6 Compactness

Given a family of sets $\{F_\alpha\} = F$, F is a *cover* of U if $U \subset \bigcup_\alpha F_\alpha$. If $(\forall F_\alpha)_F$ (F_α is an open set) then the cover is called an *open cover*.

A set U is *compact* if for *every* open covering $\{F_\alpha\}$ with $U \subset \bigcup_\alpha F_\alpha$ there *always* exists a *finite* sub-covering $\{F_1, \ldots, F_n\}$ of U such that $\bigcup_1^n F_\alpha \subset U$.

As an illustration consider $\mathfrak{R}^n$. A subset X of $\mathfrak{R}^n$ is compact iff it is closed and bounded. This means that X must have finite area and volume in n-dimensions.

A.1.7 Connectedness

A set X is *connected* if it cannot be written as $X = X_1 \cup X_2$ where X_1 and X_2 are both open sets and $X_1 \cap X_2 = \emptyset$.

A.1.8 Homeomorphisms and Topological Invariants

Let T_1 and T_2 be two topological spaces. An *homeomorphism* is a map f from T_1 to T_2:

$$f : T_1 \to T_2$$

such that f is continuous and its inverse map f^{-1} is also continuous. If there is a third topological space T_3 such that T_1 is homeomorphic to T_2 and T_2 is homeomorphic to T_3, then T_1 is homeomorphic to T_3. An homeomorphism defines an equivalence class, that of all spaces that are homeomorphic to a given topological space. If the homeomorphism f and its inverse f^{-1} are infinitely differentiable (C^∞), then f is called a *diffeomorphism*. All diffeomorphisms are homeomorphisms, but the converse is not always the case.

A *topological invariant* is a construct that does not change under homeomorphisms. They are characteristics of the equivalence class of the homeomorphism. An example of an invariant is the dimension n of $\mathfrak{R}^n$.

Homeomorphisms generate equivalence classes whose members are topological spaces. Instead, *homotopies* generate classes whose members are continuous maps. More specifically, let f_1 and f_2 be two continuous maps between the topological spaces T_1 and T_2:

$$f_1 : T_1 \to T_2,$$

$$f_2 : T_1 \to T_2.$$

Then f_1 is said to be homotopic to f_2 if f_1 can be deformed into f_2. Formally:

$$F : T_1 \times [0, 1] \to T_2, \quad F \text{ continuous}$$

and

$$F(x, 0) = T_1(x),$$

$$F(x, 1) = T_2(x).$$

This means that as the real variable t changes continuously from 0 to 1 in the interval [0, 1] the map f_1 is deformed continuously into the map f_2. Homotopy is an equivalence relation that divides the space of continuous maps from T_1 to T_2 into equivalent classes. These homotopy equivalent classes are topological invariants of the pair of spaces T_1 and T_2.

Homotopy can be used to classify topological spaces. If we identify one of the topological spaces with the n-dimensional sphere S^n, then the space of continuous maps from S^n to T, $C(S^n, T)$, can be divided into equivalence classes according to the topological space T. The equivalent classes of $C(S^n, T)$ have a group structure and form the homotopy group $\Pi_n(T)$.

It is rather straightforward to show that both compactness and connectedness are topological invariants (Nash and Sen 2011).

A.2 Manifolds

Whereas topology is the natural mathematical framework to study continuity, differential geometry is a natural framework to study differentiability. Since differentiability implies continuity, but not the other way around, differential geometry is more specific than topology. The concept of *manifold* is central to differential geometry.

A.2.1 Manifolds: Definition and Properties

A set M is a differentiable manifold if:

1. M is a topological space.
2. M is equipped with a family of pairs $\{(M_\alpha, \varphi_\alpha)\}$.
3. The M_α's are a family of open sets that cover M: $M = \bigcup_\alpha M_\alpha$. The φ_α's are homeomorphisms from M_α to open subsets O_α of $\mathfrak{R}^n$: $\varphi_\alpha : M_\alpha \to O_\alpha$.
4. Given M_α and M_β such that $M_\alpha \cap M_\beta \neq \emptyset$, the map $\varphi_\beta \circ \varphi_\alpha^{-1}$ from the subset $\varphi_\alpha(M_\alpha \cap M_\beta)$ of $\mathfrak{R}^n$ to the subset $\varphi_\beta(M_\alpha \cap M_\beta)$ of $\mathfrak{R}^n$ is infinitely differentiable (C^∞).

The family $\{(M_\alpha, \varphi_\alpha)\}$ is called an *atlas*. The individual members of the altas are *charts*. In informal language we can say that M is a space that can be covered by patches M_α which are assigned coordinates in $\mathfrak{R}^n$ by φ_α. Within each of these patches M looks like a subset of the Euclidean space $\mathfrak{R}^n$. M is not necessarily globally Euclidean or pseudo-Euclidean. If two patches overlap, then in $M_\alpha \cap M_\beta$ there are two assignations of coordinates, which can be transformed smoothly into each other. The dimension of the manifold M is the dimension n of the space $\mathfrak{R}^n$.

A manifold M is said to be Hausdorff if for any two distinct elements $x \in M$ and $y \in M$, there exist $O_x \subset M$ and $O_y \subset M$ such that $O_x \cap O_y = \emptyset$.

A given topological space M is said to be *metric* if the open sent are provided by a binary function $d(x, y)$ such that:

1. $d(x, y) \geq 0$.
2. $d(x, y) = 0$ iff $x = y$.
3. If $z \in M$, then $d(x, y) + d(y, z) = d(x, z)$.

An important property of manifolds is their *orientability*. Given a manifold M whose atlas is $\{(M_\alpha, \varphi_\alpha)\}$, M is orientable if $\det(\varphi_\beta \circ \varphi_\alpha^{-1}) > 0$ for all M_α and M_β such that $M_\alpha \cap M_\beta \neq \emptyset$. The manifold is orientable if one can define a preferred direction unambiguously.

A.2.2 Fiber Bundles

A fiber bundle is a topological space that is *locally, but not necessarily globally* the product of two spaces. All spaces that are globally products are called *trivial bundles*. Fiber bundles can be defined upon many spaces of use in physics and, in particular, in gravitation, where, as we shall see, the tangent spaces to a manifold form a bundle.

Formally, a collection (E, Π, F, G, X) is called a fiber bundle iff:

1. E is a topological space, usually called the *total space*.
2. X is a topological space, usually called the *base space*.
3. F is a topological space, usually called the *fibre*.
4. Π is an application $\Pi : E \to X$ of E onto X, called the *projection*.
5. G is a group of homeomorphisms on the fiber F.
6. There is a set of open coordinate neighborhoods $\{U_\alpha\}$ covering X, which reflects the local triviality of E. Specifically, with each U_α there is a given homeomorphism such that:

$$\varphi_\alpha : \Pi^{-1}(U_\alpha) \to U_\alpha \times F, \tag{A.1}$$

and

$$\Pi \varphi_\alpha^{-1}(x, f) = x, \quad \text{with } x \in U_\alpha, \ f \in F. \tag{A.2}$$

Let us consider a transformation of a local coordinate set $\{\varphi_\alpha, U_\alpha\}$ to another set $\{\varphi_\beta, U_\beta\}$. Let us suppose that $U_\alpha \cap U_\beta \neq \emptyset$. Then $g_{\alpha\beta}(x) = \varphi_\alpha \circ \varphi_\beta^{-1}$ is a continuous invertible map of the form

$$g_{\alpha\beta}(x) : (U_\alpha \cap U_\beta) \times F \to (U_\alpha \cap U_\beta) \times F. \tag{A.3}$$

The function $g_{\alpha\beta}(x)$ is a homeomorphism of the fiber F, called the *transition function*. The set of all these homeomorphisms for all choices of $\{\varphi_\alpha, U_\alpha\}$ form the group G. This group is called the *structure group* of the fiber bundle E.

Perhaps the simplest illustration of a fiber bundle is a Möbius strip. Such a space is illustrated in Fig. A.1. Here, the total space E is the whole Möbius strip. The

Fig. A.1 Simple example of
a fiber bundle: the Möbius
strip. See the text for
explanation

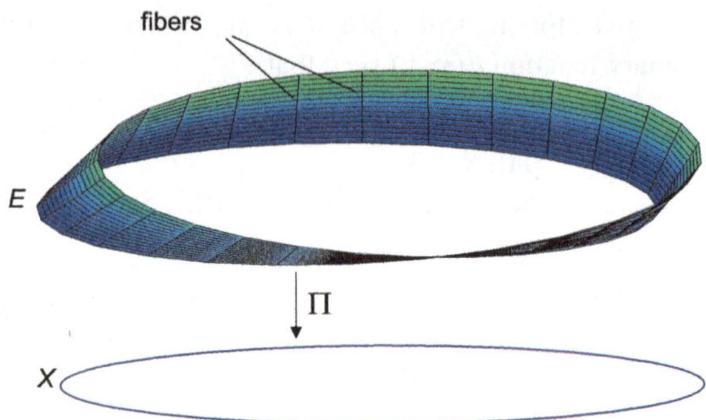

base space X is the projection onto $\Re^2$. The fiber F is a line segment on the strip
obtained from a point of $x \in X$ by $\Pi^{-1}(x)$. To an open set U_α of X corresponds a
set of fibers on E. The functions φ_α transforms $\Pi^{-1}(U_\alpha)$ into the product $U_\alpha \times F$.
We see, then, that whereas a manifold is locally like $\Re^n$ a fiber bundle is locally like
a product of topological spaces.

Manifolds and fiber bundles are related since we can define a fiber bundle formed
by all tangent spaces to a given manifold. Specifically, for each manifold M there is
a fiber bundle $T(M)$ (called *tangent bundle*) given by:

$$T(M) = \bigcup_{p \in M} T_p(M). \tag{A.4}$$

The base space of the fiber bundle is M, the fiber at any point p of M is the
tangent space $T_p(M)$, and the projection is defined by:

$$\Pi : T(M) \to M, \tag{A.5}$$

and

$$\mathbf{V} \in T_p(M) \to p. \tag{A.6}$$

The fiber $T_p(M)$ at p is a vector space of dimension n, equal to the dimension of
the manifold. Let now $p \in U_\alpha \subset M$, then

$$\varphi_\alpha : \Pi^{-1}(U_\alpha) \to U_\alpha \times \Re^n, \tag{A.7}$$

and

$$\mathbf{V} \to \left(p, a^i(p)\right), \tag{A.8}$$

where $a^i(p)$ is a coordinate assignation to p. The group structure is given by the
group of invertible $n \times n$ matrices.

The main role of the tangent bundle is to provide a domain and range for the
derivative of a smooth function. Namely, if $f : M \to M'$ is a smooth function, with
M and M' smooth manifolds, its derivative is also a smooth function.

References

T. Frankel, *The Geometry of Physics*, 3rd edn. (Cambridge University Press, Cambridge, 2012)

C. Isham, *Modern Differential Geometry for Physicists*, 2nd edn. (World Scientific, Singapore, 2005)

C. Nash, S. Sen, *Topology and Geometry for Physics* (Dover, Mineola, 2011)

Appendix B
Selected and Annotated Bibliography

The following bibliography intends to be merely orientative and by no way is complete. It reflects the authors' taste and contains some of those books we most frequently resort to in our library.

B.1 Books on General Relativity

- *Gravitation*, by C.W. Misner, K.S. Thorne and J.A. Wheeler, W.H. Freeman & Co., New York, 1973.
  Complete and detailed. Part VII is devoted to gravitational collapse and black holes.
- *General Relativity*, by R.M. Wald, The University of Chicago Press, Chicago, 1984.
  A modern introduction to General Relativity with emphasis on global techniques.
- *The Large Structure of Space-Time*, by S.W. Hawking and G.F.R. Ellis, Cambridge University Press, Cambridge, 1973.
  The classic reference on global techniques in General Relativity.
- *Introducing Einstein's Relativity*, by R. D'Inverno, Clarendon Press, Oxford, 1992.
  A complete and very useful reference on General Relativity and black holes.
- *Space-Time and Geometry*, by S. Carroll, Addison Wesley, San Francisco, 2004.
  Clear and readable introduction for undergraduate and graduate students.
- *Relativity, Second Edition*, by W. Rindler, Oxford University Press, Oxford, 2006.
  An excellent textbook.
- *General Relativity*, by M.P. Hobson, G. Efstathiou and A.N. Lasenby, Cambridge University Press, Cambridge, 2006.
  A very clear and complete introduction to mathematical aspects of General Relativity.
- *General Relativity*, by N. Straumann, Springer, Berlin, 2004.
  Complete, mathematically strong, with several astrophysical applications.

G.E. Romero, G.S. Vila, *Introduction to Black Hole Astrophysics*,
Lecture Notes in Physics 876, DOI 10.1007/978-3-642-39596-3,
© Springer-Verlag Berlin Heidelberg 2014

- *Gravity*, by J.B. Hartle, Addison Wesley, San Francisco, 2003.
  Undergraduate textbook with strong emphasis on the physical interpretations.
- *General Relativity and the Einstein's Equations*, by Y. Choquet-Bruhat, Oxford University Press, Oxford, 2009.
  A mathematically sophisticated monograph on General Relativity.
- *Introduction to General Relativity*, by L. Ryder, Cambridge University Press, Cambridge, 2009.
  Student-friendly and well-illustrated basic textbook.
- *General Relativity*, by M. Ludvigsen, Cambridge University Press, Cambridge, 1999.
  A geometric and abstract presentation of General Relativity.
- *Global Aspects in Gravitation and Cosmology*, by P.S. Joshi, Clarendon Press, Oxford, 1993.
  A good complement to Hawking and Ellis' book.
- *Numerical Relativity*, by T.W. Baumgarte and S.L. Shapiro, Cambridge University Press, Cambridge, 2010.
  Essential to the working scientist.
- *A Relativist's Toolkit*, by E. Poisson, Cambridge University Press, Cambridge, 2004.
  Faithful to the title. It is a must for every relativist.
- *Relativity on Curved Manifolds*, by F. De Felice and C.J.S. Clarke, Cambridge University Press, Cambridge, 1990.
  An advanced monograph.
- *Advanced General Relativity*, by J. Stewart, Cambridge University Press, Cambridge, 1991.
  Good reference for advanced topics such as spinors and asymptopia.
- *Rotating Fields in General Relativity*, by J.N. Islam, Cambridge University Press, Cambridge, 1985.
  Monograph fully devoted to axially symmetric solutions of Einstein's equations.
- *An Introduction to the Relativistic Theory of Gravitation*, by P. Hajicek, Springer, Heidelberg, 2008.
  Good introductory course.
- *Lecture Notes on the General Theory of Relativity*, by Ø. Grøn, Springer, Heidelberg, 2009.
  Short overview of the topic.
- *An Introduction to Relativity*, by J.V. Narlikar, Cambridge University Press, Cambridge, 2010.
  Undergraduate level. It deals with some controversial topics.
- *A First Course on General Relativity, Second Edition*, by B.F. Schutz, Cambridge University Press, Cambridge, 2009.
  Standard textbook.
- *Tensor Relativity and Cosmology*, by M. Dalarsson and N. Dalarsson, Elsevier, Amsterdam, 2005.
  An introduction which includes many explicit calculations.
- *Gravitation*, by T. Padmanabhan, Cambridge University Press, Cambridge, 2010.
  Complete, modern, with advanced topics.

- *Relativity: the General Theory*, by J.L. Synge, North-Holland Publishing Company, Amsterdam, 1960.
  Excellent book by one of the greatest relativists. Still a unique source on several topics.
- *Gravitation and Spacetime, Second Edition*, by H.C. Ohaniam and R. Ruffini, W.W. Norton & Co., New York, 1994.
  The chapter on black holes is particularly good.
- *The Theory of Space, Time and Gravitation, Second Revised Edition*, by V.A. Fock, Pergamon Press, Oxford, New York, 1964.
  A deep exposition a là Landau.
- *The Theory of Relativity, Second Edition*, by R.K. Pathria, Pergamon Press, Oxford, 1974.
  A nice oldie.
- *Space Time Matter*, by H. Weyl, Dover, New York, 1952.
  The first and still one of the best books on General Relativity (first edition 1918).
- *Theory of Relativity*, by W. Pauli, Dover, New York, 1958.
  A classic (first edition 1921).
- *The Mathematical Theory of Relativity*, by A.S. Eddington, Cambridge University Press, Cambridge, 1923.
  Classic.
- *Relativity, Thermodynamics and Cosmology*, by R.C. Tolman, Dover, New York, 1987.
  A classic and one of the few books dealing with relativistic thermodynamics. Originally published in 1934.
- *Introduction to the Theory of Relativity*, by P.G. Bergmann, Dover, New York, 1976.
  A book strongly recommended by Einstein himself. One of the few books discussing the later field theories developed by Einstein. Originally published in 1942.
- *General Theory of Relativity*, by P.A.M. Dirac, Princeton University Press, Princeton, 1996.
  The lectures given by Dirac on the topic. Originally published in 1975.
- *Principles of Relativity Physics*, by J.L. Anderson, Academic Press, New York, 1967.
  An excellent book, full of physical insight.

B.2 Books on Black Holes

- *Black Holes*, by J.-P. Luminet, Cambridge University Press, Cambridge, 1992.
  A popular thought-provoking introduction.
- *Gravity's Fatal Attraction*, by M. Begelman and M. Rees, Scientific American, New York, 1998.
  Superbly illustrated, conceptually clear.
- *Exploring Black Holes*, by E.F. Taylor and J.A. Wheeler, Addison Wesley, San Francisco, 2000.

A didactic primer.

- *Stars and Relativity*, by Y.B. Zel'dovich and I.D. Novikov, Dover, New York, 1996.
  Originally published in 1971, it was one of the first books to discuss black holes by then named "frozen stars".

- *Black Holes: the Membrane Paradigm*, by K. Thorne, R.H. Price and D.A. McDonald, Yale University Press, New Haven, 1986.
  The main reference on a much debated analogy.

- *Black Holes*, by D. Raine, E. Thomas, Imperial College Press, London, 2005.
  Good, well-written and concise.

- *Black Holes and Relativistic Stars*, by R.M. Wald (ed.), Chicago University Press, Chicago, 1998.
  An outstanding collection of original papers dedicated to the memory of S. Chandrasekhar.

- *The Mathematical Theory of Black Holes*, by S. Chandrasekhar, Oxford University Press, Oxford, 1983.
  There is a lot of material that you will only find in this book, but not recommended for the beginner.

- *Introduction to Black Hole Physics*, by P.V. Frolov and A. Zelnikov, Oxford University Press, Oxford, 2011.
  Perhaps the best book available on black holes.

- *Black Hole Gravito-Hydromagnetics, Second Edition*, by B. Punsly, Springer, Berlin, 2008.
  Strongly focused on ergospheric effects and relativistic magnetohydrodynamics.

- *Black Holes, White Dwarfs and Neutron Stars*, by S.L. Shapiro and S.A. Teukolsky, John Wiley & Sons, New York, 1983.
  Classic but a bit outdated on some of the astrophysical aspects.

- *Black Hole Physics*, by V.P. Frolov and I.D. Novikov, Kluwer Academic Publishers, Dordretch, 1998.
  As complete as expensive, but if you can afford it very worthy.

- *Black Hole Uniqueness Theorems*, by M. Heusler, Cambridge University Press, Cambridge, 1996.
  Advanced and unique in its kind.

- *Physics and Astrophysics of Neutron Stars and Black Holes, Second Edition*, by R. Giacconi and R. Ruffini, Cambridge Scientific Publishers, Cambridge, 2009.
  A rich resource of material on both Physics and Astrophysics of black holes.

B.3 Books on Related Topics in Astrophysics

- *High-Energy Radiation from Black Holes*, by C.D. Dermer and G. Menon, Princeton University Press, Princeton, 2009.
  Mostly devoted to radiative processes.

- *Relativistic Astrophysics and Cosmology*, by P. Hoyng, Springer, Heidelberg, 2006.
  Concise and insightful.

- *Compact Objects in Astrophysics*, by M. Camenzind, Springer, Berlin, 2007.
  Very complete with many applications.
- *MHD Flows in Compact Astrophysical Objects*, by V.S. Beskin, Springer, Heidelberg, 2010.
  Key reference for jets and outflows.
- *Active Galactic Nuclei*, by J.H. Krolik, Princeton University Press, Princeton, 1999.
  Covers both observational and theoretical aspects of AGN.
- *The Physics of Extragalactic Radio Sources*, by D.S. De Young, The University of Chicago Press, Chicago, 2002.
  Includes a very good discussion of the hydrodynamics of jets.
- *Quasars and Active Galactic Nuclei*, by A.K. Kembhavy and J.V. Narlikar, Cambridge University Press, Cambridge, 1999.
  Highly-recommended as an introduction to the topic.
- *High-Energy Astrophysics, Third edition*, by M.S. Longair, Cambridge University Press, Cambridge, 2011.
  Outstanding textbook.
- *Very High-Energy Cosmic Gamma Radiation*, by F.A. Aharonian, World Scientific, New Jersey, 2004.
  Broad coverage of gamma-ray astronomy.
- *Beams and Jets in Astrophysics*, by P.A. Hughes (ed.), Cambridge University Press, Cambridge, 1991.
  Very useful source of information for the researcher.
- *Theory of Black Hole Accretion Disk*, by M.A. Abramowicz, G. Björnsson and J.E. Pringle (eds.), Cambridge University Press, Cambridge, 1998.
  A menagerie of valuable reviews. We specially recommend those on ADAFs and stability in black hole binaries.
- *Accretion Power in Astrophysics, Second Edition*, by J. Frank, A. King and D. Raine, Cambridge University Press, Cambridge, 1992.
  Perhaps the most widely used book on accretion by astrophysicists.
- *Accretion*, by A. Treves, L. Maraschi and M.A. Abramowicz, World Scientific, Singapur, 1989.
  The subtitle of this book is "A Collection of Influential Papers". It delivers what it promises.
- *X-ray Binaries*, by W.H.G. Lewin, J. van Paradijs and E.P.J. van den Heuvel (eds.), Cambridge University Press, Cambridge, 1997.
  A complete introduction.
- *High-Energy Astrophysics*, by F. Melia, Princeton University Press, Princeton, 2009.
  Discusses black holes in binaries, gamma-ray bursts and supermassive black holes.
- *Relativistic Astrophysics of the Transient Universe*, by M.H.P.M. van Putten and A. Levinson, Cambridge University Press, Cambridge, 2012.
  Updated and engaging treatment of many astrophysical manifestations of black holes.

Index

G.E. Romero, G.S. Vila, *Introduction to Black Hole Astrophysics*,
Lecture Notes in Physics 876, DOI 10.1007/978-3-642-39596-3,
© Springer-Verlag Berlin Heidelberg 2014

Made in the USA
Monee, IL
07 July 2026

56552011R00188